INTRODUCTION TO

Statistical Data Analysis for the Life Sciences

CLAUS THORN EKSTRØM

HELLE SØRENSEN

CRC Press
Taylor & Francis Group
Boca Raton   London   New York

CRC Press is an imprint of the
Taylor & Francis Group an **informa** business

A CHAPMAN & HALL BOOK

CRC Press
Taylor & Francis Group
6000 Broken Sound Parkway NW, Suite 300
Boca Raton, FL 33487-2742

© 2011 by Taylor and Francis Group, LLC
CRC Press is an imprint of Taylor & Francis Group, an Informa business

No claim to original U.S. Government works

Printed in the United States of America on acid-free paper
10 9 8 7 6 5 4 3 2 1

International Standard Book Number: 978-1-4398-2555-6 (Paperback)

Library of Congress Cataloging-in-Publication Data

Ekstrøm, Claus Thorn, 1971-
    Introduction to statistical data analysis for the life sciences / Claus Thorn Ekstrøm and Helle Sørensen.
        p. cm.
    Includes bibliographical references and index.
    ISBN 978-1-4398-2555-6 (pbk. : alk. paper)
    1. Mathematical statistics--Textbooks. 2. Life sciences--Statistical methods--Textbooks. I. Sørensen, Helle, 1971- II. Title.

QA276.E38 2011
570.1'5195--dc22   2010026466

Visit the Taylor & Francis Web site at
http://www.taylorandfrancis.com

and the CRC Press Web site at
http://www.crcpress.com

Contents

Preface

We believe that a textbook in statistics for the life sciences must focus on applications and computational statistics combined with a reasonable level of mathematical rigor. In the spring of 2008 we were asked to revise and teach the introductory statistics course taken by the majority of students at the Faculty of Life Sciences at the University of Copenhagen. We searched for a textbook that could replace the earlier textbook by Skovgaard et al. (1999) but were unable to find one with the right combination of data examples, statistical theory, and computing. We decided to make our own material, and this book is the result of our efforts.

The book covers material seen in many textbooks on introductory statistics but differs from these books in several ways. First and foremost we have kept the emphasis on both data analysis and the mathematics underlying classical statistical analysis. We have tried to give the reader a feeling of being able to model and analyze data very early on and then "sneak in" the probability and statistics theory as we go along. Second, we put much emphasis on the modeling part of statistical analysis and on biological interpretations of parameter estimates, hypotheses, *etc*. Third, the text focuses on the use and application of statistical software to analyze problems and datasets. Our students should not only be able to determine *how* to analyze a given dataset but also have a computational toolbox that enables them to actually *do* the analysis — for real datasets with numerous observations and several variables.

We have used R as our choice of statistical software. R is the *lingua franca* of statistical computing; it is a free statistical programming software and it can be downloaded from http://cran.r-project.org. By introducing the students to R we hope to provide them with the necessary skills to undertake more sophisticated analyses later on in their careers. R commands and output are found at the end of each chapter so that they will not steal too much attention from the statistics, and so the main text can be used with any statistical software program. However, we believe that being able to use a software package for statistical analyses is essential for all students. Appendix B provides a short introduction to R that can be used for students with no previous experience of R. All the datasets used in the book are available from the supporting web site

http://www.statistics.life.ku.dk/isdals/

The book can be read sequentially from start to end. Some readers may prefer to have a proper introduction to probability theory (Chapter 9) before introducing statistics, inference, and modeling, and in that case Chapter 9 can be read between Chapters 1 and 2. Chapters 2 and 3 cover linear regression and one-way analysis of variance with emphasis on modeling, interpretation, estimation, and the biological questions to be answered, but without details about variation of estimates and hypothesis tests. These two chapters are meant as appetizers and should provide the readers a feeling of what they will be able to accomplish in the subsequent chapters and to make sure that the reader keeps in mind that we essentially intend to apply the theory to analyze data.

Chapters 4 to 7 cover the normal distribution and statistical inference: estimation, confidence intervals, hypothesis testing, prediction, and model validation with thorough discussions on one- and two-sample problems, linear regression, and analysis of variance. The different data types are treated "in one go" since the analyses are similar from a statistical point of view, but the different biological interpretations are also stressed.

Chapter 8 extends the theory to linear normal models (*e.g.*, multiple linear regression and two-way analysis of variance models), shows that linear regression and analysis of variance are essentially special cases of the same class of models, and more complicated modeling terms such as interactions are discussed.

Chapter 9 is a self-contained introduction to probability theory including independence and conditional probabilities.

In Chapter 10 we present the binomial distribution and discuss statistical inference for the binomial distribution. Chapter 11 is concerned with analysis of count data and the use of chi-square test statistics to test hypotheses. Emphasis is on the analysis of 2×2 tables as well as on general $r \times k$ tables. Chapter 12 is about logistic regression and thus combines aspects from linear models with the binomial distribution.

Each of these chapters contains a number of exercises related to the topic and theory of that chapter. Roughly half the exercises are supposed to be done by hand, whereas a computer should be used for the remaining ones (marked with an ® symbol). A few of the exercises include R commands and related output that can be used to answer the problems. These exercises are supposed to give the reader a possibility to get familiar with the R language and learn to read and interpret output from R without getting into trouble with the actual programming. A small number of exercises are of a more mathematical nature; *e.g.*, derivation of formulas. Such exercises are marked with an **[M]**.

Chapter 13 contains ten larger case exercises where readers are encouraged to apply their knowledge to larger datasets and learn more about important topics. We consider these exercises an important part of the book. They are suitable for self-study because the analyses are made in many small steps and much help is provided in the questions.

The book ends with three appendices. Appendix A includes an overview of inference methods. Appendix B contains an introduction to R that can be used as a starting point for readers unfamiliar with R. Finally, Appendix C contains a few statistical tables for those situations where a computer is not available to calculate the relevant tail probabilities or quantiles.

We used the book for a 7.5 ECTS course for a second/third year bachelor course with four lectures and four hours of exercises per week for eight weeks. In addition, three hours were used with one of the case exercises from Chapter 13: The students worked without instruction for two hours followed by one hour with discussion.

We are grateful to our colleagues at the Faculty of Life Sciences at the University of Copenhagen — in particular to Ib Skovgaard and Mats Rudemo, who authored an earlier textbook and who throughout the years have collected data from life science studies at the University of Copenhagen. Many thanks go to the students who participated in the "Statistical Data Analysis 1" course in 2008 and 2009 and helped improve the original manuscript with their comments.

Claus Thorn Ekstrøm
Helle Sørensen

Chapter 1

Description of samples and populations

Statistics is about making statements about a population from data observed from a representative sample of the population. A *population* is a collection of subjects whose properties are to be analyzed. The population is the complete collection to be studied; it contains all subjects of interest. A *sample* is a part of the population of interest, a subset selected by some means from the population. The concepts of population, sample, and statistical inference are illustrated in Figure 1.1.

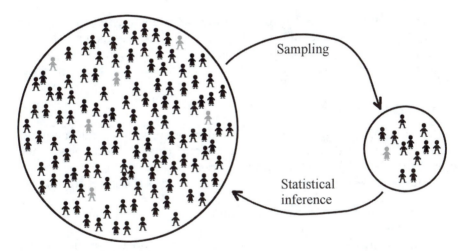

Figure 1.1: Population and sample. In statistics we sample subjects from a large population and use the information obtained from the sample to infer characteristics about the general population. Thus the upper arrow can be viewed as "sampling" while the lower arrow is "statistical inference".

A *parameter* is a numerical value that describes a characteristic of a population, while a *statistic* is a numerical measurement that describes a characteristic of a sample. We will use a statistic to infer something about a parameter.

Imagine, for example, that we are interested in the average height of a population of individuals. The average height of the population, μ, is a parameter, but it would be too expensive and/or time-consuming to measure the height of all individuals in the population. Instead we draw a random sample of, say, 12 individuals and measure the height of each of them. The average of those 12 individuals in the sample is our statistic, and if the sample is representative of the population and the sample is sufficiently large, we have confidence in using the statistic as an *estimate* or guess of the true population parameter μ. The rest of this book is concerned with methods for making inferences about population parameters based on sample statistics.

The distinction between population and sample depends on the context and the type of inference that you wish to perform. If we were to deduce the average height of the total population, then the 12 individuals are indeed a sample. If for some reason we were only interested in the height of these 12 individuals, and had no intention to make further inferences beyond the 12, then the 12 individuals themselves would constitute the population.

1.1   Data types

The type(s) of data collected in a study determine the type of statistical analysis that can be used and determine which hypotheses can be tested and which model we can use for prediction. Broadly speaking, we can classify data into two major types: categorical and quantitative.

1.1.1   Categorical data

Categorical data can be grouped into categories based on some qualitative trait. The resulting data are merely labels or categories, and examples include gender (male and female) and ethnicity (*e.g.*, Caucasian, Asian, African). We can further sub-classify categorical data into two types: *nominal* and *ordinal*.

Nominal. When there is no natural ordering of the categories we call the data nominal. Hair color is an example of nominal data. Observations are distinguished by name only, and there is no agreed upon ordering. It does not make sense to say "brown" comes before "blonde" or "gray". Other examples include gender, race, smoking status (smoker or non-smoker), or disease status.

Ordinal. When the categories may be ordered, the data are called ordinal variables. Categorical variables that judge pain (*e.g.*, none, little, heavy) or income (low-level income, middle-level income, or high-level income) are examples of ordinal variables. We know that households with low-level income earn less than households in the middle-level bracket, which in turn earn less than the high-level households. Hence there is an ordering to these categories.

It is worth emphasizing that the difference between two categories cannot be measured even though there exists an ordering for ordinal data. We know that high-income households earn more than low- and medium-income households, but not how much more. Also we cannot say that the difference between low- and medium-income households is the same as the difference between medium- and high-income households.

1.1.2 Quantitative data

Quantitative data are numerical measurements where the numbers are associated with a scale measure rather than just being simple labels. Quantitative data fall in two categories: *discrete* and *continuous*.

Discrete. Discrete quantitative data are numeric data variables that have a finite or countable number of possible values. When data represent counts, they are discrete. Examples include household size or the number of kittens in a litter. For discrete quantitative data there is a proper quantitative interpretation of the values: the difference between a household of size 9 and a household of size 7 is the same as the difference between a household of size 5 and a household of size 3.

Continuous. The real numbers are continuous with no gaps; physically measurable quantities like length, volume, time, mass, *etc.*, are generally considered continuous. However, while the data in theory are continuous, we often have some limitations in the level of detail that is feasible to measure. In some experiments, for example, we measure time in days or weight in kilograms even though a finer resolution could have been used: hours or seconds and grams. In practice, variables are never measured with infinite precision, but regarding a variable as continuous is still a valid assumption.

Categorical data are typically summarized using frequencies or proportions of observations in each category, while quantitative data typically are summarized using averages or means.

Example 1.1. Laminitis in cattle. Danscher et al. (2009) examined eight heifers in a study to evaluate acute *laminitis* in cattle after oligofructose overload. Due to logistic reasons, the 8 animals were examined at two different locations. For each of the 8 animals the location, weight, lameness score, and number of swelled joints were recorded 72 hours after oligofructose was administered. A slightly modified version of the data is shown in Table 1.1, and these data contain all four different types of data.

Location is a nominal variable as it has a finite set of categories with no specific ordering. Although the location is labeled with Roman numerals, they have no numeric meaning or ordering and might as well be renamed to "A" and "B". Weight is a quantitative continuous variable even though it is only reported in whole kilograms. The weight measurements are actual measurements on the continuous scale and taking differences between the

Table 1.1: Data on acute *laminitis* for eight heifers

Location	Weight (kg)	Lameness score	No. swelled joints
I	276	Mildly lame	2
I	395	Mildly lame	1
I	356	Normal	0
I	437	Lame	2
II	376	Lame	0
II	350	Moderately lame	0
II	331	Lame	1
II	331	Normal	0

values is meaningful. Lameness score is an ordinal variable where the order is defined by the clinicians who investigate the animals: normal, mildly lame, moderately lame, lame, and severely lame. The number of swelled joints is a quantitative discrete variable — we can count the actual number of swelled joints on each animal.   □

1.2 Visualizing categorical data

Categorical data are often summarized using tables where the frequencies of the different categories are listed. The *frequency* is defined as the number of occurrences of each value in the dataset. If there are only a few categories then tables are perfect for presenting the data, but if there are several categories or if we want to compare frequencies in different populations then the information may be better presented in a graph. A *bar chart* is a simple plot that shows the possible categories and the frequency of each category.

The *relative frequency* is useful if you want to compare datasets of different sizes; *i.e.*, where the number of observations in two samples differ. The relative frequency for a category is computed by dividing the frequency of that category by the total number of observations for the sample, n,

$$\text{relative frequency} = \frac{\text{frequency}}{n}.$$

The advantage of the relative frequency is that it is unrelated to the sample size, so it is possible to compare the relative frequencies of a category in two different samples directly since we draw attention to the relative proportion of observations that fall in each category.

A *segmented bar chart* presents the relative frequencies of the categories in a sample as a single bar with a total height of 100% and where the relative frequencies of the different categories are stacked on top of each other. The information content from a segmented bar chart is the same as from a plot of the relative frequency plot, but it may be easier to identify differences in the distribution of observations from different populations.

Example 1.2. Tibial dyschrondroplasia. *Tibial dyschondroplasia* (TD) is a disease that affects the growth of bone of young poultry and is the primary cause of lameness and mortality in commercial poultry. In an experiment 120 broilers (chickens raised for meat) were split into four equal-sized groups, each given different feeding strategies to investigate if the feeding strategy influenced the prevalence of TD:

Group A: feeding ad libitum.
Group B: 8 hours fasting at age 6-12 days.
Group C: 8 hours fasting at age 6-19 days.
Group D: 8 hours fasting at age 6-26 days.

At the time of slaughter the presence of TD was registered for each chicken. The following table lists the result:

	Group A	Group B	Group C	Group D	Total
TD present	21	7	6	12	46
TD absent	9	23	24	18	74

The difference between the relative frequencies of TD and non-TD chickens is very clear when comparing the four groups in Figure 1.2. □

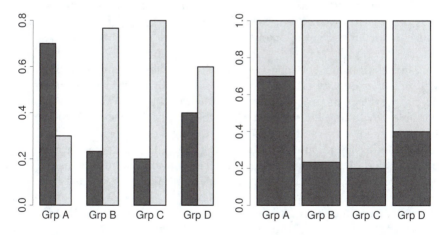

Figure 1.2: Relative frequency plot (left) for broiler chickens with and without presence of *tibial dyschondroplasia* (dark and light bars, respectively). The segmented bar plot (right) shows stacked relative frequencies of broiler chickens with and without *tibial dyschondroplasia* for the four groups.

1.3   Visualizing quantitative data

With categorical variables we can plot the frequency or relative frequency for each of the categories to display the data. The same approach works for discrete quantitative data when there are just a few different possible values, but frequency plots of each observed variable are not informative for quantitative continuous variables since there will be too many different "categories" and most of them will have a very low frequency. However, we can do something that resembles the frequency plot for categorical data by grouping the quantitative continuous data into bins that take the place of the categories used for categorical data. We then count the number of observations that fall into each bin and the resulting bins, and their related relative frequencies give the distribution of the quantitative continuous variable.

We can display the bin counts in a *histogram*, which is analogous to the bar chart, and the histograms allows us to graphically summarize the distribution of the dataset; *e.g.*, the center, spread, and number of modes in the data. The *relative frequency histogram* can be used to compare the distributions from different populations since the relative frequency histogram has the inherent feature that areas for each bar in the histogram are proportional to the probability that an observation will fall in the range covered by the bin. The shape of the relative frequency histogram will be identical to the shape of the histogram and only the scale will differ. Note that if for some reason the bin widths are not equal, then the areas of the histogram bars will no longer be proportional to the frequencies of the corresponding categories simply because wider bins are more likely to contain more observations than smaller bins.

The relationship between two quantitative variables can be illustrated with a *scatter plot*, where the data points are plotted on a two-dimensional graph. Scatter plots provide information about the relationship between the variables, including the strength of the relationship, the shape (whether it is linear, curved, or something else), and the direction (positive or negative), and make it easy to spot extreme observations. If one of the variables can be controlled by the experimenter then that variable might be considered an explanatory variable and is usually plotted on the x-axis, whereas the other variable is considered a response variable and is plotted on the y-axis. If neither the one or the other variable can be interpreted as an explanatory variable then either variable can be plotted on either axis and the scatter plot will illustrate only the relationship but not the causation between the two variables.

Example 1.3. Tenderness of pork. Two different cooling methods for pork meat were compared in an experiment with 18 pigs from two different groups: low or high pH content. After slaughter, each pig was split in two and one side was exposed to rapid cooling while the other was put through

Table 1.2: Tenderness of pork from different cooling methods and pH levels

Pig no.	pH group	Tunnel cooling	Rapid cooling
73	high	8.44	8.44
74	high	7.11	6.00
75	high	6.00	5.78
76	high	7.56	7.67
77	low	7.22	5.56
78	high	5.11	4.56
79	low	3.11	3.33
80	high	8.67	8.00
81	low	7.44	7.00
82	low	4.33	4.89
83	low	6.78	6.56
84	low	5.56	5.67
85	low	7.33	6.33
86	low	4.22	5.67
87	high	5.78	7.67
94	low	5.78	5.56
95	low	6.44	5.67
96	low	8.00	5.33

a cooling tunnel. After the experiment, the tenderness of the meat was measured. Data are shown in Table 1.2 and are from a study by Møller et al. (1987).

Figure 1.3 shows the histograms and relative frequency histograms for the high- and low-pH groups. Notice that the shapes for the low- and high-pH groups do not change from the histograms to the relative frequency histograms. The relative frequency histograms make it easier to compare the distributions in the low- and high-pH groups since the two groups have different numbers of observations.

Figure 1.4 shows the relationship of tenderness between the rapid and tunnel cooling methods for the combined data of low- and high-pH groups. Scatter plots are extremely useful as tools to identify relationships between two quantitative continuous variables. □

1.4 Statistical summaries

Categorical data are best summarized in tables like the one shown in Example 1.2 on p. 5. Quantitative data do not have a fixed set of categories, so

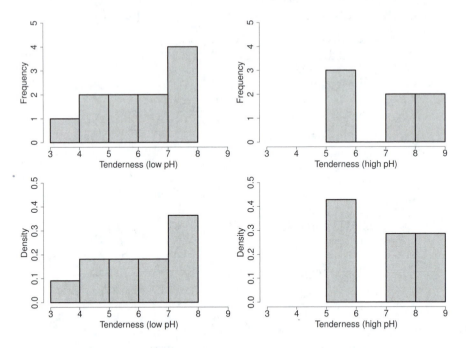

Figure 1.3: Histograms (top row) and relative frequency histograms (bottom row) for tunnel cooling of pork for low- and high-pH groups.

representing those in a table is infeasible. One work-around for this problem would be to make bins and present the frequency of each bin in the same way we make the histograms. However, information is lost by grouping data in bins and the number of bins and bin widths may have a huge influence on the resulting table. Instead it may be desirable to identify certain summary statistics that capture the main characteristics of the distribution.

1.4.1   Center and dispersion of a distribution

Often it is desirable to have a single number to describe the values in a dataset, and this number should be representative of the data. It seems reasonable that this representative number should be close to the "middle" of the data such that it best describes all of the data, and we call any such number a measure of *central tendency.*

Very different sets of data can have the same central tendency. Thus a single representative number is insufficient to describe the distribution of the data, and we are also interested in how closely the central tendency represents the values in the dataset. The *dispersion* represents how much the observations in a dataset differ from the central tendency; *i.e.,* how widely the data are "spread out".

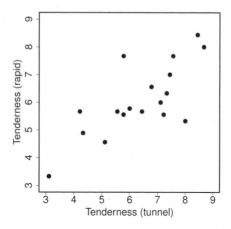

Figure 1.4: Scatter plot of tenderness for rapid cooling and tunnel cooling.

In the following we let $y_1, \ldots, y_n$ represent independent*, quantitative observations in a sample of size n from some population.

Median and inter-quartile range

We can order the observations $y_1, \ldots, y_n$ from lowest to highest and we use the following notation to represent the set of ordered observations: $y_{(1)}, \ldots, y_{(n)}$. Thus $y_{(1)}$ is the smallest value of $y_1, \ldots, y_n$, $y_{(2)}$ is the second smallest, *etc.*

The *median* of n numbers is a measure of the central tendency and is defined as the middle number when the numbers are ordered. If n is even then the median is the average of the two middle numbers:

$$\text{Median} = \begin{cases} y_{\left(\frac{n+1}{2}\right)} & \text{if } n \text{ is odd} \\ \frac{1}{2}\left[y_{(n/2)} + y_{(n/2+1)}\right] & \text{if } n \text{ is even} \end{cases} \tag{1.1}$$

The median may be used for both quantitative and ordinal categorical data.

The *range* is one measure of dispersion and it is defined as the highest value in the dataset minus the lowest value in the dataset:

$$\text{Range} = y_{(n)} - y_{(1)} \tag{1.2}$$

One weakness of the range is that it uses only two values in its calculation and disregards all other values. Two sets of data could have the same range but be "spread out" in very different fashions. Consider the following three datasets:

---

*Independence is discussed more closely on p. 78 and in Chapter 9. Roughly speaking, independence means that the observations do not provide any information about each other — *e.g.*, even if the previous observation is larger than the mean, there is no reason to believe that the next observation will be larger than the mean.

| Dataset 1 : 14, 14, 14, 14, 14, 14, 34 |
| Dataset 2 : 14, 16, 19, 22, 26, 30, 34 |
| Dataset 3 : 14, 14, 14, 34, 34, 34, 34 |

The medians for the three datasets are 14, 22, and 34, respectively. The range for each set is $34 - 14 = 20$, but the three sets are very different. The first set is not very dispersed, with the exception of the single value of 34. The second dataset has values that are more or less evenly dispersed between 14 and 34, while the last set has 3 values of 14 and 4 values of 34 with no values in between. Clearly the range does not provide a lot of the information about the spread of the observations in the dataset; *i.e.*, how far away from the center that typical values are located.

Another measure of dispersion that tries to capture some of the spread of the values is the *inter-quartile range* (IQR). The inter-quartile range is calculated as follows: We remove the top 25% and the bottom 25% of all observations and then calculate the range of the remaining values. We denote the first and third quartile as $Q1$ and $Q3$, respectively.

$$IQR = Q3 - Q1. \tag{1.3}$$

The advantage of the IQR over the range is that the IQR is not as sensitive to extreme values because the IQR is based on the middle 50% of the observations.

Generally, we can identify separate cut-off points taken at regular intervals (called *quantiles*) if we order the data according to magnitude. In the following we divide the ordered data into 100 essentially equal-sized subsets such that the xth quantile is defined as the cut-off point where x% of the sample has a value equal to or less than the cut-off point. For example, the 40th quantile splits the data into two groups containing, respectively, 40% and 60% of the data. For a finite dataset it may be impossible to obtain the exact partitions of the data for a given quantile, and instead we round up and define the xth quantile as the smallest ranked observation such that at least x% of the data have values equal or below the xth quantile[†].

The first *quartile* is defined as the 25th quantile and the third quartile is defined as the 75th quantile. The median (1.1) corresponds to the 50th quantile, so the first quartile, the median, and the third quartile split the data into 4 groups of equal size.

Boxplot

A *boxplot* (also called a *box-and-whiskers plot*) summarizes the data graphically by plotting the following five summaries of the data: minimum, first

---

[†]Several definitions of quantiles exist for finite datasets, and while the various definitions may lead to slightly different values, the differences tend to be small with large datasets. The median value for a finite dataset with an even number of observations is taken as the average value of the two middle observations even though our definition of quantile provides a slightly different value — the higher of the two middle observations.

quartile, median, third quartile, and maximum, as shown below for dataset 2 listed above:

The middle 50% of the data are represented by a box and the median is shown as a fat line inside the box. Two whiskers extend from the box to the minimum and the maximum value. The five summaries used in the boxplot present a nice overview of the distribution of the observations in the dataset, and the visualization makes it easy to determine if the distribution is symmetric or skewed.

The IQR is sometimes used to identify *outliers* — observations that differ so much from the rest of the data that they appear extreme compared to the remaining observations. As a rule-of-thumb, an outlier is an observation that is smaller than 1.5·IQR under the first quartile or larger than 1.5·IQR over the third quartile; *i.e.*, anything outside the following interval:

$$[Q1 - 1.5 \cdot IQR; Q3 + 1.5 \cdot IQR]. \tag{1.4}$$

It is often critical to identify outliers and extreme observations as they can have an enormous impact on the conclusions from a statistical analysis. We shall later see how the presence or absence of outliers is used to check the validity of statistical models.

Sometimes data are presented in a *modified boxplot*, where outliers are plotted as individual points and where the minimum and maximum summaries are replaced by the smallest and largest observations that are still within the interval $[Q1 - 1.5 \cdot IQR; Q3 + 1.5 \cdot IQR]$. That enables the reader to determine if there are any extreme values in the dataset (see Example 1.4).

Example 1.4. Tenderness of pork (continued from p. 6). If we order the 18 measurements for tunnel cooling from the pork tenderness data according to size we get

$$3.11 \quad 4.22 \quad 4.33 \quad 5.11 \quad 5.56 \quad 5.78 \quad 5.78 \quad 6.00 \quad 6.44$$
$$6.78 \quad 7.11 \quad 7.22 \quad 7.33 \quad 7.44 \quad 7.56 \quad 8.00 \quad 8.44 \quad 8.67$$

such that $y_{(1)} = 3.11, y_{(2)} = 4.22$, *etc.* There is an even number of observations in this sample, so we should take the average of the middle two observations to calculate the median; *i.e.*,

$$\text{Median} = \frac{6.44 + 6.78}{2} = 6.61.$$

The range of the observations is $8.67 - 3.11 = 5.56$.

Since there are 18 observations in the dataset, we have that the lower quartile should be observation $18 \cdot 1/4 = 4.5$. We round that value up so the lower quartile corresponds to observation 5; *i.e.*, $Q1 = 5.56$. Likewise, the upper quartile is observation $18 \cdot 3/4 = 13.5$, which we round up to 14, so $Q3 = 7.44$. Thus, the inter-quartile range is $7.44 - 5.66 = 1.88$. The modified boxplots for both tunnel and rapid cooling are shown below.

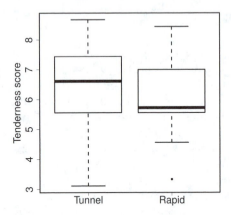

From the modified boxplots we see that the distribution of values for tunnel cooling is fairly symmetric whereas the distribution of the observations from rapid cooling is highly skewed. By placing boxplots from two samples next to each other we can also directly compare the two distributions: the tenderness values from tunnel cooling generally appear to be higher than the values from rapid cooling although there are a few very small values for tunnel cooling. We can also see from the boxplot that there is a single outlier for rapid cooling. It is worth checking the dataset to see if this is indeed a genuine observation. □

The mean and standard deviation

The mean is another measure of the center for quantitative data. Let us start by introducing some notation. Let $y_1, \ldots, y_n$ denote the quantitative observations in a sample of size n from some population. The *sample mean* is defined as

$$\bar{y} = \frac{\sum_{i=1}^{n} y_i}{n} \tag{1.5}$$

and is calculated as a regular average: we add up all the observations and divide by the number of observations. The *sample standard deviation* is a measure of dispersion for quantitative data and is defined as

$$s = \sqrt{\frac{\sum_{i=1}^{n} (y_i - \bar{y})^2}{n-1}}. \tag{1.6}$$

Loosely speaking, the standard deviation measures the "average deviation from the mean" observed in the sample; *i.e.*, the standard deviation measures how far away from the center we can expect our observations to be on average.

The *sample variance* is denoted s^2 and is simply the sample standard deviation squared:

$$s^2 = \frac{\sum_{i=1}^{n}(y_i - \bar{y})^2}{n-1}. \tag{1.7}$$

The mean and standard deviation provide more information than the median and inter-quartile range because their values utilize information from all the available observations. The mean, median, standard deviation, and inter-quartile range have the same units as the values from which they are calculated.

Example 1.5. Tenderness of pork (continued from p. 6). The mean of the tunnel cooling data is

$$\bar{y} = \frac{3.11 + 4.22 + 4.33 + 5.11 + \cdots + 7.56 + 8.00 + 8.44 + 8.67}{18} = 6.382.$$

The standard deviation becomes

$$s = \sqrt{\frac{(3.11 - 6.382)^2 + (4.22 - 6.382)^2 + \cdots + (8.67 - 6.382)^2}{18 - 1}} = 1.527.$$

Thus the mean tenderness for tunnel cooling is 6.382 and the corresponding standard deviation is 1.527 units on the tenderness scale. □

Looking at formula (1.7) for the variance, we see that it is roughly the average of the squared deviations. It would be the average if we divided the sum in (1.7) by n instead of $n-1$. The variance of the *population* (not the sample, but population) is $\sigma = \sum_i (y_i - \mu)^2 / n$, which requires knowledge about the true mean of the population, μ. We could calculate this variance if full information about the total population was available, but in practice we need to replace μ with our "best guess" of μ, which is $\bar{y}$. The sample mean $\bar{y}$ depends on the observations from the sample and will vary from sample to sample, so it is not a perfectly precise estimate of μ. We divide by $n-1$ in (1.6) and (1.7) in order to take this uncertainty about the estimate of μ into account. It can be shown that if we divide the sample variance by n we tend to underestimate the true population variance. This is remedied by dividing by $n-1$ instead. The reason is that the sum of the deviations is always zero (per construction). Hence, when the first $n-1$ deviations have been calculated, then the last deviation is given automatically, so we essentially have only $n-1$ "observations" to provide information about the deviations.

The sample mean and sample standard deviation have some nice properties if the data are transformed as shown in Infobox 1.1.

Infobox 1.1: Sample mean and standard deviation of linearly transformed data

Let $\bar{y}$ and s be the sample mean and sample standard deviation from observations $y_1, \ldots, y_n$ and let $y_i' = c \cdot y_i + b$ be a linear transformation of the y's with constants b and c. Then $\bar{y}' = c \cdot \bar{y} + b$ and $s' = |c| \cdot s$.

The results presented in Infobox 1.1 can be proved by inserting the transformed values y_i' in the formulas (1.5) and (1.6). The first part of the result tells us that if we multiply each observation, y_i, by a constant value, c, and add a constant, b, then the mean of the new observations is identical to the original mean multiplied by the factor c and added b. Thus, if we measured, say, height in centimeters instead of meters, the mean would be exactly 100 times as big for the centimeters as it would be for the measurements in meters. In addition, the standard deviation of the transformed variables, $y_1', \ldots, y_n'$, is identical to the standard deviation of the original sample multiplied by the factor c. The standard deviation of our measurements in centimeters is going to be exactly 100 times as large as the standard deviation in meters.

This is a nice property since it means that simple linear transformation will not have any surprising effects on the mean and standard deviation.

Mean or median?

The mean and median are both measures of the central tendency of a dataset, but the two measures have different advantages and disadvantages.

The median partitions the data into two parts such that there is an equal number of observations on either side of the median or that the two areas under the histogram have the same size — regardless of how far away the observations are from the center. The mean also partitions the data into two parts, but it uses the observed values to decide where to split the data. In a sense, a histogram balances when supported at the mean since both area size *and* distance from center are taken into account. Just like a seesaw, a single value far away from the center will balance several values closer to the center, and hence the percentage of observations on either side of the mean can differ from 50%.

One advantage of the median is that it is not influenced by extreme values in the dataset. Only the two middle observations are used in the calculation, and the actual values of the remaining observations are not used. The mean on the other hand is sensitive to all values in the dataset since every observation in the data affects the mean, and extreme observations can have a substantial influence on the mean value.

The mean value has some very desirable mathematical properties that make it possible to prove theorems, and useful results within statistics and inference methods naturally give rise to the mean value as a parameter estimate. It is much more problematic to prove mathematical results related to the median even though it is more robust to extreme observations. Generally the mean is used for symmetric quantitative data, except in situations with extreme values, where the median is used. The mean and standard deviation may appear to have limited use since they are only really meaningful to symmetric distributions. However, the central limit theorem from probability theory proves that sample means and estimates can indeed be considered to be symmetric regardless of their original distribution provided that the sample size is large (see Section 4.4).

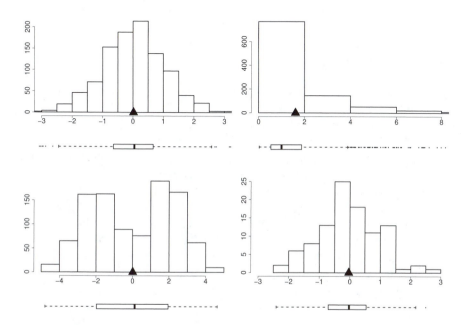

Figure 1.5: Histograms, boxplots, and means (▲) for 4 different datasets. The lower right dataset contains 100 observations — the remaining three datasets all contain 1000 observations.

Figure 1.5 shows histograms, modified boxplots, and means for four different distributions. The top-left distribution has first and third quartiles that are about the same distance from the median and the same is true for the whiskers. This in combination with the outline of the histogram all indicates that the distribution is symmetric, and we also see that the median and the mean are almost identical. The top-right distribution in Figure 1.5 is highly skewed, and we see that there is a substantial difference between the mean value and the median. Notice also from the modified boxplot that several observations are outside the outlier interval (1.4) and are plotted as points. This does not necessarily mean that we have so many extreme observations in this case since the distribution is highly skewed. The outlier interval (1.4) is defined by the IQR, and when the distribution is highly skewed the outlier interval will have difficulty identifying outliers in one direction (the non-skewed direction, or towards the left in Figure 1.5) while it may identify too many outliers in the other direction (the skewed direction, or towards the right in Figure 1.5). The bottom-left distribution is bimodal and is clearly symmetric and the bottom-right distribution is also symmetric and resembles the distribution in the upper-left panel but is a bit more ragged.

1.5   What is a probability?

Most people have an intuitive understanding of what a "probability" is, and we shall briefly cover the concept of probabilities in this section. For a more mathematical definition of probabilities and probability rules see Chapter 9.

We think of the *probability* of a random event as the limit of its relative frequency in an infinitely large number of experiments. We have already used the relative frequency approach earlier, when we discussed presentation of categorical and continuous data. When a random experiment is performed multiple times we are not guaranteed to get the exact same result every time. If we roll a die we do not get the same result every time, and similarly we end up with different daily quantities of milk even if we treat and feed each cow the same way every day.

In the simplest situation we can register whether or not a random event occurs; for example, if a single die shows an even number. If we denote this event (*i.e.*, rolling an even number with a single die) A and we let n_A be the number of occurrences of A out of n rolls, then n_A/n is the relative frequency of the event A. The relative frequency stabilizes as n increases, and the probability of A is then the limit of the relative frequency as n tends towards infinity.

Example 1.6. Throwing thumbtacks. A brass thumbtack was thrown 100 times and it was registered whether the pin was pointing up or down towards the table upon landing (Rudemo, 1979). The results are shown in Table 1.3, where '1' corresponds to "tip pointing down" and '0' corresponds to "tip pointing up". The relative frequency of the event "pin points down" as a function of the number of throws is shown in Figure 1.6.

We see from Figure 1.6 that the relative frequency varies highly when n is low but that it stabilizes on a value around 0.6 as n tends towards infinity. Hence we conclude that the probability of observing a pin pointing down when throwing a thumbtack is around 0.6 or 60%.   □

Table 1.3: Thumbtacks: 100 throws with a brass thumbtack. 1= pin points down, 0= pin points up

11001	10100	10110	01110	10011
00001	11010	11011	10011	10111
01011	11010	01001	00111	10011
11011	00111	10100	10011	11010

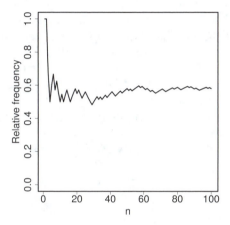

Figure 1.6: Thumbtack throwing. Relative frequency of the event "pin points down" as the number of throws increases.

1.6   R

In Example 1.3 on p. 6 we had information on tunnel cooling for 18 pigs for two different pH groups. We will use that dataset to illustrate various R functions for visualizing data and calculating summary statistics for quantitative data. Categorical data are illustrated with the *tibial dyschrondroplasia* data from Example 1.2.

We start by entering the two datasets into R:

```
> # First data on tibial dyschrondroplasia
> m <- matrix(c(21, 9, 7, 23, 6, 24, 12, 18), ncol=4)

> # Then data on cooling methods
> tunnel <- c(8.44, 7.11, 6.00, 7.56, 7.22, 5.11, 3.11, 8.67,
+ 7.44, 4.33, 6.78, 5.56, 7.33, 4.22, 5.78, 5.78, 6.44, 8.00)
> rapid <- c(8.44, 6.00, 5.78, 7.67, 5.56, 4.56, 3.33, 8.00,
+ 7.00, 4.89, 6.56, 5.67, 6.33, 5.67, 7.67, 5.56, 5.67, 5.33)
```

```
> ph <- c("hi", "hi", "hi", "hi", "lo", "hi", "lo", "hi", "lo",
+ "lo", "lo", "lo", "lo", "lo", "hi", "lo", "lo", "lo")
```

1.6.1   Visualizing data

Categorical data are visualized as bar plots, which are produced by the barplot() function in R. When the first argument to barplot() is a vector, then the default plot consists of a sequence of bars with heights corresponding to the elements of the vector. If the first argument is a matrix, then each bar is a segmented bar plot where the values in each column of the matrix correspond to the height of the elements of the stacked bar.

If we prefer the stacked bar plot to show relative frequencies then we need to divide each column in the matrix by the column sum. The prop.table() function converts the matrix to the relative frequencies given either the row or column sums. The second argument to prop.table() determines if the elements of the table are relative to the row sums (margin=1) or the column sums (margin=2). The following lines produce the two plots shown in Figure 1.2 on p. 5 and use the options besides=TRUE and names= to barplot().

```
> relfrq <- prop.table(m, margin=2)
> relfrq
       [,1]        [,2] [,3] [,4]
[1,]   0.7 0.2333333   0.2   0.4
[2,]   0.3 0.7666667   0.8   0.6
> # Make juxtaposed barplot
> barplot(relfrq,   beside=TRUE,
+ names=c("Grp A", "Grp B", "Grp C", "Grp D"))
> # Stacked relative barplot with labels added
> barplot(relfrq,   names=c("Grp A", "Grp B", "Grp C", "Grp D"))
```

We use a simple scatter plot to illustrate the relationship between two quantitative variables. The plot() function is used to produce a scatter plot, and we can add additional information to the plot by specifying the labels for the x-axis and the y-axis with the xlab and ylab options to plot(). The following command will generate the scatter plot seen in Figure 1.4 on p. 9:

```
> plot(tunnel, rapid, xlab="Tenderness (tunnel)",
+ ylab="Tenderness (rapid)")
```

plot() is a generic function in R and the output depends on the number and type of objects that are provided as arguments to the function. A scatter plot is produced when two numeric vectors are used in plot(). If only a single numeric vector is used as an argument, plot(tunnel), then all the observations for that vector are plotted with the observation number on the x-axis and the corresponding value on the y-axis.

Histograms and relative frequency histograms are both produced with the hist() function. By default the hist() function automatically groups the quantitative data vector into bins of equal width and produces a frequency histogram. We can force the hist() function to make a frequency plot or a relative frequency plot by specifying either the freq=TRUE or the freq=FALSE option, respectively. The following two commands produce the upper left histogram and lower left relative frequency histogram seen in Figure 1.3 on p. 8 and use the main= option to include a title.

```
> hist(tunnel[ph=="lo"], xlab="Tenderness (low pH)",
+ main="Histogram")              # Add title to plot
> hist(tunnel[ph=="lo"], xlab="Tenderness (low pH)",
+ freq=FALSE, main="Histogram") # Force relative frequency plot
```

The number of bins is controlled by the breaks= option to hist(). If the breaks= option is not entered, then R will try to determine a reasonable number of bins. If we include an integer value for the breaks= option, then we fix the number of bins.

```
> # Use the breaks option to specify the number of bins
> # regardless of the size of the dataset
> hist(tunnel[ph=="lo"], xlab="Tenderness (low pH)",
+ breaks = 8, main="Histogram")
```

Horizontal and vertical boxplots are produced by the boxplot() function. By default, R creates the modified boxplot as described on p. 11.

```
> boxplot(tunnel)
```

The standard boxplot where the whiskers extend to the minimum and maximum value can be obtained by setting the range=0 option to boxplot(). In addition, the boxplot can be made horizontal by including the horizontal=TRUE option.

```
> # Horiz. boxplot with whiskers from minimum to maximum value
> boxplot(tunnel, range=0, horizontal=TRUE)
```

boxplot() is a generic function just like plot() and changes the output based on the type and number of arguments. If we provide more than a single numeric vector as input to boxplot(), then parallel boxplots will be produced. Often it is easier to compare the distribution among several vectors if they are printed next to each other. The command below will produce the figure seen in Example 1.4 on p. 11.

```
> # Parallel boxplots
> boxplot(tunnel, rapid, names=c("Tunnel", "Rapid"))
```

Note that we specify the names of the different vectors. If we do not specify the names then R will label each boxplot sequentially from 1 and upwards.

1.6.2 Statistical summaries

We can use the mean(), median(), range(), IQR(), sd(), and var() functions to calculate the mean, median, range, inter-quartile range, standard deviation, and variance, respectively, for the vector of measurements.

```
> mean(tunnel)        # Calculate the mean value
[1] 6.382222
> median(tunnel)      # Calculate the median
[1] 6.61
> range(tunnel)       # Calculate the range. Low and high values
[1] 3.11 8.67
> IQR(tunnel)         # Calculate the inter-quartile range
[1] 1.7975
> sd(tunnel)          # Calculate the standard deviation (SD)
[1] 1.527075
> var(tunnel)         # Calculate the variance
[1] 2.331959
> sd(tunnel)**2       # The variance equals the SD squared
[1] 2.331959
```

Quantiles can be calculated using the quantile() function. By default, R uses a slightly different method to calculate quantiles, and to get the definition we have presented in the text we should use the type=1 option.

```
> quantile(tunnel, 0.25, type=1) # 25th quantile of tunnel data
  25%
5.56
> quantile(tunnel, c(0.10, 0.25, 0.60, 0.95), type=1)
  10%  25%  60%  95%
4.22 5.56 7.11 8.67
```

1.7 Exercises

1.1 Data types. For each of the following experiments you should identify the variable(s) in the study, the data type of each variable, and the sample size.

1. For each of 10 beetles, a biologist counted the number of times the beetle fed on a disease-resistant plant during a 4-hour period.

2. In a nutritional study, 40 healthy males were measured for weight and height as well as the weight of their food intake over a 24-hour period.

3. Seven horses were included in a 2-week study. After the first week a veterinarian measured the heart rate of each of the horses after an identification chip was inserted in its neck. At the end of the second week the veterinarian measured the heart rate after branding the horses with a hot iron.

4. The birth weight, number of siblings, mother's race, and age were recorded for each of 85 babies.

1.2 Blood pressure. Consider the following data on diastolic blood pressure (measured in mmHg) for 9 patients:

Patient	1	2	3	4	5	6	7	8	9
Blood pressure	96	119	119	108	126	128	110	105	94

1. Determine the median, the range, and the quartiles.

2. Determine the inter-quartile range.

3. Construct a boxplot of the data.

4. Calculate the mean, the standard deviation, and the variance.

5. What are the units for the mean, standard deviation, and variance?

6. How will the mean change if we add 10 mmHg to each of the measurements? How will this change the standard deviation and the variance?

7. Do you think the mean will increase, decrease, or stay roughly the same if we measure the diastolic blood pressure of more individuals? How do you think more individuals will influence the standard deviation?

1.3 Distribution of mayflies. To study the spatial distribution of mayflies (*Baetis rhodani*), researchers examined a total of 80 random 10 centimeter square test areas. They counted the number of mayflies, Y, in each square as shown:

Mayflies	4	5	6	7	8	9	10
Frequency	2	2	5	7	10	9	10
Mayflies	11	12	13	14	15	16	17
Frequency	10	8	6	4	4	2	1

1. The mean and standard deviation of Y are $\bar{y} = 10.09$ and $s = 2.96$. What percentage of the observations are within

   (a) 1 standard deviation of the mean?

   (b) 2 standard deviations of the mean?

2. Determine the total number of mayflies in all 80 squares. How is this number related to the $\bar{y}$?

3. Determine the median of the distribution.

1.4 Distribution shapes. Different words are often used to describe the overall shape of a distribution. Determine which of the following 6 phrases best matches the histograms seen in the figure below: symmetrical, bimodal and symmetrical, skewed right, skewed left, bimodal and skewed right, and uniform and symmetrical.

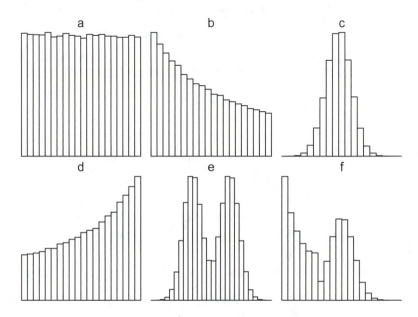

1.5 Design of experiments. Assume that it is of interest to compare the milk yield from cows that have received two different feeding strategies (A and B) to determine if the feeding strategies lead to systematic differences in the yields. Discuss the advantages and disadvantages of the following four design strategies and whether or not they can be used to investigate the purpose of the experiment.

1. Feed one cow after plan A and one cow after plan B.

2. 100 cows from one farm are fed according to plan A while 88 cows from another farm are fed according to plan B.

3. Ten cows are selected at random from a group of 20 cows and fed according to plan A while the remaining 10 cows are fed according to plan B.

4. For each of 10 twin pairs, a cow is chosen at random and fed from plan A while the other cow is fed according to plan B.

1.6 Comparison of boxplots. Consider the following comparison between the calorie content data from 10 common sandwiches from McDonald's and 9 common sandwiches from Burger King found on their respective web pages.

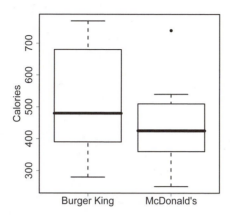

Describe the distributions (*i.e.*, the shape, center, and spread of each distribution) and how they compare to one another.

1.7 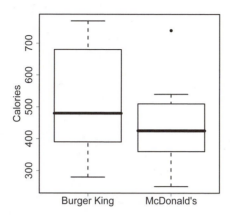 **Histograms and boxplots.** Use the following data from Rudemo (1979) on the lengths in millimeters of 20 cones from conifer (*Picea abies*).

125.1	114.6	99.3	119.1	109.6
102.0	104.9	109.6	134.0	108.6
120.3	98.7	104.2	91.4	115.3
107.7	97.8	126.4	104.8	118.8

1. Read the data into R.

2. Calculate the standard deviation and variance of the cone lengths. What is the relationship between the two numbers?

3. Use `hist()` to plot a histogram of the cone lengths. Use `boxplot()` to plot a modified boxplot. What can you discern about the cone lengths from the two figures?

4. Construct a new vector with the name `conelenm` which contains the same lengths but now measured in centimeters. How will the mean and standard deviation change after we change the units?

5. Plot a histogram and boxplot of the transformed data (use `hist()`). You can choose if you want to change the intervals on

the x-axis or if you want frequencies or relative frequencies on the y-axis. What can you say about the shape of the distribution? Is it symmetric, skewed to the right, or skewed to the left?

1.8 Which distribution? Consider the following three boxplots (1, 2, and 3) and histograms (x, y, and z). Which histogram goes with each boxplot? Explain your answer.

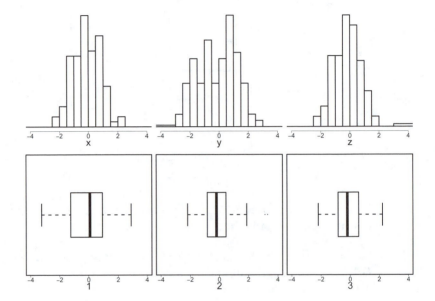

Chapter 2

Linear regression

The purpose of a data analysis is often to describe one variable as a function of another variable. The functional relationship between the two variables may in some situations be based on a well-known theoretical hypothesis, and in other situations we may have no prior knowledge about the relationship but would like to use the observed data to identify a relationship empirically.

Simple linear regression attempts to model the relationship between two quantitative variables, x and y, by fitting a linear equation to the observed data. The linear equation can be written as

$$y = \alpha + \beta \cdot x$$

where α (also called the *intercept*) is the value of y when $x = 0$ and β is the *slope* (*i.e.*, the change in y for each unit change in x); see Figure 2.1.

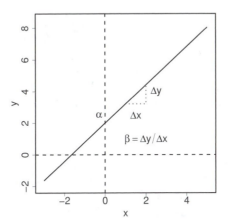

Figure 2.1: The straight line.

When we want to model the relationship between two variables we assume that one variable is the *dependent* variable (y in the linear equation) while the other is an *explanatory variable* (x in the regression formula). We want to model y as a linear function of x in the hope that information about x will give us some information about the value of y; *i.e.*, it will "explain" the value of y, at least partly. For example, a modeler might use a linear regression model to relate the heart beat frequency of frogs to the body temperature or to relate the tenderness of pig meat to the length of the meat fibers.

Example 2.1. Stearic acid and digestibility of fat. Jørgensen and Hansen (1973) examined the digestibility of fat with different levels of stearic acid. The average digestibility percent was measured for nine different levels of stearic acid proportion. Data are shown in the table below, where x represents stearic acid and y is digestibility measured in percent.

x	29.8	30.3	22.6	18.7	14.8	4.1	4.4	2.8	3.8
y	67.5	70.6	72.0	78.2	87.0	89.9	91.2	93.1	96.7

The data are plotted in Figure 2.2 together with the straight line defined by $y = 96.5334 - 0.9337 \cdot x$. In Section 2.1 it will become clear why these values are used for the parameters in the model.

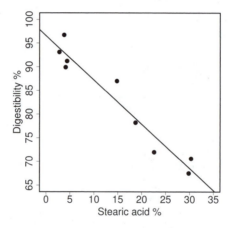

Figure 2.2: Digestibility of fat for different proportions of stearic acid in the fat. The line is $y = -0.9337 \cdot x + 96.5334$.

Figure 2.2 shows that the relationship between stearic acid and digestibility appears to scatter around a straight line and that the line plotted in the figure seems to capture the general trend of the data.

We now have a model (the straight line) for the data that enables us to give statements about digestibility even for levels of stearic acid that were *not* part of the original experiment as long as we assume that the relationship between stearic acid and digestibility can indeed be modeled by a straight line. Based on our "model" we would, for example, expect a digestibility of around 87% if we examine fat with a stearic acid level of 10%. □

For each value of x the linear equation gives us the corresponding y-value. However, most real-life data will never show a perfect functional relationship between the dependent and the explanatory variables — just as we saw in Example 2.1. Despite the linear relationship between digestibility and stearic acid, it is obvious that a straight line will never fit all the observations perfectly. Some of the observations are above the line and some are below, but the general trend matches a straight line as seen in Figure 2.2.

2.1 Fitting a regression line

Fitting a regression line means identifying the "best" line; *i.e.*, the optimal parameters to describe the observed data. What we mean by "best" will become clear in this section.

Let $(x_i, y_i), i = 1, \ldots, n$ denote our n pairs of observations and assume that we somehow have "guesstimates" of the two parameters, $\hat{\alpha}$ and $\hat{\beta}$, from a linear equation used to model the relationship between the x's and the y's. Notice how we placed "hats" over α and β to indicate that the values are not necessarily the true (but unknown) values of α and β but *estimates*. Our model for the data is given by the line

$$y = \hat{\alpha} + \hat{\beta} \cdot x.$$

For any x, we can use this model to predict the corresponding y-value. In particular, we can do so for each of our original observations, $x_1, \ldots, x_n$, to find the *predicted values*; *i.e.*, the y-values that the *model* would expect to find:

$$\hat{y}_i = \hat{\alpha} + \hat{\beta} \cdot x_i.$$

We can use these predicted values to evaluate how well the model fits to the actual observed values. This is achieved by looking at the *residuals*, which are defined as follows:

$$r_i = y_i - \hat{y}_i. \tag{2.1}$$

The residuals measure how far away each of our actual observations (y_i's) are from the expected value given a specific model (the straight line in this case). We can think of the residuals as the rest or remainder of the observed y's that are not explained by the model. Clearly, we would like to use a model that provides small residuals because that means that the values predicted by the model are close to our observations.

Example 2.2. Stearic acid and digestibility of fat (continued from p. 26). Let us for now assume that we have eyeballed the data and have found that a line defined by the parameters

$$\hat{\alpha} = 96.5334 \qquad \hat{\beta} = -0.9337$$

provides a good straight line to describe the observed data. We can then calculate the predicted value for each observed x; *e.g.*,

$$\hat{y}_1 = 96.5334 - 0.9337 \cdot 29.8 = 68.709$$

This value is slightly higher than the observed value of 67.5, and the residual for the first observation is

$$r_1 = 67.5 - 68.709 = -1.209.$$

Figure 2.3 shows a graphical representation of the residuals for all nine levels of stearic acid. □

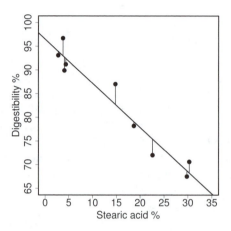

Figure 2.3: Residuals for the dataset on digestibility and stearic acid. The vertical lines between the model (the straight line) and the observations are the residuals.

Note that the residuals, r_i, measure the vertical distance from the observation to the fitted line and that positive, negative, and zero residuals correspond to observations that are above, below, and exactly on the regression line, respectively. Until now we have just assumed that it was possible to identify a straight line that would fit our observed data. Two researchers may, however, have different opinions on which regression line should be used to model a dataset; *e.g.*, one researcher suggests that $y = 1.8x + 2$ best describes the data while the other proposes $y = 1.7x + 2.3$. From our discussion so far it should be clear that it would be desirable to have a regression line where

- the residuals are small. That indicates that the regression line is close to the actual observations.

- the residual sum is zero. That means the observations are spread evenly above and below the line. If the residual sum is non-zero we could always change the intercept of the model such that the residual sum would be zero.

Different lines can yield a residual sum of zero, as can be seen in Figure 2.4 where two different regression lines are plotted for the stearic acid dataset. The solid line is defined by $y = 96.5334 - 0.9337 \cdot x$ while the dashed line is defined as $0.6 \cdot x + 74.15$. Both regression lines have residual sum zero but it is clear from the figure that the solid line is a much better model for the observed data than the dashed line. Hence, the sum of the residuals is not an adequate measure of how well a model fits the data simply because a large positive residual can be canceled by a corresponding negative residual.

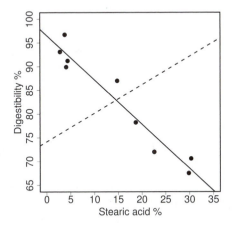

Figure 2.4: Two regression lines for the digestibility data. The solid line is defined by $y = -0.9337 \cdot x + 96.5334$ while the dashed line is defined by $y = 0.6 \cdot x + 74.15$. Both regression lines have residual sum zero.

What we need is a way to consider the magnitude of the residuals such that positive and negative residuals will not cancel each other. The preferred solution is the method of *least squares*, where the residuals are squared before they are added, which prevents positive and negative residuals from canceling each other*. One way to think about the squared residuals is that we desire a model where we have as few observations as possible that are far away from the model. Since we square the residuals we take a severe "punishment" from observations that are far from the model and can more easily accommodate observations that are close to the predicted values. Figure 2.5 shows a graphical representation of the squared residuals. The gray areas correspond to the square of the residuals so each observation gives rise to a square gray area. Instead of just looking at the sum of residuals and trying to find a model that is as close to the observations as possible, we essentially try to identify a model that minimizes the sum of the gray areas.

---

*An alternative would be to use the absolute residuals. This approach also prevents the cancellation of positive and negative residuals, but the calculus of minimizing the sum of absolute residuals (see Section 2.1.1) can be rather tricky. Another reason why the sum of squared residuals is preferred is that the corresponding estimates are identical to the estimates found by the more general maximum likelihood approach. Maximum likelihood will be discussed briefly in Section 5.2.7.

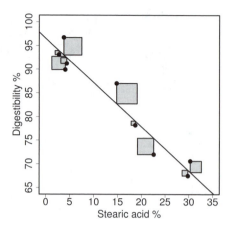

Figure 2.5: Squared residuals for the dataset on digestibility and stearic acid. Gray areas represent the squared residuals for the proposed regression line.

2.1.1   Least squares estimation

The least squares method estimates the unknown parameters of a model by minimizing the sum of the squared deviations between the data and the model. Thus for a linear regression model we seek to identify the parameters α and β such that

$$\sum_{i=1}^{n} (y_i - \alpha - \beta \cdot x_i)^2 \tag{2.2}$$

becomes as small as possible. A standard approach to find the maximum or minimum of a function is to differentiate the function and identify the parameter values for which the derivative equals zero. The partial derivatives of the function $Q(\alpha, \beta; x, y) = \sum_{i=1}^{n}(y_i - \alpha - \beta \cdot x_i)^2$ are

$$
\begin{aligned}
\frac{\partial Q}{\partial \alpha} &= \sum_{i=1}^{n} \frac{\partial}{\partial \alpha}(y_i - \alpha - \beta \cdot x_i)^2 = \sum_{i=1}^{n} 2(y_i - \alpha - \beta \cdot x_i) \cdot (-1) \\
&= -2 \cdot \sum_{i=1}^{n} y_i - n\alpha - \beta \sum_{i=1}^{n} x_i \tag{2.3}
\end{aligned}
$$

$$
\frac{\partial Q}{\partial \beta} = \sum_{i=1}^{n} 2(y_i - \alpha - \beta \cdot x_i) \cdot (-x_i) \tag{2.4}
$$

To find the minima of Q we set the partial derivatives equal to zero and solve the two equations with two unknowns, α and β. It can be shown that there is a unique minimum of (2.2), which means we can find a unique line that fits our data best.

We can summarize the results above as follows: For a linear regression

Table 2.1: Calculations for the stearic acid data

i	x_i	y_i	$(x_i - \bar{x})$	$(y_i - \bar{y})$	$(x_i - \bar{x})^2$	$(y_i - \bar{y})^2$	$(x_i - \bar{x})(y_i - \bar{y})$
1	29.8	67.5	15.21	-15.41	231.38	237.50	-234.42
2	30.3	70.6	15.71	-12.31	246.84	151.56	-193.42
3	22.6	72.0	8.01	-10.91	64.18	119.05	-87.41
4	18.7	78.2	4.11	-4.71	16.90	22.20	-19.37
5	14.8	87.0	0.21	4.09	0.05	16.72	0.86
6	4.1	89.9	-10.49	6.99	110.02	48.85	-73.31
7	4.4	91.2	-10.19	8.29	103.81	68.71	-84.46
8	2.8	93.1	-11.79	10.19	138.98	103.81	-120.12
9	3.8	96.7	-10.79	13.79	116.40	190.13	-148.77
	131.3	746.2	0.00	0.00	1028.55	958.54	-960.40

model the line that best fits the data has slope and intercept given by

$$\hat{\beta} = \frac{\sum_{i=1}^{n}(x_i - \bar{x})(y_i - \bar{y})}{\sum_{i=1}^{n}(x_i - \bar{x})^2} \tag{2.5}$$

$$\hat{\alpha} = \bar{y} - \hat{\beta} \cdot \bar{x}. \tag{2.6}$$

Keep in mind that by "best straight line" we mean the one that minimizes the residual sum of squares. Note that as a consequence of (2.6) we have that the best straight line will always go through the point $(\bar{x}, \bar{y})$ since $\bar{y} = \hat{\alpha} + \hat{\beta} \cdot \bar{x}$.

The least squares criterion is a general technique; it has uses beyond linear regression and can be used to estimate parameters for any particular model. In Sections 2.4 and 5.2 we discuss the use of least squares in more general situations.

Example 2.3. Stearic acid and digestibility of fat (continued from p. 26). The least squares estimates of the slope and intercept for the digestibility data are found by inserting the data into (2.5) and (2.6). The details are shown in Table 2.1.

The mean values of x and y are $\bar{x} = \frac{131.3}{9} = 14.5888$ and $\bar{y} = \frac{746.2}{9} = 82.9111$. Once we have those we can fill out the remaining columns in the table and finally calculate the estimated slope and intercept:

$$\hat{\beta} = \frac{-960.40}{1028.549} = -0.9337$$

$$\hat{\alpha} = 82.9111 - (-0.9337 \cdot 14.5888) = 96.5334$$

Thus the best regression line for the digestibility data is given by $y = -0.9337 \cdot x + 96.5334$.

The best line enables us to make predictions about the digestibility percentage for stearic acid levels we have not examined in the experiment as described in Example 2.1. In addition, the regression line allows us to provide statements about the change in digestibility: *"If we increase the stearic acid level by 10 percentage points we expect the digestibility to decrease by 9.33 percentage points".*   □

2.2   When is linear regression appropriate?

Throughout this chapter we have tried to fit a linear relationship between two variables x and y. The formulas for the estimates (2.5) and (2.6) make no assumption whether or not a straight line is at all appropriate to describe the relationship between the two variables. We can compute the estimates for any dataset with a pair of variables, but there are issues we should consider before we do so:

Quantitative variables. Linear regression applies only to two quantitative variables. Make sure both variables are quantitative before a linear regression is used to model the relationship between x and y.

Does the relationship appear to be linear? Is it reasonable to model the relationship between x and y as a straight line? We should always start our analysis by plotting the data and checking the overall relationship between x and y in a graph: a curvilinear relationship between x and y makes a linear regression inappropriate. In these situations we should either transform the data (see Section 2.2.1), if that is feasible, or use another model (see Chapter 8).

Influential points. Influential points are data points with extreme values that greatly affect the slope of the regression line. If we look closely at the estimation formula for the regression slope, (2.5), we can see that for linear regression, influential points are often outliers in the x-direction. If a point, x_i, is close to the mean value, $\bar{x}$, then the term $(x_i - \bar{x})$ will be close to zero in both the numerator and denominator of (2.5) and the actual value of y_i will have little importance on the regression slope. On the other hand, if x_i is far away from $\bar{x}$ then both the numerator and denominator of (2.5) will be large and the difference $(y_i - \bar{y})$ may have a large impact on the slope estimate. Figure 2.6 illustrates the effect of an influential point.

x on y or y on x? The regression of x on y is different from the regression of y on x, and we have to fit a new model with digestibility as the explanatory variable and stearic acid as the response variable if we wish to predict stearic acid levels from digestibility. If we take the original regression equality $y = \alpha + \beta \cdot x$ and solve for x we get $x = -\alpha/\beta + y/\beta$

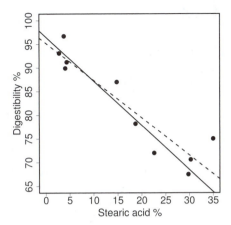

Figure 2.6: The effect of influential points on linear regression slopes. If we add a single extra point at $(35, 75)$ to the stearic acid data we will change the regression slope from -0.9337 (solid line) to -0.706 (dashed line).

— a regression line with intercept $-\alpha/\beta$ and slope $1/\beta$. However, these results are not the same ones we get when we regress x on y. The reason is that in terms of the original scatter plot, the best straight line for predicting digestibility is the one that minimizes the errors in the vertical direction. When we want the best straight line that minimizes the squared residuals of stearic acid we essentially seek to minimize residuals in the horizontal direction. Figure 2.7 shows that the regression of digestibility on stearic acid has a best fit with parameters $\hat{\alpha} = 96.5334$ and $\hat{\beta} = -0.9337$. The best fit of stearic acid on digestibility has least squares estimates of 97.6618 and -1.0020 for the intercept and slope, respectively.

In some experiments it is clear which of the variables should take the role of the response variable, y, and which variable should take the role of the explanatory variable — for example, if the researcher controls one of the variables in the experiment and records the other variable as the response. In other experiments, however, it can be unclear whether it is reasonable to model y on x or x on y. In those situations it may be more reasonable to calculate the correlation coefficient; see Section 2.3.

Interpolation is making a prediction within the range of observed values for the explanatory variable x. Interpolation may be uncertain in specific situations; for example, if the investigator has collected multiple responses from only very few values of x. In these situations there would be no way to demonstrate the linearity of the relationship between the two variables, but such situations are rarely encountered in practice. Interpolation is generally safe and is often one of the primary reasons why it is attractive to model the relationship between x and y through

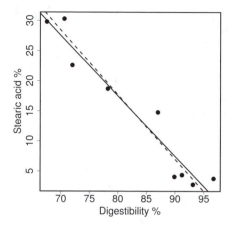

Figure 2.7: Interchanging x and y. Regression estimates of x on y cannot be determined from the regression estimates of y on x as the vertical residuals are used to fit the model for the latter while the "horizontal" residuals are needed for the former. The solid line corresponds to the regression of stearic acid percentage on digestibility while the dashed line is the ordinary regression of digestibility on stearic acid.

a known function. It enables us to predict values of y for values of x that do not exist in the sample.

Extrapolation concerns the situations where predictions are made outside the range of values used to estimate the model parameters. The prediction becomes increasingly uncertain when the distance to the observed range of values is increased as there is no way to check that the relationship continues to be linear outside the observation range. In the example with stearic acid we have no valid predictions of the digestibility when the stearic acid level is above 35%. In many experiments a linear regression may fit the data very well for some restricted intervals but may not be reasonable for the complete range of possible observed values.

2.2.1   Transformation

In Section 2.2 we discussed the usefulness of the linear regression model in situations where there appears to be a non-linear relationship between two variables x and y. In those situations the linear regression model is inappropriate. In some cases, however, we may be able to remedy the situation by transforming the response variable in such a way that the transformed data shows a linear relationship with the explanatory variable x.

Let $(x_i, y_i), i = 1, \ldots, n$ denote our n pairs of observations and assume that a straight line does not reasonably describe the relationship between x

and y. By transformation we seek a function, f, such that the transformed variables, $z_i = f(y_i)$, can be modeled as a linear function of the x's; *i.e.*,

$$z = \alpha + \beta \cdot x.$$

This is the case in the following example.

Example 2.4. Growth of duckweed. Ashby and Oxley (1935) investigated the growth of duckweed (*Lemna*) by counting the number of leaves every day over a two-week period. The data are seen in the table below:

Days	Leaves	Days	Leaves
0	100	7	918
1	127	8	1406
2	171	9	2150
3	233	10	2800
4	323	11	4140
5	452	12	5760
6	654	13	8250

We would like to model the growth of duckweed as a function of time. As always, we first plot the data (see Figure 2.8). It is obvious from the figure that a straight line is inadequate in describing the growth of the duckweed over time.

The population size of a species can often be described by an *exponential growth model* where the population size at time t is given by the formula

$$f(t) = c \cdot \exp(b \cdot t).$$

The two parameters, b and c, represent the average population increase per individual per time unit and the population size at time zero, respectively. If we take natural logarithms on both sides we get

$$\log(f(t)) = \underbrace{\log c}_{\alpha} + \underbrace{b}_{\beta} \cdot t.$$

This corresponds to a linear regression model with $\log(f(t))$ as response and t as explanatory variable.

The logarithm of the number of leaves is plotted in the bottom-left panel of Figure 2.8, and we see that a straight line fits the data almost perfectly. Thus we fit a linear regression model to the logarithmically transformed leaf count and get estimates $\hat{\alpha} = 4.4555$ and $\hat{\beta} = 0.3486$.

Finally, we back-transform these parameters to the original scale, $\hat{c} = \exp(\hat{\alpha}) = 86.099$ and $\hat{b} = \hat{\beta} = 0.3486$. If we insert these estimates in the exponential growth model we can plot the fit on the original scale, as is shown in the bottom-right panel of Figure 2.8. The interpretation of the growth rate,

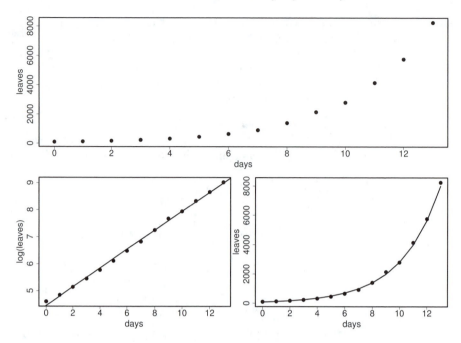

Figure 2.8: Top panel shows the original duckweed data. Bottom left shows the data and fitted regression line after logarithmic transformation and bottom right shows the fitted line transformed back to the original scale.

$\hat{b} = 0.3486$, is that if we have k leaves in our population then on average we will have $\exp(\hat{b}) \cdot k = 1.417 \cdot k$ leaves in our population the following day. □

In Example 2.4 we used the exponential growth model and transformed the data in a way that enabled us to use a straight line to model the (transformed) data. Sometimes it is also useful to transform both the response and explanatory variables (see Example 8.1 for an example of this). However, in many other situations there may not be a well-known relationship or transformation that will enable us to use a linear regression model. In those situations we will have to try different reasonable transformations and see which one fits best. Section 7.1 discusses how to compare the fit of different transformations in more detail.

2.3 The correlation coefficient

In the preceding sections on linear regression we tried to model one variable, y, as a function of x. One underlying implicit assumption in that regard is that we expect a causal relationship where the x variable directly influences the y variable. This is a reasonable assumption in many situations, especially in controlled experiments, where an investigator decides or controls the values of x. That was the case in Example 2.1, where the levels of stearic acid were chosen by the investigator. In other situations, however, we may expect an association between x and y but it may not be reasonable to say that x is the cause of y or vice versa. Examples of this include systolic and diastolic blood pressure, the height and weight of humans, tenderness of steaks and diced meat, or the size of two plants grown in the same pot.

The *sample correlation coefficient* describes the linear association between x and y and is defined as

$$\hat{\rho} = \frac{\sum_{i=1}^{n}(x_i - \bar{x})(y_i - \bar{y})}{\sqrt{(\sum_{i=1}^{n}(x_i - \bar{x})^2)(\sum_{i=1}^{n}(y_i - \bar{y})^2)}}. \tag{2.7}$$

In the numerator of (2.7) we see that we multiply the residual of x by the residual of y for each pair. Thus if both x and y deviate in the same direction from their respective means then the contribution is positive and if they deviate in different directions then the contribution will be negative.

The correlation coefficient can be interpreted as the regression slope found by regressing y on x after both variables have been scaled to have a standard deviation of 1. To be specific, let s_x and s_y be the standard deviation of the x's and y's, respectively, and set $x_i' = x_i/s_x$ and $y_i' = y_i/s_y, i = 1, \dots, n$, respectively. Then the regression slope of y' on x' is exactly $\hat{\rho}$ (see Exercise 2.9).

The denominator for the correlation coefficient (2.7) is always positive except in the extreme situation when all the x's or all the y's are identical, where the denominator is zero and the correlation coefficient is not well-defined. As a result, the sign of the correlation coefficient is identical to the sign of the regression slope (2.5) because the numerators are identical in the two formulas. Another point worth noting is that x and y enter (2.7) symmetrically, so the correlation of x and y is identical to the correlation between y and x.

The correlation is a measure of the strength of the linear relationship between the two variables and it can be shown that it is always between -1 and 1, inclusive. The value of $\hat{\rho} = 1$ occurs when the observations lie exactly on a straight line with positive slope, and $\hat{\rho} = -1$ corresponds to the situation where the observations are exactly on a straight line with negative slope. The correlation coefficient is zero when the best-fitting straight line of y on x does not depend on the observed value of the x's (*i.e.*, when $\hat{\beta} = 0$). Figure 2.9 shows correlation coefficients for different datasets.

The correlation coefficient is dimensionless since x and y are both standardized in the calculations. Thus, the value of $\hat{\rho}$ does not reflect the slope relating y to x but merely expresses the tightness of the linear relationship between x and y. It is vital to remember that a correlation, even a very strong one, does not mean we can make any conclusion about causation. Moreover,

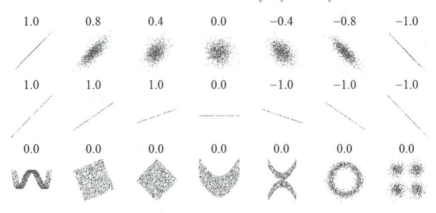

Figure 2.9: Correlation coefficients for different datasets. Note from the second row of graphs that the slope has no influence on the correlation coefficient except for the middle case where the variance of *y* is 0 so the correlation is not well-defined. The last row of graphs shows that the correlation may be zero even though the data is highly structured. (Picture courtesy of Wikipedia.)

there may be a strong non-linear relationship between x and y even though the correlation coefficient is zero.

Example 2.5. Tenderness of pork and sarcomere length. Rapid chilling of pork may result in muscle contractions which make the meat tough. In a study the average sarcomere length in the meat and the corresponding tenderness as scored by a panel of sensory judges was examined. A high score corresponds to tender meat. The data in Table 2.2 represent the average tenderness and the average sarcomere length measured on the same muscle in each of 24 pigs. The primary objective of the study was to see how closely the results are associated. The data listed in Table 2.2 and shown in Figure 2.10 are from Møller et al. (1987).

Figure 2.10 shows that there is no strong linear association between tenderness and sarcomere length but also that there is no particular non-linear relationship that describes the association. The calculated correlation coefficient for the pork tenderness data is 0.3658.   □

2.3.1   When is the correlation coefficient relevant?

Quantitative variables. The sample correlation coefficient applies only to two quantitative variables. Make sure that both variables are quantitative before the correlation coefficient between x and y is calculated.

Check linear association. Figure 2.9 illustrates how important it is to plot the data before the correlation coefficient is calculated — the associa-

Table 2.2: Data on sarcomere length and pork tenderness

Pig	Sarcomere length	Tenderness	Pig	Sarcomere length	Tenderness
1	1.621	8.44	13	1.619	7.33
2	1.529	7.11	14	1.570	4.22
3	1.500	6.00	15	1.580	5.78
4	1.549	7.56	16	1.599	6.11
5	1.458	7.22	17	1.568	7.44
6	1.454	5.11	18	1.555	3.89
7	1.405	3.11	19	1.660	7.67
8	1.506	8.67	20	1.673	8.00
9	1.580	7.44	21	1.690	8.78
10	1.635	4.33	22	1.545	5.78
11	1.648	6.78	23	1.676	6.44
12	1.574	5.56	24	1.511	8.00

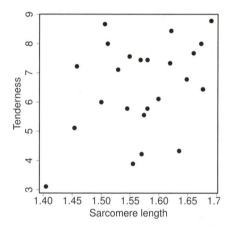

Figure 2.10: Graph of tenderness of pork and sarcomere length for 24 pigs.

tion may be highly structured, but if the relationship is non-linear the correlation coefficient may still be zero.

2.4 Perspective

We have discussed two methods to describe the linear association between n pairs of variables x and y. In linear regression we consider the explanatory variable as given and without any restrictions except it should be quantitative — in fact, the explanatory variable, x, can be fixed by the investigator and need not be continuous. The advantage of regression analysis over the correlation coefficient is that we have a model that describes the relationship of the y's given the observed x's. This functional relationship enables us to predict the values of y for values of x that we have not observed in our sample dataset.

The correlation coefficient measures the strength or "tightness" of the linear relationship and should only be used to describe the linear association between two continuous variables, x and y.

2.4.1   Modeling the residuals

The sample standard deviation (1.6) measures the average distance from the observations to their mean. In linear regression the residuals measure the distance from the observed value to the predicted value given by the regression model. We can use the same approach as for the standard deviation to calculate the standard deviation of the residuals. First we note that the sum of the residuals from a linear regression is 0:

$$
\begin{aligned}
\sum_i r_i &= \sum_{i=1}^n y_i - (\hat{\alpha} + \hat{\beta} x_i) = \\
&= \sum_{i=1}^n y_i - n(\bar{y} - \hat{\beta}\bar{x}) - \hat{\beta} \sum_{i=1}^n x_i \\
&= 0.
\end{aligned}
$$

The *residual standard deviation* is defined as

$$
s_{y|x} = \sqrt{\frac{\sum_i r_i^2}{n - 2}}, \tag{2.8}
$$

where the subscript $y|x$ is used to indicate that we refer to the standard deviation of the y's conditional of the fact that we already have taken the information from the x's into account through our model.

The residual standard deviation calculates the average distance between the observed data points and the linear predictions. Note the analogy to the standard deviation formula (1.6) except for the denominator, which is $n - 2$ instead of the $n - 1$ used in (1.6) where we had to estimate only one parameter, μ[†]. The linear regression model describes the type of relationship between x and y, but the residual standard deviation provides information

---

[†]The reason we are dividing by $n - 2$ is essentially the same as the reason we use $n - 1$ in (1.6). In linear regression we need to estimate two parameters (α and β) in order to calculate the predicted value for a given value x. Thus, we are paying a "price" in the number of observations since we have already used the same n observations to estimate both α and β before we use the data to calculate the residuals.

about the spread of our observations around the regression line. We can use that to describe the effectiveness of our prediction — if the residual standard deviation is small then the observations are generally closer to the predicted line, and they are further away if the residual standard deviation is large.

2.4.2 More complicated models

Throughout this chapter we have discussed the linear relationship between two variables. Obviously, more complicated relationships may exist. For example, y may depend on x in a quadratic way,

$$y = \beta_0 + \beta_1 \cdot x + \beta_2 \cdot x^2$$

or maybe a model for the relationship between reaction time and concentration for chemical processes (the *Michaelis-Menten kinetics*),

$$y = \frac{\beta_0 \cdot x}{\beta_1 + x},$$

turns out to be appropriate. The concept of residual least squares extends well to these more complicated models, and the same approach can be applied to find parameter estimates in these situations.

Let $f(x; \beta_1, \ldots, \beta_p)$ describe the relationship between x and y through a known function f with parameters $\beta_1, \ldots, \beta_p$. Mathematically, the least squares criterion that is minimized to obtain the parameter estimates is

$$Q(\beta_1, \ldots, \beta_p) = \sum_{i=1}^{n} [y_i - f(x_i; \beta_1, \ldots, \beta_p)]^2.$$

In other words, we need to identify estimates $\hat{\beta}_1, \ldots, \hat{\beta}_p$ such that Q is minimized. For some models, the least squares minimization can be done analytically using calculus, but for non-linear models the minimization is almost always done using iterative numerical algorithms.

Notice that it is possible to place restrictions on the parameters in an existing model and still use the least squares criterion to estimate the remaining parameters. For example, if we want to place a restriction on a linear regression that the straight line should go through the origin $(0,0)$, we would require that $\alpha = 0$. Thus, we would use the least squares criterion to minimize

$$Q(\beta_1, \ldots, \beta_p) = \sum_{i=1}^{n} [y_i - (0 + \beta \cdot x_i)]^2 = \sum_{i=1}^{n} [y_i - \beta \cdot x_i]^2.$$

This model has only one parameter, β, and the slope estimate is

$$\hat{\beta} = \frac{\sum_i x_i y_i}{\sum_i x_i^2},$$

which is different from the estimate provided in (2.5) since we here fit the best straight line that goes through the origin.

2.5 R

Fitting a linear regression model in R is done using the versatile linear model function, lm(). If we consider the data from the stearic acid example (Example 2.1 on p. 26) used throughout this chapter, we can fit a linear regression line with the code below:

```
> stearic.acid <- c(29.8, 30.3, 22.6, 18.7, 14.8, 4.1, 4.4,
+ 2.8, 3.8)
> digest <- c(67.5, 70.6, 72.0, 78.2, 87.0, 89.9, 91.2,
+ 93.1, 96.7)
> lm(digest ~ stearic.acid)

Call:
lm(formula = digest ~ stearic.acid)

Coefficients:
 (Intercept)   stearic.acid
    96.5334       -0.9337
```

In the call to lm() we specify the statistical model digest ~ stearic.acid, which can be interpreted in the following way: digest is modeled as a linear function of stearic.acid. By default, R interprets numerical vectors — in our case, both digest and stearic.acid — as quantitative variables, which is one of the requirements for a linear regression model. See Sections 5.1.2 and 8.3.1 for more information on model formulae.

The output from lm() shows the estimated parameters from the model. Here we have two parameters: the intercept and the slope. The estimated intercept is found under Intercept to be 96.5334 and the slope, −0.9337, is listed under stearic.acid since that is the covariate or variable name related to the parameter.

R automatically includes an intercept parameter in the model even though we did not specify it explicitly in the model formula. If we wish to model a linear relationship *without* an intercept (*i.e.*, a line going through the origin $(0,0)$), then we should specifically request that by including a -1 term in the formula,

```
> lm(digest ~ stearic.acid -1)

Call:
lm(formula = digest ~ stearic.acid - 1)

Coefficients:
stearic.acid
       3.371
```

Note how the estimated slope changes substantially from -0.9337 to 3.371 when we force the regression line to go through $(0, 0)$.

The function `predict()` calculates the predicted values (see p. 27) from a given model; *i.e.*, the y-values that we would expect to see if the model is correct (denoted $\hat{y}$ in the text). Likewise, the `resid()` function is used to extract the residuals, $y_i - \hat{y}_i$, from an estimated model.

```
> model <- lm(digest ~ stearic.acid) # Save lm result as model
> predict(model)                      # Compute predicted values
       1        2        3        4        5        6        7
68.70786 68.24099 75.43080 79.07240 82.71399 92.70502 92.42490
       8        9
93.91889 92.98515
> digest - predict(model)             # Calculate residuals
         1          2          3          4          5
-1.2078638  2.3590070 -3.4308034 -0.8723956  4.2860121
         6          7          8          9
-2.8050230 -1.2249006 -0.8188871  3.7148545
> resid(model)                        # Get residuals from resid
         1          2          3          4          5
-1.2078638  2.3590070 -3.4308034 -0.8723956  4.2860121
         6          7          8          9
-2.8050230 -1.2249006 -0.8188871  3.7148545
> sum(resid(model)**2)                # Sum of squared residuals
[1] 61.76449
```

The `plot()` function can be used to illustrate the relationship between the two quantitative variables, as described in Section 1.6. The `abline()` function adds a straight line to an existing plot, and we can use `abline()` to illustrate the estimated linear relationship between the two variables, as in Figure 2.2.

```
> plot(stearic.acid, digest)          # Make scatter plot
> abline(model)                       # Add straight line to plot
                                      # from previous lm output
```

In the duckweed example (Example 2.4 on p. 35) we had to transform the response variable (number of leaves) in order to fit the exponential growth model using a linear regression model. We can either make a new variable that contains the logarithm of the number of leaves or ask R to transform the response variable directly in the call to `lm()`.

```
> days <- seq(0, 13)
> leaves <- c(100, 127, 171, 233, 323, 452, 654, 918, 1406,
+ 2150, 2800, 4140, 5760, 8250)
> lm(log(leaves) ~ days)

Call:
```

```
lm(formula = log(leaves) ~ days)
```

```
Coefficients:
(Intercept)            days
     4.4555          0.3486
```

Note that the two estimates, the intercept $\hat{\alpha}$ = 4.4555 and the slope $\hat{\beta}$ = 0.3486, are estimated on the logarithmic scale and should be back-transformed to get them on the original scale of the exponential growth model.

In R, the correlation between two quantitative variables is calculated with the cor() function.

```
> cor(digest, stearic.acid)
[1] -0.9672452
```

2.6   Exercises

2.1 Heart rate of frogs. The relationship between heart rate and temperature for frogs was examined. Nine frogs were chosen at random and placed in nine different climate chambers where the temperature could be controlled. Their heart rate was measured subsequently and the data are shown below:

Temperature (°C)	Heart rate (beats/minute)
2	5
4	12
6	10
8	13
10	22
12	23
14	30
16	27
18	32

1. Fit a linear regression model by hand that explains the heart rate as a function of temperature. What are the estimates of the parameters?

2. Assume you acquire a 10th frog and place it in a climate chamber at 14 degrees. What heart rate would you expect the frog to have?

2.2 [M] Derivation of least squares estimates for linear regression.
Solve the two equations with two unknowns, α and β, defined by
the partial derivatives for the linear regression model, (2.3) and (2.4);
i.e.,

$$-2 \cdot \sum_{i=1}^{n} y_i - n\alpha - \beta \sum_{i=1}^{n} x_i = 0$$

$$\sum_{i=1}^{n} 2(y_i - \alpha - \beta \cdot x_i) \cdot (-x_i) = 0.$$

Show that the two estimates found by solving the equations are indeed a minimum of the sum of the squared residuals.

[Hint: Calculate the second-order partial derivatives and verify that they are positive.]

2.3 ⓇⒷ **Cherry trees.** For each of 31 cherry trees the diameter (in inches), height (in feet), and volume (in cubic feet) were measured (Ryan Jr. et al., 1985). Data are part of the R distribution and can be accessed as the data frame `trees` after issuing the command `data(trees)`. It is of interest to examine the relationship between diameter (called girth in the dataset), height, and volume.

1. Use `data(trees)` to get the data into R. Print the dataset to make sure that it contains exactly 31 cherry trees.

2. Make three plots that show volume against height, volume against diameter, and height against diameter. You can use the function `plot()` for this.

When exploring data it is always a good idea to view them graphically. Keep the following four points in mind when you look at these graphs:

Direction. Is the direction of the points generally positive (the y-values increase with increasing x-values)? Negative (y-values decrease with increasing x-values)? Or is there no obvious direction?

Shape. Do the observations follow a straight line or curve?

Strength. Are the observations scattered closely or widely around the general form of the data? Are they close to a line or curve or does it look more like an unstructured "cloud of points"?

"Strange" observations. Are any of the points vastly different from the rest of the data? These outliers could stem from data-entry errors and it might be a good idea to examine these more closely.

3. Describe the direction, shape, strength, and any strange observations for the three plots made in question 2.

4. Fit a model that describes the volume as a linear function of diameter (use `lm()`). What is the estimated slope of the fitted regression line? What is the estimated intercept?

5. Plot the fitted regression line in the same plot as the original data (use `abline()`).

6. Calculate the correlation coefficient between volume and height, volume and diameter, and diameter and height (use the function `cor()`). Do the results correspond to what you would expect from the plots?

7. Calculate the correlation between height and volume. Is that different from the correlation between volume and height? Why/why not?

8. We will now examine the influence of a single extreme value on the correlation coefficient. Locate the observation in the dataset that corresponds to the largest value of `volume`. Change that value to 35. How will this change influence the correlation between `volume` and `diameter` and between `volume` and `height`?

9. How will the changed value influence the estimates of the slope and intercept?

2.4 Correlation coefficient. Recall the definition of the correlation coefficient. Examine the definition and conclude which effect it will have on the correlation coefficient if we

1. add a constant, c, to all the x-values?

2. multiply a constant, k, to all the x-values?

2.5 ℝ Chocolate Chip Cookies. The correlation measures the *linear* relationship between two variables, and low correlation does not necessarily mean that there is no association between the two variables. Consider the dataset below which shows the relationship between the quality of chocolate chip cookies (on a predefined scale where a high value corresponds to high quality) and the baking temperature (in degrees Celsius).

Temperature	60	90	120	150	180	210	240	270	300
Score	2	8	13	18	18	17	14	6	3

1. Read the data into R.

2. Calculate the correlation coefficient.

3. Plot the quality score against the temperature. Is there any relationship on the plot?

2.6 [M] Least squares estimation with fixed slope. Assume you want to fit a straight line which has a fixed regression slope of 1. Derive the least squares estimate of the intercept by solving the partial derivative of the least squares criterion. How would this estimate change if we had a requirement that the regression slope should be fixed at 2?

2.7 Correlation graphs. Consider the four graphs shown in the figure below. The correlation coefficients for the four graphs are -0.86, -0.25, 0.17, and 0.70. Determine which graph corresponds to each of the four correlation coefficients.

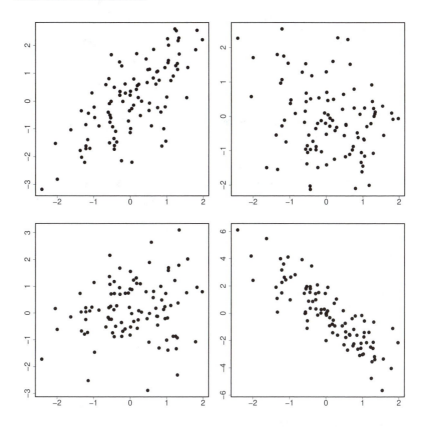

2.8 Biomass in soil. In ecological research it is often of interest to quantify the biomass in the upper layers of soil. One commonly used measure of biomass is the weight of live animals, FW (fresh weight), in the respective layer. Some practical difficulties exist, however, when the FW is to be measured in the field. Instead, animals collected from the soil are dried and the dry matter, DM, is used to determine the FW.

Ten individuals of the species *Isotoma notabilis* were collected and their dry matter and fresh weights were recorded.

Dry matter	Fresh weight	Dry matter	Fresh weight
1.379	1.993	1.436	2.238
2.479	3.811	1.657	2.370
1.514	2.454	2.112	3.287
1.226	1.883	1.864	2.467
1.975	3.466	1.902	2.632

1. Estimate the correlation coefficient between DM and FW. Does the data show any relationship between FW and DM?

2. Specify the linear regression model for FW as a function of DM and estimate the parameters.

3. In previous experiments with another species the investigators found that the fresh weight, FW, could be predicted well as a constant k multiplied with DM; *i.e.*, FW $\approx k$ DM. Transform the model by taking natural logarithms on both sides of this relationship. Fit this model to the data.

   Compare the two models to the observed data graphically and discuss if the transformed model predicts the FW from the DM just as well as the model in question 2.

2.9 [M] Correlation coefficient and standardized variables. Assume we have observed n pairs of observations, $(x_i, y_i), i = i, \ldots, n$. Let s_x and s_y be the standard deviation of the x's and the y's, respectively, and set $x_i' = x_i/s_x$ and $y_i' = y_i/s_y, i = 1, \ldots, n$.

Show that the regression slope of y' on x' is exactly identical to $\hat{\rho}$.

Chapter 3

Comparison of groups

The purpose of an analysis is often to compare different groups of data. Suppose, for example, that a meat scientist wants to examine the effect of three different storage conditions on the tenderness of meat. For that purpose 24 pieces of meat have been collected and allocated into three storage (or treatment) groups, each of size eight. The allocation is chosen at random. In each group all eight pieces of meat are stored under the same conditions, and after some time the tenderness of each piece of meat is measured. The main question is whether the different storage conditions affect the tenderness: are the observed differences between the groups due to a real effect — which we would find again if we repeated the experiment — or due to random variation? And if there are differences in meat tenderness, how large are they?

In this chapter we illustrate the setup with examples and introduce some notation. If only two groups are compared then we often talk about comparison of two samples, whereas the term *one-way analysis of variance* (or *one-way ANOVA*) is used if there are three or more groups. The terminology from one-way analysis of variance of course also applies when there are two groups.

3.1 Graphical and simple numerical comparison

An analysis should start with a graphical inspection of the data, whenever possible. Boxplots are particularly useful to reveal much of the important information from the data: Which differences can we expect to find in the analysis? How large is the variation in the data? Is the variation roughly the same for all groups? It also seems reasonable to consider the data from each group separately and compare statistical summaries computed from each of them.

First, consider the situation where the data consist of two samples; that is, two sets of measurements corresponding to two different groups. The two groups may correspond to different populations (such as men and women or different stocks) or to two different treatments.

Example 3.1. Parasite counts for salmons. An experiment with two difference salmon stocks, from River Conon in Scotland and from River Ätran in

Sweden, was carried out as follows (Heinecke et al., 2007). Thirteen fish from each stock were infected and after four weeks the number of a certain type of parasites was counted for each of the 26 fish with the following results:

Stock	No. of parasites												
Ätran	31	31	32	22	41	31	29	40	41	39	36	17	29
Conon	18	26	16	20	14	28	18	27	17	32	19	17	28

The purpose of the study was to investigate if the number of parasites during an infection is the same for the two salmon stocks.

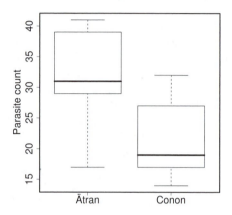

Figure 3.1: Boxplot for the two salmon samples.

Parallel boxplots for the two samples are shown in Figure 3.1, and the mean and sample standard deviations are computed to

$$\text{Ätran:}\quad \bar{y}_1 = 32.23, \quad s_1 = 7.28$$
$$\text{Conon:}\quad \bar{y}_2 = 21.54, \quad s_2 = 5.81.$$

Here we have used formulas (1.5) and (1.6) for each sample separately. Notice how we use subscripts to distinguish between the two samples.

The summary statistics and the boxplots tell the same story: the observed parasite counts are generally higher for the Ätran group compared to the Conon group, indicating that Ätran salmons are more susceptible to parasites. The result might be different, though, if we repeated the experiment and used new samples of salmons. The purpose of the statistical analysis is to clarify whether the observed difference between $\bar{y}_1$ and $\bar{y}_2$ is caused by an actual difference between the stocks or by random (sampling) variation.　□

It is not surprising that comparison between two groups is closely tied to the difference $\bar{y}_1 - \bar{y}_2$ between sample means. It is essential, though, to realize that the sample means, and hence their difference, would be different if we repeated the experiment and obtained two new samples. We need to take into account the natural variation of sample means: How much do the sample means $\bar{y}_1$ and $\bar{y}_2$ vary from sample to sample? Is the observed deviation between $\bar{y}_1$ and $\bar{y}_2$ caused by a difference between the corresponding populations or did it just occur by chance?

We now turn to an example where six groups are to be compared.

Example 3.2. Effect of antibiotics on dung decomposition. An experiment with dung from heifers was carried out in order to explore the influence of antibiotics on the decomposition of dung organic material (Sommer and Bibby, 2002). As part of the experiment, 36 heifers were divided into six groups. All heifers were fed a standard feed, but antibiotics of different types were added to the feed for heifers in five of the groups. No antibiotics were added for heifers in the remaining group (the control group). For each heifer, a bag of dung was dug into the soil, and after eight weeks the amount of organic material was measured for each bag. The data is listed below. Notice that only four bags were usable for the spiramycin group.

Control	α-Cyper-methrin	Enro-floxacin	Fenben-dazole	Ivermectin	Spiramycin
2.43	3.00	2.74	2.74	3.03	2.80
2.63	3.02	2.88	2.88	2.81	2.85
2.56	2.87	2.42	2.85	3.06	2.84
2.76	2.96	2.73	3.02	3.11	2.93
2.70	2.77	2.83	2.85	2.94	
2.54	2.75	2.66	2.66	3.06	

Figure 3.2 shows two plots of the data: the observations together with group means and the total mean (left panel) and parallel boxplots (right panel). The amount of organic material appears to be lower for the control group compared to any of the five types of antibiotics, suggesting that decomposition is generally inhibited by antibiotics. However, there is variation from group to group (between-group variation) as well as a relatively large variation within each group (within-group variation). The within-group variation seems to be roughly the same for all types, except perhaps for spiramycin, but that is hard to evaluate because there are fewer observations in that group. The boxplots (and the scattering around the group means) are reasonably symmetric.

The sample means and the sample standard deviations are computed for each group separately and listed in Table 3.1. Of course, we find the same indications as we did in the boxplots in Figure 3.2: On average the amount of organic material is lower for the control group than for the antibiotics groups, and except for the spiramycin group the standard deviations are roughly the same in all groups. □

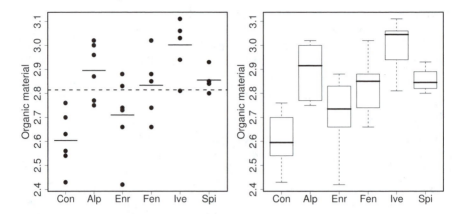

Figure 3.2: Graphical presentation of the antibiotics data. Left panel: data points with group sample means (solid line segments) and the total mean of all observations (dashed line). Right panel: parallel boxplots.

Table 3.1: Group means and group standard deviations for the antibiotics data

Antibiotics	n_j	$\bar{y}_j$	s_j	s_j^2
Control	6	2.603	0.119	0.0141
α-Cypermethrin	6	2.895	0.117	0.0136
Enrofloxacin	6	2.710	0.162	0.0262
Fenbendazole	6	2.833	0.124	0.0153
Ivermectin	6	3.002	0.109	0.0120
Spiramycin	4	2.855	0.054	0.0030

3.2   Between-group variation and within-group variation

When there are only two groups it is quite obvious that the difference in means, $\bar{y}_1 - \bar{y}_2$, is a reasonable measure of the difference between the groups, but also that we have to take into account the natural variation within the samples in order to investigate whether the observed value could just be due to chance.

When there are three or more groups there are several such pairwise differences, which, in a suitable way, should be put together. It turns out to be important to distinguish between two types of variation: *between-group variation* and *within-group variation*. The terminology will be made more precise in Section 6.3.1. At this point it is enough to understand it graphically; hence, consider the left part of Figure 3.2.

Between-group variation refers to differences between the groups; for example, deviation between the different treatments in the antibiotics example (p. 51) or between stocks in the salmon example (p. 49). It is illustrated by the vertical differences between the group means (horizontal line segments) and the total mean (dashed line).

Within-group variation refers to the variation in each of the groups; that is, within the particular treatment groups or within each of the two salmon samples. It is illustrated by the vertical deviations between the observations and their corresponding group means.

A large between-group variation is an indication of differences between the group means, but if the within-group variation is also large, then the observed differences may be due to random variation. In other words, with a large within-group variation we could get quite different results if we repeated the experiment. Hence, we will have to take both types of variation into account.

It is the distinction between different sources of variation that has given analysis of variance (ANOVA) its name.

3.3 Populations, samples, and expected values

Assume for a moment that we are interested in comparing men's and women's blood pressure. There are two distinct populations, the male and the female, and we are interested in comparing the average blood pressure in the male population, denoted α_m, and the average blood pressure in the female population, denoted α_f. The population averages α_f and α_m may also be interpreted as *expected values*: α_f is the blood pressure we would expect for a random woman if we had no further information about her. Similarly for α_m. In order to compare the average blood pressure levels in the male and the female populations we would draw a sample of men and a sample of women, measure their blood pressure, and use the observed blood pressure measurements to infer about the population averages.

The setup is similar for the salmon data in Example 3.1 (p. 49). There are two populations, the Ätran and the Conon salmon stocks, and a sample of 13 fish was drawn from each population and used in the experiment.

The situation is slightly different for the antibiotics experiment (Example 3.2, p. 51), though. Recall that 36 heifers were drawn at random and allocated randomly to six treatment groups. Hence, the groups are created by intervention on a sample from a single population (the population of heifers), but again we will use the results from the sample to infer about properties of the population. Imagine that we treated all heifers in the population with spiramycin, say, and let $\alpha_{spiramycin}$ denote the average amount of organic material for all the heifers. Then we are interested in comparing $\alpha_{spiramycin}$ to

$\alpha_{control}$, the population average if all heifers were untreated. Similarly for the other antibiotics types. We can interpret the α's as expected values: $\alpha_{spiramycin}$ is the amount of organic material that we would expect for a random heifer treated with spiramycin if we have no further information about it.

No matter how the groups are generated — as samples from different populations or as experimental groups — we wish to compare the expected values (the α's) across groups. In order to make this comparison we have to take the variation within the groups into account. This variation is described by the standard deviations in the groups.

3.4   Least squares estimation and residuals

Consider the situation with n observations split into k groups. Label the groups 1 through k and let $\alpha_1, \ldots, \alpha_k$ denote the expected values in the k groups. The question is how to estimate $\alpha_1, \ldots, \alpha_k$ from the data. We will use the least squares criterion for the estimation; *cf.* Section 2.1.1. In order to understand what that means in the present context we need a bit of notation.

Let $g(i)$ denote the group for observation i. Then $g(i)$ has one of the values $1, \ldots, k$ and the expected value for y_i is $\alpha_{g(i)}$. Notice that, with this notation, the sample mean and sample standard deviation in group j are given by

$$\bar{y}_j = \frac{1}{n_j} \sum_{i:g(i)=j} y_i,$$

$$s_j = \sqrt{\frac{1}{n_j - 1} \sum_{i:g(i)=j} (y_i - \bar{y}_j)^2}, \qquad (3.1)$$

for $j = 1, \ldots, k$. Here the sum $\sum_{i:g(i)=j}$ means the sum over all observations i which have $g(i) = j$; that is, all observations that belong to group j.

Example 3.3. Effect of antibiotics on dung decomposition (continued from p. 51). For the antibiotics data we have $n = 34$ observations split into $k = 6$ groups. Let group 1 denote the control group, group 2 the α-cypermethrin group, *etc.* If the observations are ordered group-wise, with the six control observations first and the four spiramycin observations last, then

$$g(1) = \cdots = g(6) = 1, \qquad g(31) = \cdots = g(34) = 6$$

and

$$\bar{y}_1 = \frac{1}{n_1} \sum_{i:g(i)=1} y_i = \frac{1}{n_1} \sum_{i=1}^{6} y_i = \frac{1}{6}(2.43 + \cdots + 2.54) = 2.603.$$

As an alternative we could have used the original names as the labels for the groups:

$$g(1) = \cdots = g(6) = \text{control}, \qquad g(31) = \cdots = g(34) = \text{spiramycin},$$

and similar for the other groups. □

Just as in the linear regression case (Section 2.1.1), we will estimate $\alpha_1, \ldots, \alpha_k$ by least squares. The deviations between the ith observation and its expected value is $y_i - \alpha_{g(i)}$, so the sum of squared deviations is

$$Q(\alpha_1, \ldots, \alpha_k) = \sum_{i=1}^{n} \left(y_i - \alpha_{g(i)} \right)^2. \tag{3.2}$$

This is a function of the parameters $\alpha_1, \ldots, \alpha_k$. The least squares estimates $\hat{\alpha}_1, \ldots, \hat{\alpha}_k$ are the values of $\alpha_1, \ldots, \alpha_k$ that make this function as small as possible. In Exercise 3.3 we show that the solution is

$$\hat{\alpha}_j = \bar{y}_j, \quad j = 1, \ldots, k.$$

In other words, we simply use the sample means from the observations belonging to group j to estimate the expected value (population mean) for group j. This is hardly surprising.

The residual corresponding to the ith observation is given by the distance from the observation to its group mean,

$$r_i = y_i - \hat{\alpha}_i = y_i - \bar{y}_{g(i)}. \tag{3.3}$$

The residual variance and residual standard deviation are defined as

$$s^2 = \frac{1}{n-k} \sum_{i=1}^{n} r_i^2, \quad s = \sqrt{s^2} = \sqrt{\frac{\sum_{i=1}^{n} r_i^2}{n-k}} \tag{3.4}$$

In particular, the residual variance, s^2, is the average squared residual, except that we divide by $n - k$ rather than n. Recall the definition (3.1) of the sample standard deviation; in particular, the denominator $n_j - 1$. If we add up the denominators for each group, we get exactly $n - k$:

$$\sum_{j=1}^{k} (n_j - 1) = \sum_{j=1}^{k} n_j - k = n - k.$$

Moreover, notice that s is defined in the same way as in the linear regression setup; see (2.8). The only difference is the denominator: it is $n - 2$ in the linear regression setup because we have estimated two parameters (α and β), but it is $n - k$ in the one-way ANOVA setup because we have estimated k parameters ($\alpha_1, \ldots, \alpha_k$).

The residual variance s^2 can also be computed as a weighted average of the group variance estimates, s_j^2, as follows (see Exercise 3.4):

$$s^2 = \frac{1}{n-k} \sum_{j=1}^{k} (n_j - 1)s_j^2. \tag{3.5}$$

Note that s_j^2 is assigned the weight $n_j - 1$, the denominator in (3.1). s^2 is called the *pooled sample variance*. Similarly, $s = \sqrt{s^2}$ is called the *pooled sample standard deviation*. It a measure of the standard deviation within the groups since it measures deviations from the group means. Notice that this interpretation makes sense only if the population standard deviations are the same for all groups.

Example 3.4. Effect of antibiotics on dung decomposition (continued from p. 51). The sum of squared residuals is computed as

$$\sum_{i=1}^{n} r_i^2 = (2.43 - 2.603)^2 + (2.63 - 2.603)^2 + \cdots + (2.93 - 2.855)^2$$
$$= 0.4150$$

where the group means are taken from Table 3.1. Using definition (3.4) we thus get the residual variance and residual standard deviation

$$s^2 = \frac{0.4150}{34 - 6} = 0.01482; \quad s = \sqrt{0.01482} = 0.1217.$$

If we use (3.5) and the sample variances s_j^2 from Table 3.1 we get the same result:

$$s^2 = \frac{1}{34 - 6} (5 \cdot 0.0141 + 5 \cdot 0.0136 + \cdots + 3 \cdot 0.0030) = 0.01482.$$

Recall that the pooled sample variance is a weighted average of the group variances. The pooled standard deviation, however, is not a weighted average of the groups standard deviations, but s will always be between the smallest and the largest s_j.   □

3.5   Paired and unpaired samples

We now return to a situation where the purpose is to compare two groups. However, the data structure is different from that of Section 3.1, since we now consider data consisting of pairs of observations that naturally belong to each other. *Paired samples* occur, for example, if two measurements are collected for each subject in the sample under different circumstances (treatments), or if measurements are taken on pairs of related observational units such as twins. In dietary studies with two diets under investigation, for example, it is common that the subjects try one diet in one period and the other diet in another period. Measurements are collected in both periods, and by looking at the differences between the measurements from the same subject corresponding to the two diets, the variation between subjects, which is often substantial, is reduced.

Example 3.5. Equine lameness. An experiment on equine lameness was carried out on eight horses (Jensen et al., 2009). The measurement is a symmetry score; *i.e.*, a value with information about the symmetry of the gait pattern. The symmetry score was measured twice for each horse while it was trotting: once when the horse was healthy and once when lameness on a forelimb was induced with a specially designed horseshoe. The investigators wanted to examine if the symmetry score for a horse changes due to lameness. The results were as follows:

Horse	Lame	Healthy	Difference
1	4.3541	−0.9914	5.3455
2	4.7865	1.4710	3.3155
3	6.1945	1.2459	4.9486
4	10.7383	0.4024	10.3359
5	3.3007	0.0325	3.2682
6	4.8678	−0.6396	5.5074
7	7.8965	0.7246	7.1719
8	3.9338	0.0604	3.8734

This is an example of paired samples, as there are two observations from each horse. Rather than using the original measurements, we will use the pairwise differences: We compute the difference in symmetry score between the lameness measurement and the healthy measurement for each horse. This gives us a single sample consisting of eight measurements. If lameness does not change the symmetry score we would expect the difference in symmetry score to vary around zero. The difference turns out to be positive for all horses in the study, but the statistical analysis should assess whether this is due to an actual change in symmetry score due to lameness or due to random variation. □

It is important to distinguish paired samples from unpaired — or independent — samples, because different methods of analysis are appropriate. For unpaired samples like the salmon data (Example 3.1, p. 49), we impose an assumption of independence between all observations. We will be more precise about the terminology of independence in Section 4.2.1, but loosely speaking it means that the observations do not share information.

For the lameness data, however, it is most likely that each horse has its own level of symmetry which may (or may not) change due to lameness. In other words, if a horse trots more symmetrically than another while healthy, then it probably does so, too, when it has a lame limb. Hence, two observations from the same horse share information, namely information about the general gait pattern for that particular horse, and it would not be reasonable to assume independence between two observations from the same horse. As a consequence, the between-group variation is confused with the within-group variation, making this terminology inappropriate for paired data.

By considering the differences between the healthy and the lameness measurement there is only one measurement per horse, so there is no independence problem. Moreover, we suspect that the physical proportions of the horse and the horse's general gait pattern influence the score in each condition but imagine that they more or less cancel out when we compute the difference.

It is sometimes possible to design both kinds of experiments in order to compare different groups. By using paired samples we hope to eliminate random variation that is not related to the matter of investigation. In the lameness example we could have used 16 horses, say, and used half of them in the healthy condition and half of them for lameness measurements. Then we would be in the setup of independent samples, and a large variation between horses would mask the difference between the lameness conditions. Similarly for the dietary example mentioned in the beginning of this section: We hope to eliminate the large variation between persons and emphasize the diet effect by looking at differences between the two diets for the subjects in the study. We sometimes say that we use the experimental units (horses, subjects, plants) as their own controls.

Typically fewer subjects are needed in order to estimate differences with a certain (pre-specified) precision in a paired study compared to an unpaired study. Hence, there are also practical/economical advantages of paired studies. It is not always possible, though, to design paired studies. Sometimes it is only possible to measure once on each subject; for example, because the animal is killed or the plant material destroyed during the course of measurement. A paired version of the salmon experiment (Example 3.1, p. 49) is also hard to imagine as this would require salmons from different stocks to be paired.

3.6   Perspective

In this chapter we have been concerned with three kinds of data structures:

Two independent samples where the samples correspond to two different groups or treatments and can be assumed to be independent.

Independent samples where the samples correspond to k different groups or treatments and can be assumed to be independent.

Paired samples where the observations consist of pairs of measurements, with the observations in a pair corresponding to two different groups or treatments.

The first case is a special case of the second, but we emphasize it anyway, for two reasons. First, it is very important to distinguish two independent samples from paired samples because different analysis methods are appropriate. Second, as we shall see, there are several more issues to consider in the analysis when there are three or more groups, compared to the two-sample case.

The setup with independent samples corresponds to a *one-way analysis of variance*, or *one-way ANOVA*. It is called "analysis of variance" because different sources of variation are compared (*cf.* Section 3.2) and "one-way" because only one factor — the treatment or grouping — is varied in the experiment.

The primary goal of a one-way ANOVA with k groups is to compare $\alpha_1, \ldots, \alpha_k$ representing the expected values in the groups (or the average response level in the corresponding populations). So far we have estimated the α's but we have not discussed the precision of the estimates, and thus we cannot say if the observed differences represent real differences or if they have occurred just by chance: if we repeated the experiment many times, would we observe similar differences or something quite different? We will discuss the actual statistical analysis in Chapters 5 and 6.

The most important assumption for the analysis is that of similar standard deviations in the groups. A rule-of-thumb says that the group standard deviations are "similar enough" if the ratio between the largest and the smallest group sample standard deviation is not greater than 2 (Samuels and Witmer, 2003), but the robustness depends on the sample sizes. Results are more robust for large and equally sized samples than for smaller samples or samples that differ much in size.

It is also important to consider the information obtained from the box-plots and recall that the standard deviation is only meaningful for data that are at least roughly symmetric. Moreover, the group sample sizes should be taken into account as the standard deviations are only very imprecisely estimated for small samples. For the antibiotics data, for example, the group standard deviations are quite similar except for the Spiramycin group. This group is very small, though, so we should not trust the sample estimate too much.

Sometimes the assumption of homogeneous variation is clearly not fulfilled. In particular, if the observed values are very different across the groups, it quite often occurs that there is larger variation in groups with large observed values compared to groups with small observed values. This can often be "fixed" by considering a transformation of the data: we seek a function h such that the transformed values $z_i = h(y_i)$ have homogeneous variance across the groups. This is similar to the transformation approach in Section 2.2.1 on linear regression, and the logarithmic transformation is often very useful in the current context too. See Case 2, Part II (p. 355) for an example.

Very often more than one explanatory factor is of interest in a scientific experiment. The effects of such factors should be taken into account simultaneously, leading to multi-way analysis of variance. The case with two factors — the *two-way analysis of variance* or *two-way ANOVA* — is treated in detail in Sections 8.2 and 8.4.1.

3.7   R

In this section we describe how to compute group specific means and standard deviations as well as the pooled standard deviation, and how to make R distinguish between linear regression and one-way ANOVA. Additional information and details will appear in later chapters.

Consider the antibiotics data (Example 3.2, p. 51) and assume that the data frame `antibio` contains the data:

```
> antibio
        type  org
1   Ivermect 3.03
2   Ivermect 2.81
3   Ivermect 3.06
4   Ivermect 3.11
5   Ivermect 2.94
6   Ivermect 3.06
7    Alfacyp 3.00
.
.
.
34   Control 2.54
> attach(antibio)
```

The `attach()` command makes it possible to use the variables `type` and `org` with reference to the data frame.

Parallel boxplots are produced as described in Section 1.6 with the `boxplot()` command. The following command produces a figure similar to the right one in Figure 3.2:

```
> boxplot(org~type)              # Parallel boxplots
```

3.7.1   Means and standard deviations

The workhorse for ANOVA is the lm() function, just as for linear regression. The model is fitted with lm(), and values like group means and the pooled standard deviation are easily extracted.

```
> lm(org~type-1)

Call:
lm(formula = org ~ type - 1)

Coefficients:
  typeAlfacyp    typeControl   typeEnroflox   typeFenbenda
        2.895          2.603          2.710          2.833
  typeIvermect   typeSpiramyc
        3.002          2.855
```

The vector, org, on the left-hand side of ~ in the call to lm() is modeled as a function of the vector on the right-hand side (type). The -1 means that the model is fitted "without intercept". We will explain this in more detail in Section 5.5.2 — for now, just notice that the output will then list the group means (*cf.* Table 3.1).

Additional information can be extracted with the summary() function:

```
> modelAntibio1 <- lm(org~type-1) # Save lm result as an object
> summary(modelAntibio1)          # Estimates etc.

Call:
lm(formula = org ~ type - 1)

Residuals:
     Min       1Q    Median       3Q      Max
-0.29000 -0.06000   0.01833  0.07250  0.18667

Coefficients:
              Estimate Std. Error t value Pr(>|t|)
typeAlfacyp    2.89500    0.04970   58.25   <2e-16 ***
typeControl    2.60333    0.04970   52.38   <2e-16 ***
typeEnroflox   2.71000    0.04970   54.53   <2e-16 ***
typeFenbenda   2.83333    0.04970   57.01   <2e-16 ***
typeIvermect   3.00167    0.04970   60.39   <2e-16 ***
typeSpiramyc   2.85500    0.06087   46.90   <2e-16 ***
---
Signif. codes:  0 '***' 0.001 '**' 0.01 '*' 0.05 '.' 0.1 ' ' 1
```

```
Residual standard error: 0.1217 on 28 degrees of freedom
Multiple R-squared: 0.9985,Adjusted R-squared: 0.9981
F-statistic:  3034 on 6 and 28 DF,  p-value: < 2.2e-16
```

As you can see, R now provides a large output. In later chapters we shall learn how to read and use output from such a summary() call. At this point, the important thing is to recognize the estimates for α's in the 'Estimate' column in the 'Coefficients' part of the output, and the pooled sample standard deviation (Residual standard error: 0.1217) in line three from the bottom.

Finally, note that the sample standard deviations for each of the groups can be computed with the sd() function as explained in Section 1.6. For example, for the control group:

```
> sd(org[type=="Control"])     # sd of control measurements
[1] 0.1187715
```

The sd() commands compute the standard deviation of the org variable, but only for those values for which the type variable is equal to "control"; that is, only for the six control measurements. Notice that two equality signs, ==, are needed for conditions like this. Similarly, the mean of the control measurements could be computed with

```
> mean(org[type=="Control"])   # Mean of control measurements
[1] 2.603333
```

3.7.2 Factors

The type variable in the antibio data frame has string values rather than numerical values; *e.g.*, Control rather than a number. In that case, R automatically recognizes that type is a categorical variable — or a factor — and fits a one-way ANOVA model in the lm(org~type) call.

The variable could as well, however, have been coded with numerical values, say 1 through 6, corresponding to the typeNum vector:

```
> typeNum <- rep(1:6, times=c(6,6,6,4,6,6))
> typeNum
 [1] 1 1 1 1 1 1 2 2 2 2 2 2 3 3 3 3 3 3 4 4 4 4 5 5 5 5 5 5 6 6
[31] 6 6 6 6
```

This is just a matter of renaming the groups and should not alter the analysis.

But what would happen if we used the typeNum variable instead of the original type in the lm() call? Then R would fit a linear regression; that is, model the amount of organic material as a linear function of the typeNum values. This does not make sense at all: the values 1–6 are *labels* only; the values themselves are completely arbitrary — they could be interchanged, for example — and should not be used in the analysis.

In other words, it is extremely important that R knows how to interpret the variable entered to the right of the ~ in the model formula: as a categorical variable (factor) or as a numerical variable. If the variable has string values (as type does), then R automatically interprets it as a factor. If it has numerical values (as typeNum does), then as default, R interprets it numerically, but we can change it with the factor() command and use factor(type) in the analysis:

```
> factor(typeNum)                    # typeNum as factor
 [1] 1 1 1 1 1 1 2 2 2 2 2 2 3 3 3 3 3 3 4 4 4 4 5 5 5 5 5 5 6 6
[31] 6 6 6 6
Levels: 1 2 3 4 5 6
> newModel <- lm(org~factor(typeNum))  # One-way ANOVA fit
```

Note the difference between the typeNum and the factor(typeNum) variables: in the output of factor(typeNum) it is emphasized that the values 1 through 6 are to be interpreted as levels of a factor. The two model objects modelAntibio1 and newModel are identical.

3.8 Exercises

3.1 Between-group and within-group variation. Consider a one-way ANOVA setup with three groups of size 10; that is, $k = 3$, $n_1 = n_2 = n_3 = 10$. Parallel boxplots are shown for three different cases in Figure 3.3. Use the graphs to answer the following questions:

1. In which plot is the *between-group* variation SS_{grp} the smallest? In which plot is it the largest?

2. In which plot is the *within-group* variation SS_e the smallest? The largest?

3. For each plot: do you believe that the data allow us to conclude that there is a difference between the groups? Why? We will be able to answer this question much more precisely in Chapters 5 and 6 — at this point you should just think about which features are important for answering such a question.

3.2 ℝ **Tartar for dogs.** A dog experiment was carried out in order to examine the effect of two treatments on the development of tartar. Apart from the two treatment groups there was also a control group. Twenty-six dogs were used and allocated to one of the three groups, denoted control (standard feed), P_2O_7 (pyrosulphate added to the feed), and HMP (hexametaphosphate added to the feed). After four

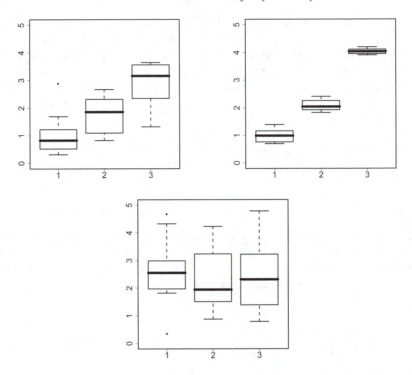

Figure 3.3: Boxplots for Exercise 3.1.

weeks each dog was examined and the development of tartar was summarized by an index taking into account the spread of tartar on the teeth as well as the thickness of the tartar. The data is given in the following table:

	Index for tartar					mean	sd
Control	0.49 1.36	1.05 1.55	0.79 1.66	1.35 1.00	0.55	1.089	0.423
P_2O_7	0.34 0.94	0.76 0.22	0.45 1.07	0.69 1.38	0.87	0.747	0.370
HMP	0.34 0.45	0.05 0.71	0.53 0.95	0.19	0.28	0.438	0.291

Answer the first questions without the use of R.

1. Convince yourself that this is a one-way ANOVA setup. What is k? What are the n_j's? Describe the g-function that relates the observations to groups: what is $g(1)$, $g(2)$, *etc.*? [Hint: See Example 3.3 on p. 54.]

2. Parameters α_{Control}, $\alpha_{\text{P}_2\text{O}_7}$, and α_{HMP} are associated with the groups. What is their interpretation? How are they estimated?

3. Compute the pooled standard deviation, s.

Use R for the remaining questions. The external dataset `tartar.txt` has two variables, `treat` and `index`.

4. Read the data into R and make parallel boxplots.

5. Use the `lm()` function to compute the group means. Check that you get the same as in the table above.

6. Use the `summary()` function to compute s. Make sure that you get the same result as in question 3.

3.3 [M] Least squares estimation. Recall the sum (3.2) of squared residuals and that the least squares estimates $\hat{\alpha}_1, \ldots, \hat{\alpha}_k$ are the values that make f as small as possible. It is convenient to split the sum into two and write

$$Q(\alpha_1, \ldots, \alpha_k) = \sum_{j=1}^{k} \sum_{i:g(i)=j} \left(y_i - \alpha_{g(i)} \right)^2.$$

Here the outer sum has a term for each group, and the inner sum has a term for each observation from the group in question.

1. Make sure you understand the above expression for f.

2. Explain why it suffices to consider each of the terms

$$f_j(\alpha_j) = \sum_{i:g(i)=j} \left(y_i - \alpha_{g(i)} \right)^2$$

on its own.

3. Show that the derivative of f_j is given by

$$f_j'(\alpha_j) = \frac{\partial f_j}{\partial \alpha_j} = 2n_j\alpha_j - 2 \sum_{i:g(i)=j} y_i \qquad (3.6)$$

and solve the equation $f_j'(\alpha_j) = 0$ for each $j = 1, \ldots, k$.

4. Explain why the solution is a minimum point for f_j.

5. What is the conclusion regarding the least squares estimates $\hat{\alpha}_1, \ldots, \hat{\alpha}_k$?

3.4 [M] Pooled residual variance. Recall definition (3.4) of the pooled residual variance s^2, and prove that (3.5) holds. [Hint: Split the sum $\sum_{i=1}^{n}$ into two sums $\sum_{j=1}^{k} \sum_{i:g(i)=j}$, as in Exercise 3.3, and use expression (3.1) for s_j.]

3.5 Experiments from your own field. Think of a situation from your own subject area where the relevant question is that of comparing two groups of data.

"Invent" two different experiments: one corresponding to comparison of two independent samples and one corresponding to paired samples.

3.6 Data structures. In order to investigate the fat intake for kids at different ages three experiments were suggested. For each of them, explain what is the resulting type of data structure.

1. The relative fat content in the feed intake was registered for 25 boys at age 10 years and for another 25 boys at age 12 years.

2. The relative fat content in the feed intake was registered for 25 boys, first when they were 10 years old and later when they were 12 years old.

3. The relative fat content in the feed intake was registered for 25 boys at age 10, 25 boys at age 12, 25 girls at age 10, and 25 girls at age 10.

3.7 ⓡ Parasite counts for salmons. Use the data from Example 3.1 (p. 49) for this exercise.

1. Use mean() and sd() to compute the sample means and the sample standard deviation for the Ätran sample and the Conon sample, respectively. Check that you get the same numbers as in Example 3.1.

2. Use formula (3.5) to compute the pooled sample standard deviation.

3. Use lm() to compute the group means and the pooled sample standard deviation. Check the results with those from the first two questions.

Chapter 4

The normal distribution

Statistical models describe the systematic behavior as well as the random variation in a precise manner. In this chapter we introduce the normal distribution as a tool to describe random variation.

So far we have focused on the average or expected behavior of the observations. In linear regression, we looked at the expected value of y for a given value of x, and in the one-way ANOVA setting, we modeled the expected value for an observation from group j. The average behavior is described by the *systematic* (or fixed) part of a model. However, we have also emphasized that there is variation around these expected values — not all y's have the same value even though they have the same corresponding value of x (linear regression), and not all observations within a group are identical (one-way ANOVA) — and that we need to take this random variation into account.

In situations where the *response variable* — the variable we wish to make a model for — is continuous, then the *normal distribution* or *Gaussian distribution* is by far the most important distribution for that purpose. The normal distribution is not useful for categorical data, and in Chapters 10 and 12 we will work with the binomial distribution for binary data.

The normal distribution is important for several reasons: First — and for our purpose most important — the Gaussian distribution turns out to describe many types of data very well, not the least biological data. This is partly due to "the central limit theorem" (CLT) stating that averages are approximately normally distributed, (almost) no matter the properties of the original variables (see Section 4.4). Second, the Gaussian distribution has very nice mathematical properties, which makes the analysis exact (not relying on approximations) and rather simple.

The Gaussian distribution is named after the German mathematician and physicist Carl Friedrich Gauss (1777—1855).

4.1   Properties

Suppose in the following that we have continuous quantitative data $y_1, \ldots, y_n$. You may for example think about observations of wheat yield, of weight gain during a period on a diet, of the concentration of some hormone in blood samples, or of the weight of crabs, as in the following example.

Example 4.1. Crab weights. The weights in grams of 162 crabs at a certain age were recorded as part of a larger experiment at the Royal Veterinary and Agricultural University in Denmark (Skovgaard, 2004). The sample mean and standard deviation are given by

$$\bar{y} = 12.76 \text{ grams}, \quad s = 2.25 \text{ grams}$$

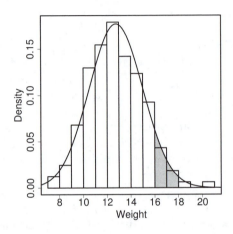

Figure 4.1: Histogram for the crab weight data together with the density for the normal distribution with mean $\bar{y} = 12.76$ and standard deviation $s = 2.25$.

Figure 4.1 shows a relative frequency histogram of the observations, together with the graph of the function

$$f(y) = \frac{1}{\sqrt{2\pi \cdot 2.25^2}} \exp\left(-\frac{1}{2 \cdot 2.25^2}(y - 12.76)^2\right).$$

As will be explained below, the function f is called the density for the normal distribution with mean 12.76 and standard deviation 2.25. The point is that the curve approximates the histogram quite well. This means that the function f is a useful tool for description of the variation of crab weights. □

4.1.1   Density, mean, and standard deviation

Histograms of continuous quantitative data were discussed in Section 1.3: The interval that contains the observations is split into subintervals and for each subinterval a rectangle is drawn with a height reflecting the number of observations in that subinterval. If the histogram is normalized such that the

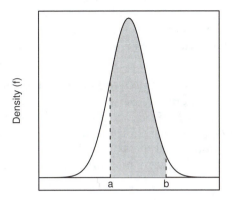

Figure 4.2: Illustration of the relation (4.2). The gray area corresponds to $\int_a^b f(y)\,dy$ and is interpreted as the probability of a random observation falling between a and b.

areas of the rectangles sum to 1, then each area represents the relative frequency of observations in the corresponding subinterval. Hence, if the sample is large, we interpret the area of a rectangle in the histogram as the probability of a random observation falling within the corresponding subinterval.

If the sample is not too small, the histogram often looks quite smooth — just as we saw it in Figure 4.1 — and it is natural to approximate it with a smooth curve. The *density* for the normal distribution corresponds to a particular type of such a smooth curve; namely, the curve given by the function

$$f(y) = \frac{1}{\sqrt{2\pi\sigma^2}} \exp\left(-\frac{1}{2\sigma^2}(y-\mu)^2\right), \quad -\infty < y < \infty. \tag{4.1}$$

Here, $\mu \in \mathbb{R}$ and $\sigma > 0$ are fixed numbers — the mean and the standard deviation, respectively (more about this shortly). For the crab weight data we used $\mu = 12.76$ and $s = 2.25$.

The interpretation of the density is similar to that of the histogram: *for an interval (a, b)* — not necessarily one of the subintervals from the histogram, but any interval — *the area under the curve from a to b is the probability that a random observation falls within the interval.* This is illustrated in Figure 4.2 where the area of the gray region is to be interpreted as the probability that a random observation falls somewhere between a and b. In particular, it is more likely that a random observation falls close to a y-value with large density $f(y)$ compared to a y-value with a small density $f(y)$. The total area under the density function is 1.

Mathematically, the area is written as an integral, so the relationship between the probability and area can be written as

$$P(a < Y < b) = \int_a^b f(y)\,dy, \quad a \le b. \tag{4.2}$$

Here Y represents a random observation and $P(a < Y < b)$ denotes the probability that such a random observation has a value between a and b. The right-hand side is the area under the density curve over the interval from a to b.

Assume that $y_1, \ldots, y_n$ are drawn according to the density (4.1). For any such sample we can compute the sample mean $\bar{y}$ and the sample standard deviation s; cf. (1.5) and (1.6). As n increases, the sample mean will approach μ and the sample standard deviation will approach σ. This explains why μ and σ are called the mean and the standard deviation of the normal distribution. Mathematically, the mean and variance are defined as integrals (see Exercise 4.9).

We say that a variable Y is normally distributed — or Gaussian — with mean μ and standard deviation σ if (4.2) is true for any $a \leq b$ where f is defined by (4.1). Then we write $Y \sim N(\mu, \sigma^2)$. Notice that we follow the tradition and use the variance σ^2 rather than the standard deviation σ in the $N(\mu, \sigma^2)$ notation.

Example 4.2. Crab weights (continued from p. 68). We already noticed from Figure 4.1 that the normal density approximates the histogram quite well for the crab weight data, so it seems reasonable to describe the variation of crab weights with the $N(12.76, 2.25^2)$ distribution. Then the probability that a random crab weighs between 16 and 18 grams is

$$\int_{16}^{18} \frac{1}{\sqrt{2\pi \cdot 2.25^2}} \exp\left(-\frac{1}{2 \cdot 2.25^2}(y - 12.76)^2\right) dy.$$

This is the area of the gray region in Figure 4.1, and turns out to be 0.065; see Example 4.4 (p. 77). Ten of the 162 crab weights are between 16 and 18 grams, corresponding to a relative frequency of $10/162 = 0.062$. The relative frequency and the probability computed in the normal distribution are close if the normal distribution describes well the variation in the sample, as in this example. Otherwise they can be very different.   □

Recall that the density f is determined by the numbers — or parameters — μ and σ. Figure 4.3 shows the density for four different values of (μ, σ); namely, the densities for $N(0, 1)$, $N(0, 4)$, $N(2, 1)$, and $N(-2, 0.25)$. Note that all normal densities are "bell shaped" and that they decrease quickly to zero as y moves away from μ.

By inspection of the definition (4.1) or the figure, it is clear that the density has the following properties:

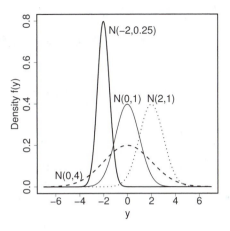

Figure 4.3: Densities for four different normal distributions.

Infobox 4.1: Properties of the density for the normal distribution

Symmetry f is symmetric around μ, so values below μ are just as likely as values above μ.

Center f has maximum value for $y = \mu$, so values close to μ are the most likely to occur.

Dispersion The density is "wider" for large values of σ compared to small values of σ (for fixed μ), so the larger σ the more likely are observations far from μ.

Of course, the interpretations of μ and σ as the mean and standard deviation fit perfectly well with the center and dispersion interpretations in the infobox.

Finally, two remarks on the relation (4.2) between probabilities and areas under the density curve. First, probabilities are non-zero for any interval since $f(y) > 0$ for all y, but single points have probability zero:

$$P(Y = a) = \int_a^a f(y)\, dy = 0$$

for any a. In particular, normal probabilities for open, closed, and half-open intervals are the same,

$$P(a < Y < b) = P(a \leq Y < b) = P(a < Y \leq b) = P(a \leq Y \leq b) \qquad (4.3)$$

and equal to $\int_a^b f(y)\, dy$. Second, the total area under the curve, $\int_{-\infty}^{\infty} f(y)\, dy$, is 1, corresponding to a total probability mass of 1 on $(-\infty, \infty)$. Actually, the density is normalized with $1/\sqrt{2\pi\sigma^2}$ exactly for this to be true.

4.1.2   Transformations of normally distributed variables

The normal distribution has many nice mathematical properties. One is that sums of normally distributed variables are normally distributed, another is that normality is retained for linear transformations of a normally distributed variable. In particular, it is possible to standardize every normally distributed variable so it is $N(0,1)$ distributed. The details are given in the following infobox.

Infobox 4.2: Transformation of normally distributed variables

(a) If Y_1 and Y_2 are *independent* and if $Y_1 \sim N(\mu_1, \sigma_1^2)$ and $Y_2 \sim N(\mu_2, \sigma_2^2)$, then

$$Y_1 + Y_2 \sim N(\mu_1 + \mu_2, \sigma_1^2 + \sigma_2^2).$$

In particular, the mean and standard deviation of $Y_1 + Y_2$ are given by $\mu_1 + \mu_2$ and $\sqrt{\sigma_1^2 + \sigma_2^2}$, respectively.

(b) If $Y \sim N(\mu, \sigma^2)$ and a and b are known real numbers, then the transformed variable $V = a + bY$ has a normal distribution,

$$V = a + bY \sim N(a + b\mu, b^2\sigma^2).$$

In particular, the mean and standard deviation of $a + bY$ are given by $a + b\mu$ and $|b|\,\text{sd}(Y)$, respectively.

(c) If $Y \sim N(\mu, \sigma^2)$ then $Z = \frac{Y - \mu}{\sigma} \sim N(0,1)$.

The third property is a special case of the second — with $a = -\mu/\sigma$ and $b = 1/\sigma$. The alert reader may have noticed that we assumed *independence* between Y_1 and Y_2 for the first property. Loosely speaking, independence between Y_1 and Y_2 means that Y_1 and Y_2 do not share any information: observing one of the variables does not provide information about the other (see also Sections 4.2.1 and 9.3).

In fact, $Y_1 + Y_2$ is normally distributed with mean $\mu_1 + \mu_2$ even if Y_1 and Y_2 are not independent — independence is only required in order for the variance formula to be correct. The variation of a sum changes whether or not the terms in the sum contain information about each other.

Also note that the formulas for the mean and variance are closely related to the corresponding formulas for the sample mean and sample standard deviation; see Infobox 1.1 (p. 14). Actually, the formulas for the mean and standard deviation in properties (a) and (b) hold for any variables, not only variables with a normal distribution; but remember that independence is required for the variance of the sum to be the sum of the variances.

Example 4.3. Crab weights (continued from p. 68). For the crab weight data, we concluded that the normal distribution with mean 12.76 and standard deviation 2.25 approximately describes the distribution of crab weights (Y). The properties from Infobox 4.2 are then illustrated as follows:

1. The total weight of two (independent) crabs is normally distributed with mean $12.76 + 12.76 = 25.52$ and variance $2.25^2 + 2.25^2 = 10.125$, corresponding to a standard deviation of $\sqrt{10.125} = 3.18$.

2. If the weight is given in kilograms instead of grams and (for some weird reason) a weight of 100 grams = 0.1 kilograms is added, then the total weight in kilograms is $V = 0.1 + Y/1000$, and V has a normal distribution with mean $0.1 + 12.76/1000 = 0.1128$ and standard deviation $2.25/1000 = 0.00225$; *i.e.*, $V \sim N(0.1128, 0.00225^2)$.

3. The variable $Z = (Y - 12.76)/2.25$, measuring the standardized deviation from the mean, is $N(0,1)$ distributed.

$\square$

Now, assume that $y_1, \ldots, y_n$ are independent and that each of them is $N(\mu, \sigma^2)$ distributed. Extending property (a) from Infobox 4.2 to a sum of n terms rather than two, we see that the sum $\sum_{i=1}^{n} y_i$ is normally distributed with mean $n\mu$ and variance $n\sigma^2$. From Infobox 4.2(b), with $a = 0$ and $b = 1/n$, it then follows that $\bar{y}$ has a normal distribution with mean μ and variance σ^2/n:

Infobox 4.3: Distribution of sample mean

If $y_1, \ldots, y_n$ are independent and each $y_i \sim N(\mu, \sigma^2)$, then

$$\bar{y} = \frac{1}{n} \sum_{i=1}^{n} y_i \sim N(\mu, \sigma^2/n). \tag{4.4}$$

In words: An average of variables which all have the same normal distribution is itself normally distributed. The mean is the same as that of the original variables, but the standard deviation is divided by $\sqrt{n}$, where n is the number of variables in the average. The result will prove extremely important in Section 4.2 (and the subsequent chapters), and we will discuss its implications and interpretations at that point.

4.1.3 Probability calculations

The normal distribution with mean 0 and standard deviation 1, $N(0,1)$, is called the *standard normal distribution* and has density

$$\phi(y) = \frac{1}{\sqrt{2\pi}} \exp\left(-\frac{1}{2}y^2\right).$$

Note that this function is important enough that it has its own name, ϕ. Infobox 4.2(c) shows that probabilities regarding a $N(\mu, \sigma^2)$ variable can be computed in the $N(0, 1)$ distribution:

$$P(a < Y < b) = P\left(\frac{a - \mu}{\sigma} < \frac{Y - \mu}{\sigma} < \frac{b - \mu}{\sigma}\right)$$

$$= P\left(\frac{a - \mu}{\sigma} < Z < \frac{b - \mu}{\sigma}\right) \qquad (4.5)$$

where $Z \sim N(0, 1)$. This is useful because it implies that we only need to be able to compute probabilities in the $N(0, 1)$ distribution, no matter the mean and standard deviation in the distribution of Y.

Hence, consider for a moment a $N(0, 1)$ distributed variable Z. The probability that a random observation of Z falls within a certain interval is computed as the integral of its density over the interval in question. Hence, for the probability that Z is at most z, we should integrate up to z:

$$P(Z \leq z) = \int_{-\infty}^{z} \frac{1}{\sqrt{2\pi}} e^{-\frac{1}{2}y^2} \, dy.$$

The integral cannot be solved explicitly, but certainly numerically. The result is usually denoted $\Phi(z)$; that is,

$$\Phi(z) = P(Z \leq z) = \int_{-\infty}^{z} \frac{1}{\sqrt{2\pi}} e^{-\frac{1}{2}y^2} \, dy.$$

Φ is called the *cumulative distribution function* (cdf) of Z or the cumulative distribution function of $N(0, 1)$. Note that the definition implies that the density function is the derivative of the cdf,

$$\frac{d\Phi(z)}{dz} = \Phi'(z) = \phi(z).$$

Figure 4.4 shows the graphs of ϕ and Φ. Notice that $\Phi(z)$ is always in the interval $(0, 1)$ and that Φ is increasing. This also follows from the definition of $\Phi(z)$ as $P(Z \leq z)$. The dashed lines in the graph correspond to $z = -1.645$ and $z = 1.645$, respectively. These values are selected because

$$\Phi(-1.645) = P(Z \leq -1.645) = 0.05$$
$$\Phi(1.645) = P(Z \leq 1.645) = 0.95.$$

Since the points -1.645 and 1.645 have zero probability, see (4.3), it follows that

$$P(Z > 1.645) = 0.05$$
$$P(-1.645 < Z < 1.645) = 0.90. \qquad (4.6)$$

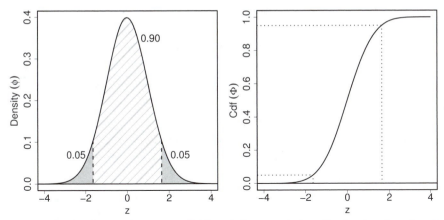

Figure 4.4: The density function ϕ (left) and the cumulative distribution function Φ (right) for $N(0,1)$. The dashed lines correspond to $z = \pm 1.645$. Each gray-shaded region has area (probability) 0.05, whereas the dashed-shaded region has area (probability) 0.90.

The probabilities are illustrated in the left part of Figure 4.4 with gray-colored and dashed-shaded regions.

Probabilities concerning an $N(0,1)$ distributed variable are easily expressed in terms of the cdf:

$$P(a < Z < b) = \int_a^b \phi(y)\,dy = \int_{-\infty}^b \phi(y)\,dy - \int_{-\infty}^a \phi(y)\,dy = \Phi(b) - \Phi(a).$$

If this is combined with (4.5), then we see that all probabilities in the $N(\mu,\sigma)$ distribution can be computed from Φ:

$$P(a < Y < b) = \Phi\left(\frac{b - \mu}{\sigma}\right) - \Phi\left(\frac{a - \mu}{\sigma}\right) \qquad (4.7)$$

Recall from (4.3) that we could replace $<$ with $\leq$ (in one or both places) and still get the same probability.

As mentioned above, there is no explicit expression for Φ. A few selected values are listed in the left-hand side of Table 4.1 for illustration, but generally values $\Phi(z)$ are looked up in statistical tables (see Appendix C.2) or computed by a computer program (see Section 4.5.1) or a pocket calculator.

The right-hand side of Table 4.1 lists important quantiles: for example, since $\Phi(1.645) = 0.95$, there is a 95% chance that a $N(0,1)$ variable is less than 1.645. We say that 1.645 is the 95% *quantile* of the standard normal distribution.

Table 4.1: Selected values of Φ, the cumulative distribution function for the standard normal distribution

z	$\Phi(z)$	z	$\Phi(z)$
-3.000	0.0013	1.282	0.9000
-2.000	0.0228	1.645	0.9500
-1.000	0.1587	1.960	0.9750
0.000	0.5000	2.326	0.9900
1.000	0.8413	2.576	0.9950
2.000	0.9772	3.090	0.9990
3.000	0.9987	3.291	0.9995

4.1.4   Central part of distribution

It is often of interest to answer questions like: what is the central 90% (say) of the distribution; that is, in which interval will we expect the central 90% of our observations to occur? In other words — and due to the symmetry of $N(\mu, \sigma^2)$ around μ — we are looking for k such that the interval $(\mu - k, \mu + k)$ has probability 90%.

We already found the answer for $N(0, 1)$: Recall from (4.6), Table 4.1, or Figure 4.4 that if $Z \sim N(0, 1)$ then

$$P(-1.645 < Z < 1.645) = 0.90.$$

Hence, the interval from -1.645 to 1.645 contains the central 90% of the distribution.

Because of (4.5), this is useful for the general $N(\mu, \sigma^2)$ distribution, too. If $Y \sim N(\mu, \sigma^2)$ then

$$P(\mu - 1.645\sigma < Y < \mu + 1.645\sigma)$$

$$= P\left(\frac{\mu - 1.645\sigma - \mu}{\sigma} < \frac{Y - \mu}{\sigma} < \frac{\mu + 1.645\sigma - \mu}{\sigma} \right)$$

$$= P\left(-1.645 < \frac{Y - \mu}{\sigma} < 1.645 \right)$$

$$= 0.90.$$

In other words: for *any normal distribution* $N(\mu, \sigma^2)$ *it holds that the interval from* $\mu - 1.645 \cdot \sigma$ *to* $\mu + 1.645 \cdot \sigma$ *defines the central 90% of the distribution*. The interval is often written as $\mu \pm 1.645 \cdot \sigma$. Notice that we should use the 95% quantile 1.645 in order to get the central 90% of the observations. This is what was illustrated in the left part of Figure 4.6.

Similar computations are made for other percentages or other quantiles. Some important examples are listed in Table 4.2 and illustrated in Figure 4.5.

Table 4.2: Intervals and corresponding probabilities for $N(\mu, \sigma^2)$

Interval	$\mu \pm 1.645 \cdot \sigma$	$\mu \pm 1.960 \cdot \sigma$	$\mu \pm 2.576 \cdot \sigma$	$\mu \pm \sigma$	$\mu \pm 3 \cdot \sigma$
Probability	0.90	0.95	0.99	0.68	0.997

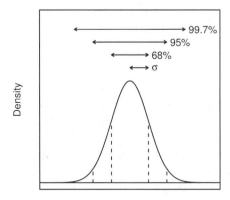

Figure 4.5: Density for $N(\mu, \sigma^2)$ with probabilities for intervals $\mu \pm \sigma$, $\mu \pm 2\sigma$, and $\mu \pm 3\sigma$.

For example, we would expect the interval as $\mu \pm 1.96\sigma$ to contain approximately 95% of the observations, since the 97.5% quantile in $N(0,1)$ is 1.96. This corresponds to a commonly used rule-of-thumb that roughly 95% of the observations are within a distance of 2 times the standard deviation from the mean. The "rule" is used for other distributions, too, but the distribution should look somewhat like the normal distribution; in particular, it should not be too asymmetric.

Example 4.4. Crab weights (continued from p. 68). If we assume that crab weights are normally distributed with mean 12.76 and standard deviation 2.23, then we would expect 95% of all crabs weights to be in the interval $12.76 \pm 1.96 \cdot 2.25$; that is, between 8.35 and 17.17.

With the cumulative distribution function we are also able to compute the probability from Example 4.1. We get

$$
\begin{aligned}
P(16 < Y < 18) &= P\left(\frac{16 - 12.76}{2.25} < Z < \frac{18 - 12.76}{2.25}\right) \\
&= P(1.44 < Z < 2.33) \\
&= \Phi(2.33) - \Phi(1.44) \\
&= 0.990 - 0.925 \\
&= 0.065,
\end{aligned}
$$

where the Φ values are looked up in a table or computed with a computer or pocket calculator. If we accept that the crabs in our sample are representative, we can conclude that there is a 6.5% chance of observing a weight between 16 and 18 grams for a randomly selected crab from the population.   □

4.2   One sample

Consider observations $y_1, \ldots, y_n$ and assume that we are interested in their expected value; that is, the value we would expect to see for a random observation of this type. In particular we may think of $y_1, \ldots, y_n$ as the values for a sample drawn from some population, then the expected value is the average value in the population. Hence, we are not interested in the sample as such, rather in it being representative of the population, and we use the sample to infer information about the population.

For example, the crabs from Example 4.1 (p. 68) were collected in order to obtain knowledge on the weight of crabs in general: What is the average crab weight (in the population of crabs)? What can be thought of as "normal" weights for a crab? How much should the weight differ in order for it to be "unusual"?

We use the sample mean $\bar{y}$ as an estimate of the population mean. However, we would get (slightly) different sample means for different samples, so how much can we "trust" the one we got? The larger the sample, the more trustworthy — or the more precise — the sample mean; but how precise is that? The point is that if the population values can be assumed to be distributed according to a normal distribution, and if the sample elements are drawn at random from the population, then we can answer these questions!

4.2.1   Independence

In the following we will always assume that observations in a sample are independent. Formally, *independence* of random variables $Y_1, \ldots, Y_n$ means that the probability that $Y_1 \leq a_1$ at the same time as $Y_2 \leq a_2$ etc. can be computed as the product of probabilities for each Y_i for all $a_1, \ldots, a_n$:

$$P(Y_1 \leq a_1, \ldots, Y_n \leq a_n) = P(Y_1 \leq a_1) \cdots P(Y_n \leq a_n). \qquad (4.8)$$

As explained below Infobox 4.2, it means that each observation brings completely new information to the dataset. Let us illustrate by an example.

Example 4.5. Sampling of apple trees. Suppose that we are collecting data for a study on yield from apple trees and plan to select 20 trees from a population of 1000 trees. If all trees are numbered from 1 to 1000, and we let a computer draw 20 random numbers from 1 to 1000, then we can assume that the yields from the corresponding apple trees are independent.

On the other hand, if we sample 10 trees in the above manner, and furthermore for each of these trees select a "twin" tree which is very much alike, then the trees within a pair hold information about each other, simply because the "twins" are selected to be alike. Hence the yields are *not* independent. A third strategy might be to sample 10 trees as above and for each tree select a tree which is very different (in size, say). Again, the corresponding yields will not be independent.

Finally, assume that the 1000 apple trees grew in 10 different orchards and that two trees were drawn from each orchard. We would expect trees from the same orchard to be more similar than trees from different orchards, because they share the same environment (soil, rainfall, *etc.*). Hence, trees from the same orchard share some information and it is not reasonable to assume that they are independent. □

4.2.2 Estimation

Assume that $y_1, \ldots, y_n$ are independent observations from a normal distribution with mean μ and standard deviation σ. We say that $y_1, \ldots, y_n$ are *iid.* $N(\mu, \sigma^2)$, where iid. means independent and identically distributed.

The mean μ and the standard deviation σ are unknown values (population values), so we use our sample to compute their estimates. The natural estimates are the sample mean and the sample standard deviation,

$$\hat{\mu} = \bar{y}, \quad \hat{\sigma} = s.$$

Recall from (4.4) that $\bar{y}$ has a normal distribution with mean and standard deviation given by

$$E(\hat{\mu}) = E(\bar{y}) = \mu, \quad sd(\hat{\mu}) = sd(\bar{y}) = \frac{1}{\sqrt{n}}\sigma. \tag{4.9}$$

From a statistical point of view the interpretation is the following:

Infobox 4.4: Statistical properties of sample means

The sample mean is an unbiased estimate *Sample means on average "hit" the right population mean.* That is, taking a (very) large number of different samples, the mean of the sample means will be μ.

The sample mean is a consistent estimate *Sample means get more and more precise as the sample size increases,* as the standard deviation of $\hat{\mu}$ decreases.

Unbiasedness and consistency are both very desirable properties. The first one tells us that the sample mean on average gives us the correct expected value (population mean). The second tells us that we can improve the precision of our estimate of μ by increasing the number of observations; in particular, make the standard deviation of $\hat{\mu}$ as small as we wish by taking a large enough sample.

The distribution of $\bar{y}$ is illustrated in Figure 4.6. For the left plot we simulated 1000 samples of size 10 from $N(0, 1)$, and computed the sample mean (mean of ten values) for each sample. The plot shows a histogram of the 1000 sample means together with the density for the $N(0, 1/10)$ distribution. For the right plot we have used sample size 25 instead and the density is that of $N(0, 1/25)$. We see that the normal densities fit the histograms very well, and clearly the distribution gets more narrow as n increases.

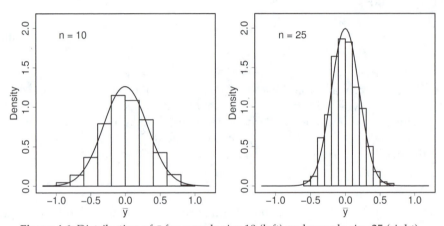

Figure 4.6: Distribution of $\bar{y}$ for sample size 10 (left) and sample size 25 (right).

4.3   Are the data (approximately) normally distributed?

For many applications it is important that the distribution of a certain variable is approximately a normal distribution, so we must carry out some kind of model validation. It would only rarely be correct to say that a certain variable is *exactly* distributed according to a normal distribution (and we would never be able to check it), but luckily the statistical methods work well as long as the normality assumption holds reasonably well. Usually the assumption of a normal distribution is validated graphically, as illustrated in the following.

4.3.1 Histograms and QQ-plots

If the sample is large enough that it makes sense to make a histogram, then we may compare the histogram of the observations to the normal density with mean and standard deviation equal to the sample mean and sample standard deviation. We did so for the crab weight data from Example 4.1 (p. 68) in Figure 4.1 and concluded that the normal density approximated the histogram quite well, suggesting that crab weights are approximately normally distributed.

Another relevant plot is the *QQ-plot*, or quantile-quantile plot, which compares the sample quantiles to those of the normal distribution. The QQ-plot is shown for the crab weights in Figure 4.7. If data are $N(\mu, \sigma^2)$ distributed, the points in the QQ-plot should be scattered around the straight line with intercept μ and slope σ, so we compare the points with the straight line with the estimated parameters: intercept $\bar{y}$ and slope s. We see that there are no serious deviations from the straight line relationship.

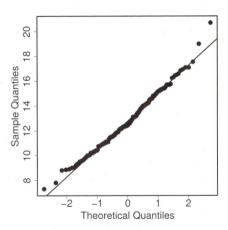

Figure 4.7: QQ-plot of the crab weight data together with the line with intercept equal to $\bar{y} = 12.76$ and slope equal to $s = 2.25$.

The idea behind the QQ-plot is the following: Assume first that we have a sample $z_1, \ldots, z_n$ and that we want to check if the values could come from the $N(0, 1)$ distribution. Let $z_{(j)}$ denote the jth smallest observation among $z_1, \ldots, z_n$ such that $z_{(1)} < z_{(2)} < \cdots < z_{(n)}$. These observations split the interval from $-\infty$ to $+\infty$ into $n + 1$ parts. Each interval between two $z_{(j)}$'s is ascribed probability $1/n$ and the intervals $(-\infty, z_{(1)})$ and $(z_{(n)}, +\infty)$ are ascribed probability $1/(2n)$ each. Let u_j be the $N(0, 1)$ quantile corresponding to the accumulated probabilities up to $z_{(j)}$; i.e., let u_j be the $(j - 0.5)/n$ quantile of $N(0, 1)$. Now, if the z_i's are $N(0, 1)$ distributed, then the sample quantiles and the $N(0, 1)$ quantiles should be similar: $z_{(j)} \approx u_j$ for all $j = 1, \ldots, n$. Therefore, if we plot the ordered observations against the corresponding

$N(0,1)$ quantiles, the points should be scattered around the straight line with intercept zero and slope one. Returning to the original sample $y_1, \ldots, y_n$, recall that $y_i = \mu + \sigma z_i$ where $z_i \sim N(0,1)$ if $y_i \sim N(\mu, \sigma^2)$. Hence, a plot of the ordered observations against the corresponding $N(0,1)$ quantiles should show points scattered around a straight line, and the intercept and slope of that line can be thought of as estimates of μ and σ.

4.3.2   Transformations

It is not unusual that the histogram or the QQ-plot reveals that the assumption of a normal distribution is unreasonable. This can have several reasons.

For example, imagine that there are observations corresponding to two different groups, for example men and women or children and adults, and that the values are quite different for the two groups. A histogram of the complete dataset would show a distribution with two modes, one for each group, and the QQ-plot would be "S-shaped". In this case a reasonable model is that of *two normal distributions*, not a common normal distribution for both groups. This is illustrated by the BMR variable in Example 4.6 below (Figure 4.9).

Another problem may be that the data are severely skewed (asymmetric). This is quite common for data with positive values only: very large values are possible and do perhaps occur, whereas negative values are impossible. In these cases log-transformation of the data may sometimes solve the problem. This means that we consider $\log(y_1), \ldots, \log(y_n)$ as our observations instead of the original $y_1, \ldots, y_n$ and check if the log-transformed data are normally distributed. This is illustrated by the vitamin A variable in Example 4.6 below (Figure 4.8).

Other transformations may of course also be useful. Notice that computations with the normal distribution are proper concerning the transformed variable, not the original one. This means that probability computations involving the original variable should be rephrased in terms of the transformed variables and computed on this scale. Case 2 (p. 353) illustrates this point for a particular dataset.

Example 4.6. Vitamin A intake and BMR. The food intake by Danish people was studied in a comprehensive survey by Haraldsdottir et al. (1985), including data from 2224 persons. The left panels of Figure 4.8 show a histogram and the QQ-plot for the intake of vitamin A for men (1079 values). The data are clearly skewed to the right and certainly *not* normally distributed. The right panels of the figure show the same figures for the log-transformed values of the data. The density for the normal distribution with mean and standard deviation equal to the sample mean and sample standard deviation for the logarithmic values (7.48 and 0.44, respectively) is plotted together with the histogram. The normal density approximates the histogram very well and expect for a few points the QQ-plot shows a straight line relationship, so it seems reasonable to assume that the log-intake is normally distributed.

Notice that we used the natural logarithm, but we could also have used the logarithmic function with base 10, say. Since transformed values corresponding to different bases are linear transformations of one another, the result would be equivalent.

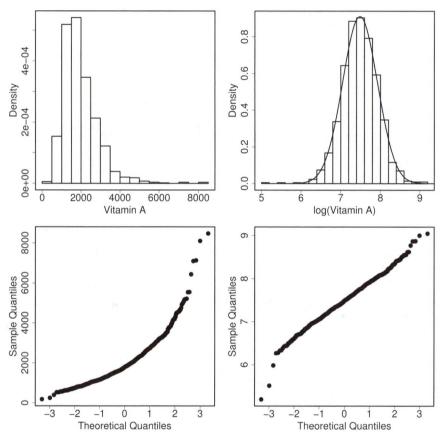

Figure 4.8: Histograms and QQ-plots of the intake of vitamin A for 1079 men: original values to the left, log-transformed values to the right.

Many other variables were registered in the same study. Histograms of a variable called BMR, related to the basal metabolic rate, are shown in Figure 4.9. The two plots in the top — for men and women, respectively — show reasonable (although not perfect) agreement between the histogram and the corresponding normal density curve. Note that the centers are different, around 7.5 and 5.5, respectively. The lower left histogram includes all data, and a QQ-plot is shown in the lower right figure. The distribution is clearly bimodal and the normal approximation is thus not appropriate.

In conclusion, the normal approximation is reasonable for the BMR variable only when we have taken into account the difference between men and women.   □

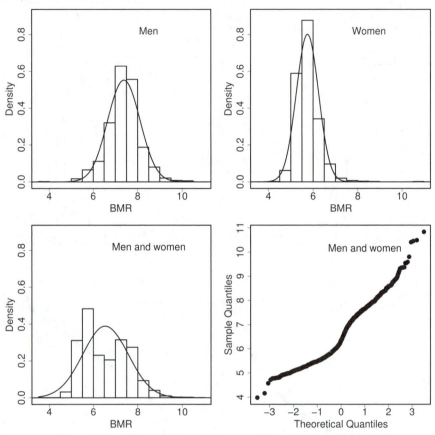

Figure 4.9: Histograms of the BMR variable for 1079 men (upper left), 1145 women (upper right), and all 2224 persons (bottom left), and QQ-plot for all 2224 persons (bottom right).

4.3.3   The exponential distribution

The normal distribution is by far the most important one for quantitative data, but it is not always appropriate and there are many other distributions. In this section we will briefly introduce the *exponential distribution*. It is often used for data measuring the time until some event occurs.

The exponential distribution with rate (or parameter) $\gamma > 0$ has density

$$g(y) = \gamma e^{-\gamma y}, \quad y > 0.$$

Notice that the density is defined only for $y > 0$, so the exponential distribution should be used only for variables that are always positive. The density has the same interpretation as for the normal distribution: If Y has an exponential distribution, then the probability that Y is in the interval (a, b) for $0 \le a < b$ is equal to the area under the density curve over the interval. We thus get

$$P(a < Y < b) = \int_a^b g(y)\, dy = \int_a^b \gamma e^{-\gamma y} = [-e^{-\gamma y}]_a^b = e^{-\gamma a} - e^{-\gamma b}. \quad (4.10)$$

The mean of the exponential distribution with rate γ is $1/\gamma$ (see Exercise 4.10), so for data $y_1, \ldots, y_n$ it is natural to estimate γ by $1/\bar{y}$.

Example 4.7. Interspike intervals for neurons. A study of the membrane potential for neurons from guinea pigs was carried out by Lansky et al. (2006). The data consists of 312 measurements of interspike intervals; that is, the length of the time period between spontaneous firings from a neuron. The histogram in the left panel of Figure 4.10 shows that the distribution is highly asymmetric, so the normal distribution is not applicable.

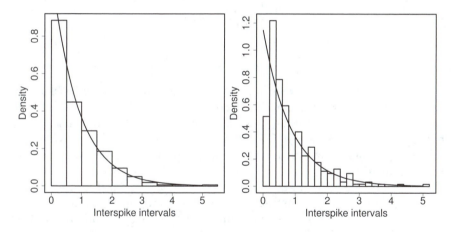

Figure 4.10: Histograms for 312 interspike intervals with different resoulutions together with the density for the exponential distribution with rate 1.147.

The mean for the interspike data is 0.872, so we estimate γ by $\hat{\gamma} = 1/0.872 = 1.147$. The corresponding density is added to the histogram in the left part of Figure 4.10, and it seems that it fits quite well to the histogram, meaning that the interspike intervals are well described by the exponential distribution. However, a closer look at the data reveals that this is not completely true. The right panel in Figure 4.10 is a histogram for the same data but with a finer grid. The density is the same as in the left panel. The plot shows that there are too few measurements very close to zero compared to what we would expect from the exponential distribution. □

4.4   The central limit theorem

Recall from Section 4.2.2 that the sample mean has a normal distribution,

$$\bar{y} \sim N(\mu, \sigma^2/n)$$

if $y_1, \ldots, y_n$ are iid. from $N(\mu, \sigma^2)$. That is, if the sample is drawn from a normal distribution, then the sample mean is normally distributed, too.

The *central limit theorem* (CLT) states that *the mean of independent variables drawn from the same distribution is approximately normally distributed* as long as the sample size is large — (almost) no matter the distribution of the original variables.

Infobox 4.5: Central limit theorem (CLT)

If $y_1, \ldots, y_n$ are iid. variables with mean μ and standard deviation σ, then

$$\bar{y} = \frac{1}{n} \sum_{i=1}^{n} y_i \overset{\text{app.}}{\sim} N(\mu, \sigma^2/n)$$

if n is large enough.

The central limit theorem is quite astonishing — there is no *a priori* reason to believe that averages/means have the same distribution type regardless of the distribution of the original variables — and shows the importance of the normal distribution: probabilities about an average can be (approximately) computed in the normal distribution, no matter how the original observations are distributed.

We will not go into details about the precise mathematical formulation of the central limit theorem. Instead, the following examples illustrate the theorem for y_i's with only two possible values (binary variables) and for variables with a skewed distribution. The binary distribution is as far from the normal distribution as one can imagine, but the distribution of the mean is still close to normal for large sample sizes.

Example 4.8. Central limit theorem for binary variables. Assume that $y_1, \ldots, y_n$ are independent and either one or zero, with probabilities p and $1 - p$, respectively:

$$y_i = \begin{cases} 1 & \text{with probability } p \\ 0 & \text{with probability } 1 - p \end{cases}$$

Then the possible values for the sum $\sum_{i=1}^{n} y_i$ are $0, 1, \ldots, n$. Actually, the sum follows a binomial distribution, which we will get back to in Chapter 10. The mean $\bar{y} = \frac{1}{n} \sum_{i=1}^{n} y_i$ is the relative frequency of ones. Recall from Section 1.5

that $\bar{y}$ stabilizes around p as n increases. The central limit theorem tells us more than that; namely, that the distribution is approximately a normal distribution.

We have not talked about mean, variance, and standard deviation for binary variables, but let us do so now. The mean of y_i, or the expected value, is

$$\mathrm{E}y_i = 1 \cdot p + 0 \cdot (1 - p) = p.$$

This is a weighted average of the possible values (zero and one), with weight equal to the corresponding probabilities. The variance can be defined as the expected squared difference between y_i and its mean. Since

$$(y_i - \mathrm{E}y_i)^2 = (y_i - p)^2 = \begin{cases} (1 - p)^2 & \text{with probability } p \\ (0 - p)^2 = p^2 & \text{with probability } 1 - p \end{cases}$$

the expected value is

$$\begin{aligned} \mathrm{Var}(y_i) &= \mathrm{E}(y_i - p)^2 \\ &= (1 - p)^2 \cdot p + p^2 \cdot (1 - p) \\ &= p + p^3 - 2p^2 + p^2 - p^3 \\ &= p - p^2 \\ &= p(1 - p). \end{aligned}$$

Again the expected values is computed as a weighted average of the possible values, this time $(1 - p)^2$ and p^2. The standard deviation is

$$\mathrm{sd}(y_i) = \sqrt{\mathrm{Var}(y_i)} = \sqrt{p(1 - p)}.$$

The central limit theorem thus states that the relative frequency of ones — the mean $\bar{y}$ — is approximately distributed according to the normal distribution with mean p and standard deviation $\sqrt{p(1 - p)/n}$ for n large enough:

$$\bar{y} \overset{\text{app.}}{\sim} N\left(p, \frac{p(1 - p)}{n}\right).$$

The approximation is clearly bad for small n. For $n = 3$, for example, the possible values of $\bar{y}$ are 0, 1/3, 2/3, and 1, and $N(p, p(1 - p)/3)$ is a bad approximation. The situation is illustrated in Figure 4.11 for $p = 0.5$ and two different sample sizes: $n = 10$ and $n = 100$. For the left panel we simulated 1000 samples of size 10, computed the sample mean for each of the 1000 samples, and plotted the histogram of the sample means as well as the $N(0.5, 0.025)$ density. Already for sample size 10 the approximation to the normal distribution is quite good. The right panel corresponds to $n = 100$, where the histogram and the normal density are very close. $\qquad \square$

Example 4.9. Central limit theorem for a bimodal distribution. Assume that

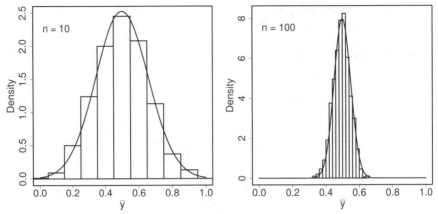

Figure 4.11: Illustration of the central limit theorem for 0/1-variables with $p = 0.5$. Histograms of the sample means for 1000 simulated samples of size 10 (left) and 100 (right) compared to the corresponding normal density curve.

the concentration of a substance in cow milk is normally distributed with mean 100 and standard deviation 5 (in some unit) for healthy cows but normally distributed with mean 40 and standard deviation 10 for cows with a certain disease. A drug can be used for treatment which brings back the concentration to the level for healthy cows. For 10% of diseased cows, however, the drug does not work and the level of concentration is not increased.

The distribution of the concentration for a random cow under treatment is illustrated in the upper left panel of Figure 4.12. The histogram is based on 2000 simulated values. The distribution is clearly bimodal, with 90% of the observations around 100 and 10% around 40. The distribution has mean 94, variance 356.5, and standard deviation 18.9, and the solid line is the density for $N(94, 356.5)$. This density of course fits very badly with the histogram, but as we shall see, the central limit theorem still applies.

For the upper right part of Figure 4.12 we simulated $y_1, \ldots, y_5$ from the bimodal distribution and computed $\bar{y}$ for the simulated data; *i.e.*, $n = 5$. We repeated this 2000 times and plotted the histogram of the 2000 sample means together with the density for $N(94, 356.5/5)$. The histogram has several peaks and is still very different from the normal density. In the lower panels we did the same, but now for $n = 25$ (left) and $n = 100$ (right); so each $\bar{y}$ is an average of 25 and 100 values, respectively. The normal densities are changed accordingly. For $n = 100$ the sample distribution for $\bar{y}$ is almost indistinguishable from the normal distribution, just as postulated by the central limit theorem for when n is large enough.   □

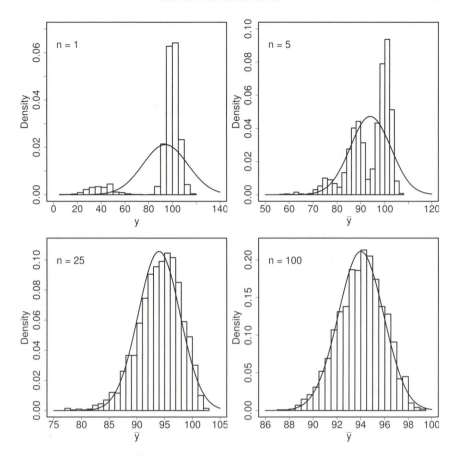

Figure 4.12: Sample distribution of the sample mean $\bar{y}$ of n observations from a bimodal distribution. The bimodal distribution is illustrated in the upper left panel (corresponding to $n = 1$). The curves are normal densities.

4.5 R

4.5.1 Computations with normal distributions

The functions pnorm(), qnorm(), dnorm() are useful for computations with normal distributions. They give the cumulative distribution function (p for probability), the quantiles (explaining the q), and the density (explaining the d), respectively. Let us start with pnorm() yielding values of the cumulative distribution function (cdf). For example, let $Y \sim N(4, 9)$. Then the probability $P(Y \leq 2)$ can be computed with pnorm() in several different ways:

```
> pnorm(2, mean=4, sd=3)   # P(Y<=2) in N(4,9)
```

```
[1] 0.2524925
> pnorm(2,4,3)                # P(Y<=2) in N(4,9)
[1] 0.2524925
> pnorm(-2/3,0,1)             # P(Z<=-2/3) in N(0,1)
[1] 0.2524925
> pnorm(-2/3)                 # P(Z<=-2/3) in N(0,1)
[1] 0.2524925
```

In the first command the first argument to pnorm() is the value 2 since we want to compute $P(Y \leq 2)$, and the second and third arguments denote the mean and the standard deviation, respectively, in the distribution of Y. Note that the third argument is the standard deviation, σ, not the variance, σ^2. Since

$$P(Y \leq 2) = P\left(Z \leq \frac{2-4}{3}\right) = P(Z \leq -2/3),$$

the probability is similarly computed as pnorm(-2/3,0,1), the third command. Finally, the default values of the mean and standard deviation are zero and one, respectively, so the fourth command is equivalent to the third.

Similarly, the probability in Example 4.4 (p. 77) may be computed in different ways:

```
> pnorm(18, 12.76, 2.25) - pnorm(16, 12.76, 2.25)
[1] 0.06500122
> pnorm(2.33) - pnorm(1.44)
[1] 0.06503062
```

Quantiles for the standard normal distribution are computed with the qnorm() function, and values of ϕ, the density function for the standard normal distribution, are computed with dnorm(). For example, the 95% and the 97.5% quantiles are 1.645 and 1.960 (*cf.* Table 4.1 on p. 76), and $\phi(0)$ is close to 0.4 (*cf.* Figure 4.3 on p. 71):

```
> qnorm(0.95)     # 95% quantile of N(0,1)
[1] 1.644854
> qnorm(0.975)    # 97.5% quantile of N(0,1)
[1] 1.959964
> dnorm(0)        # Density of N(0,1) evaluated at y=0
[1] 0.3989423
```

Like pnorm, the functions qnorm and dnorm can be used for normal distributions with non-zero mean and non-unit standard deviation by supplying the mean and standard deviation as extra arguments. For example, for the $N(4,9)$ distribution:

```
> qnorm(0.975,4,3)    # 97.5% quantile of N(4,9)
[1] 9.879892
> dnorm(1,4,3)        # Density of N(4,9) at y=1
[1] 0.08065691
```

4.5.2 Random numbers

It is sometimes useful to be able to draw random numbers from the normal distribution. This is possible with the rnorm() function (r for random). As above, the mean and standard deviation should be specified if the target distribution is not the standard normal distribution:

```
> rnorm(5)              # Five random draws from N(0,1)
[1]    2.4093946   1.0126402 -1.9877058   0.3397944 -2.3190924
> rnorm(2,4,3)
[1] 1.995428 3.960494  # Two random draws from N(4,9)
```

4.5.3 QQ-plots

QQ-plots are easily constructed with the qqnorm() function. Let the vector wgt contain the 162 measurements from Example 4.1 (p. 68). Then the command

```
> qqnorm(wgt)
```

produces a plot with points similar to those in Figure 4.7 on p. 81. If we want to compare the graph with a normal distribution with *known* mean and *known* standard deviation, then a line with intercept equal to the mean and slope equal to the standard deviation might be added with the abline() function. Alternatively, the qqline() function adds the straight line corresponding to the normal distribution with the same 25% and 75% quantiles as the sample quantile values.

```
> abline(12.8, 2.25)   # Compare to N(12.8, 2.25^2)
> qqline(wgt)          # Comp. to N with same 25%, 75% quantiles
```

4.6 Exercises

4.1 Histograms for normally distributed variables. For a sample $y_1, \ldots, y_n$ from the normal distribution with mean μ and standard deviation σ, we expect 95% of the observations to fall within the interval $(\mu - 2\sigma, \mu + 2\sigma)$; that is, in the interval

$$"\text{mean} \pm 2 \cdot \text{standard deviation}". \qquad (4.11)$$

This follows from Section 4.1.4, but if you did not read that far yet you can just use the result.

Histograms are shown for two samples, each of size 1000. The data are drawn from the normal distribution, but with two different means and standard deviations. For each histogram, use (4.11) to give a (rough) estimate of the mean and the standard deviation.

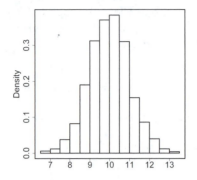 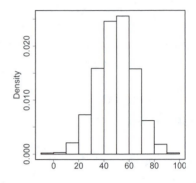

4.2 Transformations and probability calculations. Recall the vitamin A data from Example 4.6 (p. 82). The investigation showed that the distribution of the log-transformed vitamin A intake is well described by the normal distribution with mean 7.48 and standard deviation 0.44. We used the natural logarithm.

1. Compute the probability that a random man has a vitamin A intake between 3000 and 4000.

   [Hint: Rewrite the condition in terms of the logarithmic vitamin A intake.]

2. Calculate the central 90% area of the distribution of the vitamin A intake.

   [Hint: Use the results from Section 4.1.4. On which scale do the results apply?]

4.3 ℝ Investigation of normally distributed variables by simulation. In this exercise we will try to get a feeling for variables with a normal distribution. We will do so by simulation: we make R draw samples from the normal distribution and examine some of their properties.

First, let us generate a single sample with R and examine it.

1. Try the command

   ```
   y <- rnorm(10, mean=5, sd=1.5)
   ```

   and print y on the screen. y is a sample of 10 observations from the normal distribution with mean 5 and standard deviation 1.5; that is, from $N(5, 1.5^2) = N(5, 2.25)$.

2. Make a sample y1 of size 1000 from $N(5, 2.25)$. Compute the sample mean, the sample standard deviation, and the sample variance of y1. How does this compare to the theoretical values

used for simulation (mean 5, standard deviation 1.5, and variance 2.25)?

3. Use the command hist(y1, freq=F) to make a histogram where the total area of the bars is one. Afterwards use the commands

```
x <- seq(0,10,0.1)
lines(x, dnorm(x,5,1.5))
```

The first command defines a vector x with values $0, 0.1, \ldots, 10$. In the second command the density for the $N(5, 1.5^2)$ distribution is computed at each of the values in x, and the density is superimposed on the existing graph. Explain what you see.

Second, let us see what happens if we add observations from two samples of the above type.

4. Make another sample, y2, of size 1000 from the normal distribution with mean 8 and standard deviation 1, $N(8, 1)$. Then define z by the command z <- y1+y2.

   Check that z is the element-by-element sum of y1 and y2. This means: the first value in z is the sum of the first value in y1 and the first element in y2, *etc.*

5. Think of the values in y1 as the weights of 1000 apples drawn at random and of the values in y2 as the weights of 1000 oranges drawn at random. Then, what is the interpretation of the values in z?

6. Compute the sample mean, sample standard deviation, and sample variance of z. What is the relation between these values and the means, standard deviations, and variances of y1 and y2? [Hint: The mean part is the simplest, and here the relation holds exactly. Look at the variances before the standard deviations and note that the relation you are looking for is only approximate.]

7. Make a histogram of z. Superimpose the graph of the density of the normal distribution with mean and standard deviation from question 6 (as in question 3). Does z look like a sample from a normal distribution?

8. Try to formulate the results in more general terms: What is the distribution of the sum of two independent normally distributed variables? In particular, what is the mean and the standard deviation of the sum? You may take a look at Infobox 4.2 to get inspiration.

Third, let us see what happens with linear transformations of normally distributed variables. Recall that y1 is a sample from the normal distribution with mean 5 and standard deviation 1.5.

9. Define v <- 5+2*y1. Compute the sample mean and sample standard deviation of v. Explain the results.

   Check — with a histogram or a QQ-plot — if v seems to be a sample from the normal distribution with this mean and standard deviation.

10. Try to formulate the general result: What is the distribution of $v = a + b \cdot y$ if y has a normal distribution with mean μ and standard deviation σ? In particular, what is the mean and standard deviation of v? You may take a look at Infobox 4.2 again.

11. Finally, define u <- (y1-5)/1.5. What is the distribution of u? [Hint: Use the result from question 10. Compare to Infobox 4.2 once again.]

4.4 **Yield of butter fat.** The yearly yield of butter fat from a cow of a certain breed is assumed to be normally distributed with mean 200 kg and standard deviation 20 kg.

You may use the following R-output for the computations. Recall that pnorm is the cdf, Φ, for the standard normal distribution.

```
> pnorm(0.5)
[1] 0.6914625
> pnorm(-0.5)
[1] 0.3085375
> pnorm(-2.25)
[1] 0.01222447
> pnorm(2.25)
[1] 0.9877755
> pnorm(1.25)
[1] 0.8943502
> pnorm(1.768)
[1] 0.9614695
> pnorm(-2.5)
[1] 0.006209665
```

1. Make a rough drawing of the density for the normal distribution with mean 200 and standard deviation 20.

2. For each of the "events" below: Compute the probability that it occurs and illustrate it on your density drawing.

   (a) The yearly yield of butter fat from a cow is less than 190 kg.

   (b) The yearly yield of butter fat from a cow is more than 245 kg.

   (c) The yearly yield of butter fat from a cow is between 190 kg and 245 kg.

Now, consider *two* cows drawn at random and the total yearly yield from them.

3. What is the distribution of the total yearly butter fat yield from the two cows? Do you use any assumptions on the relationship between the yield from the two cows?

4. Compute the probability that the total yield is larger than 450.

5. Finally, consider a sample of 16 cows and compute the probability that the total yearly butter fat yield from these 16 cows exceeds 3000 kg.

   [Hint: What is the distribution of the total yield?]

4.5 ℞ Length of gestation period for horses, evaluation of QQ-plots. The length of the gestation period (the period from conception to birth) was registered for 13 horses. The observed number of days were

339 339 339 340 341 340 343 348 341 346 342 339 337

Use R to answer the following questions.

1. Read the data into R and compute the mean and the standard deviation.

2. Make a histogram of the data and explain why this is not very useful in order to evaluate if the normal distribution is appropriate for these data.

3. Make a QQ-plot of the data. Do you believe that it is reasonable to use a normal distribution for the data?

Experience is needed to evaluate QQ-plots (and histograms). We will try to obtain such experience by looking at *simulated data*; that is, we use the computer to generate samples from a normal distribution and make the corresponding QQ-plots. This will give an idea about how different QQ-plots can be — even for data that we know are drawn from the normal distribution.

4. Use the following command to draw a sample of size 13 from the normal distribution with mean 341.08 days and standard deviation 3.07:

   ```
   ysim <- rnorm(13, mean=341.08, sd=3.07)
   ```

5. Make a QQ-plot of the simulated data.

6. Repeat questions 4 and 5 ten times (say) in order to get a feeling of how different QQ-plots may appear even when the data are indeed normally distributed. With the simulated plots in mind, do you now think it is reasonable to use a normal distribution for the gestation data?

7. Simulate a sample of size 1000 from the normal distribution with mean 341.08 and standard deviation 3.07 and make the QQ-plot for these data. Repeat it a few times. How much do these QQ-plots differ, compared to the QQ-plots from questions 4 and 5? Discuss what your findings imply for validation of the normality assumption.

4.6 Length of gestation period for horses (continued). Use the gestation data from Problem 4.5 to answer the following questions. The sample mean is 341.08 and the sample standard deviation is 3.07. In the following you can assume that the length of gestation in a population of horses is normally distributed with this mean and standard deviation.

1. Compute the probability that a horse drawn at random from the population has a gestation period of length between 337 and 343 days (both numbers included).

   You can use a table or the following output from R:

   ```
   > pnorm(-1.329)
   [1] 0.09192398
   > pnorm(0.625)
   [1] 0.7340145
   ```

2. Compute the probability that a horse drawn at random from the sample — not the population — has a gestation period of length between 337 and 343 days (both numbers included).

3. Explain the difference between the two probabilities from questions 1 and 2.

4. Compute an interval where we would expect to find the gestation lengths for 95% of the horses in the population.

4.7 Ⓡ **Packs of minced meat.** In meat production, packs of minced meat are specified to contain 500 grams of minced meat. A sample of ten packs was drawn at random and the recorded weights (in grams) were

$$496.1 \quad 501.7 \quad 494.3 \quad 475.9 \quad 511.2$$
$$502.4 \quad 492.5 \quad 500.6 \quad 489.5 \quad 465.7$$

1. What is the probability that a pack drawn at random *from the sample* contains less than 500 grams?

2. Read the data into R and compute the sample mean ($\bar{y}$) and the sample standard deviation (s).

3. Make a QQ-plot of the data. Is it reasonable to describe the distribution of weights with a normal distribution?

For the next questions, assume that the weight of packs from the production is normally distributed with mean and standard deviation equal to $\bar{y}$ and s from question 2.

4. Compute the probability that a pack drawn at random *from the production* contains less than 500 g. Explain the difference between this probability and the probability from question 1.

5. Compute the probability that a pack drawn at random from the production contains between 490 and 510 grams.

6. Would it be unusual to find a pack with less than 480 grams of meat?

7. Compute the probability that the total weight of two packs of minced meat drawn at random from the production is less than 1000 grams.
   [Hint: Use Infobox 4.2.]

4.8 **Machine breakdown.** Consider a machine and assume that the number of hours the machine works before breakdown is distributed according to the exponential distribution with rate 0.002; *cf.* Section 4.3.3.

1. Use formula (4.10) to compute the probability that the machine breaks down after 100 hours but before 1000 hours of work.

2. Compute the probability that the machine breaks down before 800 hours of work.
   [Hint: Which values of a and b are appropriate in (4.10)?]

3. Compute the probability that the machine works at least 500 hours before it breaks down.

4.9 **[M] Mean and variance for the normal distribution.** Consider a normally distributed variable Y with density f given in (4.1); that is,

$$f(y) = \frac{1}{\sqrt{2\pi\sigma^2}} \exp\left(-\frac{1}{2\sigma^2}(y - \mu)^2\right), \quad -\infty < y < \infty.$$

The expected value or mean of Y is defined as the average value where each possible value y between $-\infty$ and $+\infty$ is assigned the weight $f(y)$:

$$E(Y) = \int_{-\infty}^{\infty} yf(y)\,dy.$$

1. Show that $\int_{-\infty}^{\infty}(y-\mu)f(y)\,dy = 0$.
   [Hint: Use a substitution $v = y - \mu$ and the fact that f is symmetric around μ.]

2. Show that $E(Y) = \mu$.

The variance of Y is defined

$$\text{Var}(Y) = \int_{-\infty}^{\infty} (y-\mu)^2 f(y)\,dy$$

3. Explain why $\text{Var}(Y)$ is "the expected or average squared deviation from the mean".

4. Show that $\text{Var}(Y) = \sigma^2$.
   [Hint: Differentiate f twice and show that

$$f''(y) = f(y)\frac{(y-\mu)^2}{\sigma^4} - f(y)\frac{1}{\sigma^2}$$

and move some terms around.]

4.10 [M] The exponential distribution. Assume that Y is distributed according to the normal distribution; *cf.* Section 4.3.3. The mean or average value of Y is defined as

$$EY = \int_0^{\infty} yg(y)\,dy = \int_0^{\infty} y\gamma e^{-\gamma y}\,dy.$$

Show that $EY = 1/\gamma$.

[Hint: Use integration by parts.]

Chapter 5

Statistical models, estimation, and confidence intervals

Statistical inference allows us to draw general conclusions from the available data. Chapters 1–3 introduced data types, linear regression, and one-way ANOVA (comparison of groups) without details about statistical inference. Chapter 4 introduced the normal (Gaussian) distribution. We are now ready to combine the two parts: if the variation can be described by the normal distribution, then we can carry out proper statistical analyses; in particular, compare groups and infer about the association between two variables.

First of all we need a *statistical model*. Given a statistical model we can discuss *estimation, confidence intervals, hypothesis testing,* and *prediction* — the topics of this and the following two chapters. We will focus on the class of *normal linear models*. These are models where the variable of interest (the response) depends linearly on some background variables (the explanatory variables) and where the random deviations are assumed to be independent normally distributed. Normal linear models are also called *Gaussian linear models,* or just *linear models*. The linear regression model, the one-way ANOVA model, and the model for a single sample are the simplest linear models and are covered in this and the next two chapters. In Chapter 8 we extend the basic models to more complicated regression and ANOVA models.

Linear models are useful for continuous variables (for example, physically measurable quantities like weight, height, concentrations), but not for categorical data because the normal distribution is not appropriate for such data. However, the principles concerning estimation, confidence intervals, hypothesis tests, *etc.*, are fundamental concepts for all statistical analyses, regardless of the data type.

5.1   Statistical models

A *statistical model* describes the outcome of a variable, the *response variable,* in terms of *explanatory variables* and random variation. For linear models the response is a quantitative variable. The explanatory variables describe the expected values of the response and are also called covariates or predictors. An explanatory variable can be either *quantitative*, with values having an explicit numerical interpretation as in linear regression, or *categorical*, corresponding to a grouping of the observations as in the ANOVA setup. When a categorical variable is used as an explanatory variable it is often called a *factor*. For example, *sex* would be a factor whereas *age* would be a quantitative explanatory variable (although it could also be grouped into age categories and thus used as a factor).

The explanatory variables explain part of the variation in the response variable, but not all the variation. The remaining variation is biological variation, which we cannot — or do not want to — describe in terms of explanatory variables. This is reflected in the statistical model consisting of two parts: the fixed part and the random part. The fixed or systematic part describes how the response variable depends on the explanatory variables. In other words, it describes *the systematic change in the response as the explanatory variables change*. The random part describes the variation from this systematic behavior.

Example 5.1. Stearic acid and digestibility of fat (continued from p. 26). The digestibility percent is the response and the level of stearic acid is a quantitative explanatory variable. The fixed part of the model explains how the digestibility percent depends on the level of stearic acid, and for this we will of course use the linear relationship $y = \alpha + \beta \cdot x$ from Chapter 2. The random part explains how the observed values differ from this linear relationship; that is, how the points in an (x, y)-plot of the data scatter around a straight line.   □

Example 5.2. Effect of antibiotics on dung decomposition (continued from p. 51). The amount of organic material is the response and the antibiotic type is the explanatory variable, a factor. The fixed part of the model describes how the amount of organic material differs between antibiotic types, whereas the random part describes how the amount of organic material differs between heifers getting the same antibiotic type.   □

In both examples there is only one explanatory variable. In general there can be more than one, and they could be of different types (quantitative variables and factors). More on this in Chapter 8. On the other hand, there are no explanatory variables for the crab weight data in Example 4.1 (p. 68). The response is the weight of the crabs, but there is no additional information about the crabs.

5.1.1   Model assumptions

In the linear regression setup, the response y is a linear function of the explanatory variable x plus a remainder term describing the random variation,

$$y_i = \alpha + \beta x_i + e_i, \quad i = 1, \ldots, n.$$

The intercept and the slope (α and β) are parameters. The values are not known to us, but the data hold information about them. In the one-way ANOVA setup with k groups, the group means $\alpha_1, \ldots, \alpha_k$ are parameters, and we write

$$y_i = \alpha_{g(i)} + e_i, \quad i = 1, \ldots, n,$$

where $g(i) = x_i$ is the group that corresponds to y_i; *cf.* Section 3.3.

In general, let

$$y_i = \mu_i + e_i, \quad i = 1, \ldots, n.$$

Here, μ_i includes the information from the explanatory variables and is a function of parameters and explanatory variables. The function should be linear as a function of the parameters. To be specific, assume that there is only one explanatory variable x, let $\theta_1, \ldots, \theta_p$ be p *parameters*, and let f be the function that describes μ_i as a function of the explanatory variables and the parameters,

$$\mu_i = f(x_i; \theta_1, \ldots, \theta_p). \tag{5.1}$$

We call $\theta_1, \ldots, \theta_p$ the *mean parameters* or the *fixed effect parameters*. The notation suggests that there is only one explanatory variable x, but f could easily depend on several variables; *e.g.*, $\mu_i = f(x_{i1}, x_{i2}; \theta_1, \ldots, \theta_p)$.

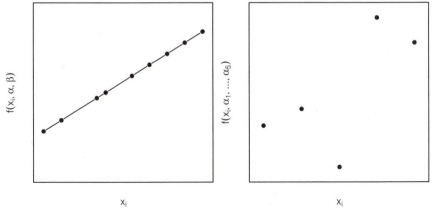

Figure 5.1: Illustration of the linear regression model (left) and the one-way ANOVA model for $k = 5$ (right).

In linear regression $p = 2$, (θ_1, θ_2) correspond to (α, β) and $f(x_i; \alpha, \beta) = \alpha + \beta \cdot x_i$, as illustrated in the left panel of Figure 5.1. The points represent values of x_i found in the dataset, but f is defined for all possible values of x_i so we can interpolate between the points. For the one-way ANOVA with k groups we have $p = k$, θ_j corresponds to α_j, and $f(x_i; \alpha_1, \ldots, \alpha_k) = \alpha_{g(i)} = \alpha_{x_i}$. This is illustrated for $k = 5$ in the right panel of Figure 5.1. Here the different values of x_i correspond to five different groups and interpolation is not meaningful — the values of x_i need not even be quantitative. For the

ANOVA model there are no restrictions for the function values, whereas for the linear regression the function values are forced to be on a straight line.

We have not yet described the *remainder terms* $e_1, \ldots, e_n$. They are assumed to be independent and $N(0, \sigma^2)$ distributed; in particular, to have mean zero. Independence of $e_1, \ldots, e_n$ is equivalent to independence of $y_1, \ldots, y_n$. It follows from Infobox 4.2 that

$$y_i \sim N(\mu_i, \sigma^2),$$

so μ_i is the *mean* or the *expected value* of y_i and σ is the standard deviation of y_i. The standard deviation σ describes the "typical" deviation between the response and its expected value and is also a parameter in the model. Notice that the mean is allowed to differ between observations (due to the explanatory variables) but that the standard deviation is assumed to be the same for all observations. Let us summarize.

Infobox 5.1: Linear model (version 1)

A linear model for $y_1, \ldots, y_n$ has the following properties.

Mean: The mean of y_i, denoted μ_i, is a function of the explanatory variables and unknown parameters. This function must be linear in the parameters.

Variance homogeneity: All $y_1, \ldots, y_n$ have the same standard deviation σ.

Normality: $y_1, \ldots, y_n$ are normally distributed, $y_i \sim N(\mu_i, \sigma^2)$.

Independence: $y_1, \ldots, y_n$ are independent.

Notice that only the first assumption describing the specific dependence on the explanatory variables differs among the different models (linear regression, one-way ANOVA, *etc.*). The last three assumptions are the same for all linear models, and we may write them as assumptions on $e_1, \ldots, e_n$ instead:

Infobox 5.2: Linear model (version 2)

A linear model for $y_1, \ldots, y_n$ can be written on the form

$$y_i = \mu_i + e_i, \quad i = 1, \ldots, n,$$

where μ_i is a function of parameters and explanatory variables (linear in the parameters), and where the remainder terms $e_1, \ldots, e_n$ are independent and normally distributed, all with mean zero and standard deviation σ.

Notice that μ_i is assumed to depend *linearly* on the parameters. As an example of a *non-linear model*, consider $\mu_i = \alpha \cdot \exp(\beta \cdot x_i)$, where μ_i depends exponentially on β; see also Example 2.4 (p. 35). We will not discuss this issue any further at this point; just note that not all models are linear models.

The model assumptions have to be validated when we use the model for our data. This is usually done by examination of the residuals in a manner resembling the model validation approach we saw in Section 7.1.

5.1.2   Model formulas

The actual computations for the data analysis are almost always carried out with statistical software programs. The precise syntax differs from program to program, but a statistical model is usually specified through a *model formula* like

$$y = x \tag{5.2}$$

where y is the response variable and x is the explanatory variable. For example, the model formulas for the digestibility data from Example 2.1 would read

digest = acid

and the model formula for the antibiotics data from Example 3.2 would read

organic = type

The exact translation of a model formula to a statistical model depends on the type of explanatory variable: if x is a quantitative variable the model formula corresponds to a linear regression; if x is a factor then the model formula corresponds to a one-way ANOVA. Evidently, it is therefore essential that the explanatory variable is coded correctly, so the program "knows" which analysis to perform. In particular, the model formula for the model "fitted without intercept":

$$y = x - 1$$

has completely different meaning and implications for the two model types. This will be discussed below for each of the models and in Section 5.5.

Model formulas provide an elegant specification of statistical models as they describe exactly what we are aiming at: how does the explanatory variable influence the response? This will prove particularly useful for complicated model structures that involve two or more explanatory variables; see Chapter 8. The remaining assumptions (variance homogeneity, normality, independence) are often thought of as technical assumptions. They are necessary as descriptions of the random variation, but we are rarely interested in them for their own sake. Moreover, they are the same throughout the analysis, so they can be suppressed in the model formula specification of the model.

5.1.3   Linear regression

Consider the situation where the explanatory variable x is a quantitative variable. The *linear regression model* assumes that

$$y_i = \alpha + \beta \cdot x_i + e_i \quad i = 1, \ldots, n, \tag{5.3}$$

where $e_1, \ldots, e_n$ are independent and $N(0, \sigma^2)$ distributed. Or, equivalently, the model assumes that $y_1, \ldots, y_n$ are independent and

$$y_i \sim N(\alpha + \beta \cdot x_i, \sigma^2), \quad i = 1, \ldots, n. \tag{5.3*}$$

The parameters of the model are α, β, and σ. The slope parameter β is the expected increment in y as x increases by one unit, whereas α is the expected value of y when $x = 0$. This interpretation of α does not always make biological sense as the value zero of x may be nonsense. More generally, α moves the line with slope β up or down.

The remainder terms $e_1, \ldots, e_n$ represent the vertical deviations from the straight line. The assumption of variance homogeneity means that the typical size of these deviations is the same across all values of x. In particular, the "average" vertical distance is the same for small and large x-values, and roughly 95% of the observations are assumed to be within distance 2σ from the regression line.

When x is a quantitative variable, then the model formula y = x is translated to the model given by (5.3). Note that the intercept is implicitly understood and does not have to be specified. Rather, if one wants the model with zero intercept ($\alpha = 0$),

$$y_i = \beta \cdot x_i + e_i, \quad i = 1, \ldots, n$$

corresponding to a straight line through $(0,0)$, then this should be stated explicitly in the model formula,

$$y = x - 1.$$

Notice how the linear regression model focuses on the effect of x on y. No statistical model is assumed for the x's — the values are treated as known, fixed values — and we say that the model is *conditional on* x. A more complicated (two-dimensional) model would be appropriate if we were interested in the variation of x as well.

5.1.4   Comparison of groups

For the situation with k different groups, we let $x_i = g(i)$ denote the group of the ith observation; *cf.* Section 3.4. The *one-way ANOVA model* assumes that

$$y_i = \alpha_{g(i)} + e_i, \quad i = 1, \ldots, n, \tag{5.4}$$

where $e_1, \ldots, e_n$ are independent and $N(0, \sigma^2)$ distributed. Or, equivalently, that $y_1, \ldots, y_n$ are independent and

$$y_i \sim N(\alpha_{g(i)}, \sigma^2), \quad i = 1, \ldots, n. \tag{5.4*}$$

In other words, it is assumed that there is a normal distribution for each group, with *means that are different from group to group* and given by the α's but with *the same standard deviation in all groups* (namely, σ) representing the *within-group variation*.

The parameters of the model are $\alpha_1, \ldots, \alpha_k$ and σ, where α_j is the expected value (or the population average) in the jth group. In particular, we are often interested in the *group differences* $\alpha_j - \alpha_l$, since they provide the relevant information if we want to compare the jth and the lth group.

When x is a factor or categorical variable then the model formula $y = x$ is translated to the model given by (5.4). Sometimes, the factor has character values such as `Control`, `ivermect`, *etc.*, as for the antibiotics data (Example 3.2, p. 51), and then there is no unambiguity as to which model the model formula refers to. However, sometimes groups are numbered 1 through k, say. Then it is vital that these numbers are interpreted as labels of the groups, not as actual numbers since the computer program will then carry out a linear regression analysis.

The case with *two unpaired samples*, corresponding to two groups, is a special case of the one-way ANOVA setup with $k = 2$ if it is reasonable to assume that the standard deviation is the same for both samples. That is, $y_i \sim N(\alpha_1, \sigma^2)$ if it belongs to group 1 and $y_i \sim N(\alpha_2, \sigma^2)$ if it belongs to group 2.

5.1.5 One sample

For the crab weight data in Example 4.1 (p. 68) the crab weight is the response variable. There is no explanatory variable (although there could have been one such as sex or age). We assume that $y_1, \ldots, y_n$ are independent with

$$y_i \sim N(\mu, \sigma^2), \quad i = 1, \ldots, n. \tag{5.5}$$

All observations are assumed to be sampled from the same distribution, so this is the *one sample* case from Section 4.2. Equivalently, we may write

$$y_i = \mu + e_i, \quad i = 1, \ldots, n, \tag{5.5*}$$

where $e_1, \ldots, e_n$ are independent and $N(0, \sigma^2)$ distributed. The parameters are μ and σ^2, where μ is the expected value (or population mean) and σ is the average deviation from this value (or the population standard deviation).

We will use the model formula

$$y = 1$$

for the one sample model, indicating the constant mean for all observations.

Notice that the model for a single sample is also used for paired samples where the measurements are related to each other in pairs. As described in Section 3.5, the difference between the two measurements within the pair is used as the response for the ith subject and treated as a single sample. To be specific, if (y_{i1}, y_{i2}) denotes the measurements for pair i, then the analysis is carried out on $d_i = y_{i2} - y_{i1}$. If the deviation between the two sampling situations is believed to be better described on a multiplicative scale, then it might be more appropriate to analyze $f_i = y_{i2}/y_{i1}$. In both situations the two measurements within a pair are combined to a single observation.

5.2   Estimation

Recall that we are considering statistical models of the type

$$y_i = \mu_i + e_i \quad i = 1, \ldots, n,$$

where $e_1, \ldots, e_n$ are independent and $N(0, \sigma^2)$ distributed, and μ_i depends on an explanatory variable (or several variables) through unknown parameters, as described by the function f,

$$\mu_i = f(x_i; \theta_1, \ldots, \theta_p).$$

We are interested in the values of the parameters, and we will use the data to compute their estimates. Different samples lead to different estimates, and so we need to be concerned with the variability and precision of the estimates: we shall talk about the *standard error* of the estimates and about *confidence intervals*, which are ranges of values that — in a certain sense — are in accordance with the data. Notice that σ is also a parameter in the model. Often it is not of much interest in itself, but it is important for the precision of the estimates of the θ's.

5.2.1   Least squares estimation of the mean parameters

Estimation means "finding the values of the parameters that make the model fit the data the best". Hence, we need to define what it should mean that a model fit the data well or not so well, and we will do so by means of squared residuals, which leads to *least squares estimation* (LS estimation). We already carried out least squares estimation for linear regression and one-way ANOVA in Chapters 2 and 3 — now comes the general idea for linear models.

Recall that for given values of the parameters $\theta_1, \ldots, \theta_p$, the *expected value* of y_i is

$$f(x_i; \theta_1 \ldots, \theta_p).$$

If the model fits well to the data, then the observed values y_i and the expected values $f(x_i; \theta_1, \ldots, \theta_p)$ should all be "close" to each other. Hence, the sum of squared deviations is a measure of the model fit:

$$Q(\theta_1, \ldots, \theta_p) = \sum_{i=1}^{n} (y_i - f(x_i; \theta_1, \ldots, \theta_p))^2.$$

The least squares estimates $\hat{\theta}_1, \ldots, \hat{\theta}_p$ are the values for which $Q(\theta_1, \ldots, \theta_p)$ is the smallest possible. Expressions for the estimators are obtained by computing the partial derivatives of f with respect to θ_j and solving the equations $\frac{\partial f}{\partial \theta_j} = 0$ for $j = 1, \ldots, p$. It turns out that the solution actually corresponds to a minimum point (rather than a maximum point or a saddle point).

5.2.2 Estimation of the standard deviation σ

Inserting the estimates, we get the *predicted values* or *fitted values*, the *residuals*, and the *residual sum of squares*:

$$\hat{y}_i = \hat{\mu}_i = f(x_i; \hat{\theta}_1, \ldots, \hat{\theta}_p), \quad r_i = y_i - \hat{y}_i, \quad SS_e = \sum_{i=1}^{n} (y_i - \hat{y}_i)^2 \qquad (5.6)$$

The fitted value, $\hat{\mu}_i$ or $\hat{y}_i$, is the value we would expect for the ith observation if we repeated the experiment, and we can think of the residual r_i as a "guess" for e_i since $e_i = y_i - \mu_i$. Therefore, it seems reasonable to use the residuals to estimate the variance σ^2:

$$s^2 = \hat{\sigma}^2 = \frac{SS_e}{n - p} = \frac{1}{n - p} \sum_{i=1}^{n} (y_i - \hat{y}_i)^2. \qquad (5.7)$$

The number $n - p$ in the denominator is called the *residual degrees of freedom*, or *residual df* (denoted df_e), and s^2 is often referred to as the *residual mean squares*, the residual variance, or the *mean squared error* (denoted MS_e): $s^2 = MS_e = SS_e / df_e$.

The standardization with $n - p$ ensures that the expected value of s^2 is exactly σ^2. In other words, if we repeated the experiment many times and computed s^2 for every dataset, then the average of these estimates would approach σ^2. Notice that $SS_e = Q(\hat{\theta}_1, \ldots, \hat{\theta}_p)$, which is the smallest possible value of Q. Hence, the least squares estimates $\hat{\theta}_1, \ldots, \hat{\theta}_p$ are those values that yield the smallest possible variance estimate.

The standard deviation σ is estimated as the residual standard deviation; *i.e.*, as the square root of s^2:

$$\hat{\sigma} = s = \sqrt{s^2} = \sqrt{\frac{SS_e}{df_e}} \qquad (5.8)$$

Notice that we have already met the estimated standard deviation in this form several times: in (1.6) for a single sample with $p = 1$; in (2.8) for the linear regression case with $p = 2$; and in (3.4) for the one-way ANOVA case with $p = k$. In the one-way ANOVA case s^2 may also be computed as a weighted average of the sample standard deviations computed from each group; see (3.5).

5.2.3 Standard errors and distribution of least squares estimates

The observed data result in estimates $\hat{\theta}_1, \ldots, \hat{\theta}_p$ of the parameters. Another sample from the population, or a replication of the experiment, would result in different data and hence different estimates, so the following questions arise: How much can we trust the estimates? How much different could they have been? And more generally, what are their properties? For linear models these questions can be answered very precisely, because the distribution of the least squares estimates is known.

Consider for a moment the one sample setup from Sections 4.2.2 and 5.1.5: $y_1, \ldots, y_n$ are independent and all $N(\mu, \sigma^2)$ distributed. The only mean parameter is the common mean, μ, which is estimated by $\hat{\mu} = \bar{y}$. From (4.4) we know that $\bar{y}$ has a normal distribution with mean equal to the true (but still unknown) value of μ and variance equal to σ^2 divided by n:

$$\hat{\mu} = \bar{y} \sim N(\mu, \sigma^2/n) \tag{5.9}$$

As pointed out in Infobox 4.4, this means that $\hat{\mu}$ is unbiased ("hits" the correct value on average, if we repeated the experiment many times) and consistent (becomes more precise as the sample size increases since the variance decreases). Furthermore, it has a normal distribution, so we know how to make probability computations concerning the estimate.

The point is that the same properties hold for the least squares estimates in all linear models! Consider the model from Section 5.1.1,

$$y_i = \mu_i + e_i, \quad i = 1, \ldots, n, \tag{5.10}$$

where $e_1, \ldots, e_n$ are iid. $N(0, \sigma^2)$ and μ_i depends (linearly) on the parameters $\theta_1, \ldots, \theta_p$. Furthermore, let $\hat{\theta}_1, \ldots, \hat{\theta}_p$ denote the least squares estimates. Then each $\hat{\theta}_j$ is normally distributed with mean equal to the true value θ_j (so $\hat{\theta}_j$ is unbiased) and a variance that decreases to zero as the sample size increases (so $\hat{\theta}_j$ is consistent). If the variance is denoted $k_j \cdot \sigma^2$, we thus write

$$\hat{\theta}_j \sim N(\theta, k_j \cdot \sigma^2). \tag{5.11}$$

The constant k_j depends on the model and the data structure — but not on the observed y-values. In particular, the value of k_j could be computed even before the experiment was carried out. In the one sample case, we see from (5.9) that the constant k_j is simply $1/n$; in particular, it decreases when n

increases. The same is true for other linear models: k_j decreases as the sample size increases.

Taking the square root of the variance in (5.11) and replacing the unknown value σ by its estimate s, we get the *standard error* of $\hat{\theta}_j$. In other words, $SE(\hat{\theta}_j)$ is the (estimated) standard deviation of $\hat{\theta}_j$,

$$SE(\hat{\theta}_j) = s\sqrt{k_j} \qquad (5.12)$$

We have focused on the distribution of each $\hat{\theta}_j$ — one at a time. In fact, this does not tell the full story, since the estimates in many cases are correlated, such that the variation of one estimate contains information about the variation of another estimate. If we combine parameter estimates we must take this correlation into account, but we will not go into details about this in this book.

5.2.4 Linear regression

Consider the linear regression model

$$y_i = \alpha + \beta \cdot x_i + e_i \quad i = 1, \ldots, n, \qquad (5.13)$$

where $e_1, \ldots, e_n$ are independent and $N(0, \sigma^2)$ distributed. As already derived in Chapter 2, the least squares estimates are

$$\hat{\beta} = \frac{\sum_{i=1}^n (x_i - \bar{x})(y_i - \bar{y})}{\sum_{i=1}^n (x_i - \bar{x})^2}, \quad \hat{\alpha} = \bar{y} - \hat{\beta} \cdot \bar{x}.$$

In order to simplify formulas, define SS_x as the denominator in the definition of $\hat{\beta}$:

$$SS_x = \sum_{i=1}^n (x_i - \bar{x})^2.$$

It then turns out that

$$\hat{\beta} \sim N\left(\beta, \frac{\sigma^2}{SS_x}\right), \quad \hat{\alpha} \sim N\left(\alpha, \sigma^2\left(\frac{1}{n} + \frac{\bar{x}^2}{SS_x}\right)\right).$$

In particular, when the sample size increases, then SS_x increases and the variances of $\hat{\beta}$ and $\hat{\alpha}$ decrease — thus the estimates get more and more precise, as we would expect (and as claimed above).

The standard errors are

$$SE(\hat{\beta}) = \frac{s}{\sqrt{SS_x}}, \quad SE(\hat{\alpha}) = s\sqrt{\frac{1}{n} + \frac{\bar{x}^2}{SS_x}},$$

where

$$s = \sqrt{\frac{SS_e}{df_e}} = \sqrt{\frac{1}{n-2}\sum_{i=1}^{n}(y_i - \hat{\alpha} - \hat{\beta} \cdot x_i)^2}$$

is the estimated standard deviation for a single observation y_i.

In regression analysis we often seek to estimate the expected value for a particular value of x, which is not necessarily one of the x-values from the dataset. Let x_0 be such an x-value of interest. The expected value of the response is denoted μ_0; that is, $\mu_0 = \alpha + \beta \cdot x_0$. It is estimated by

$$\hat{\mu}_0 = \hat{\alpha} + \hat{\beta} \cdot x_0.$$

It turns out that

$$\hat{\mu}_0 \sim N\left(\alpha + \beta \cdot x_0, \sigma^2 \left(\frac{1}{n} + \frac{(x_0 - \bar{x})^2}{SS_x}\right)\right),$$

so the standard error is

$$SE(\hat{\mu}_0) = s\sqrt{\frac{1}{n} + \frac{(x_0 - \bar{x})^2}{SS_x}}. \tag{5.14}$$

In particular, $\hat{\mu}_0$ is an unbiased and consistent estimate of μ_0.

Example 5.3. Stearic acid and digestibility of fat (continued from p. 26). We already computed estimates for the digestibility data in Example 2.3 (p. 31):

$$\hat{\alpha} = 96.5334, \quad \hat{\beta} = -0.9337$$

so for an increase in stearic acid level of 1 percentage point we will expect the digestibility to decrease by 0.93 percentage points.

Moreover, $SS_e = 61.7645$, so

$$s^2 = \frac{61.7645}{9-2} = 8.8234, \quad s = \sqrt{8.8234} = 2.970,$$

and $\bar{x} = 14.5889$ and $SS_x = 1028.549$, so

$$SE(\hat{\beta}) = \frac{2.970}{\sqrt{1028.549}} = 0.0926, \quad SE(\hat{\alpha}) = 2.970 \cdot \sqrt{\frac{1}{9} + \frac{14.5889^2}{1028.549}} = 1.6752.$$

Finally, if we consider a stearic acid level of $x_0 = 20\%$, then we will expect a digestibility percentage of

$$\hat{\mu}_0 = 96.5334 - 0.9337 \cdot 20 = 77.859,$$

which has standard error

$$SE(\hat{\mu}_0) = 2.970 \cdot \sqrt{\frac{1}{9} + \frac{(20 - 14.5889)^2}{1028.549}} = 1.1096.$$

$\square$

5.2.5 Comparison of groups

Consider the one-way ANOVA model

$$y_i = \alpha_{g(i)} + e_i, \quad i = 1, \ldots, n,$$

where $g(i)$ denotes the group corresponding to the ith observation and $e_1, \ldots, e_n$ are independent and $N(0, \sigma^2)$ distributed. In Section 3.4 we saw that the least squares estimates for the group means $\alpha_1, \ldots, \alpha_k$ are simply the group averages: $\hat{\alpha}_j = \bar{y}_j$. Since the observations from a group constitute a sample on their own, the distribution follows from the one sample case:

$$\hat{\alpha}_j = \bar{y}_j = \frac{1}{n_j} \sum_{i:g(i)=j} y_i \sim N\left(\alpha_j, \frac{1}{n_j}\sigma^2\right).$$

The corresponding standard errors are given by

$$\mathrm{SE}(\hat{\alpha}_j) = s\sqrt{\frac{1}{n_j}} = \frac{s}{\sqrt{n_j}}.$$

In particular, mean parameters for groups with many observations (large n_j) are estimated with larger precision than mean parameters with few observations — as should of course be the case.

In the above formulas we used the estimated standard deviation $s = \sqrt{s^2}$ for a single observation. In the ANOVA setup, s^2 is given by

$$s^2 = \frac{SS_e}{df_e} = \frac{1}{n-k} \sum_{i=1}^{n} (y_i - \hat{y}_i)^2 = \frac{1}{n-k} \sum_{i=1}^{n} (y_i - \hat{\alpha}_{g(i)})^2,$$

which we recognize as the pooled variance estimate (3.4).

In the one-way ANOVA case we are very often interested in the *differences* or *contrasts* between group levels rather than the levels themselves. For example, how much larger is the expected response in the treated group compared to the control group? Hence, we are interested in quantities $\alpha_j - \alpha_l$ for two groups j and l. Naturally, the estimate is simply the difference between the two estimates,

$$\widehat{\alpha_j - \alpha_l} = \hat{\alpha}_j - \hat{\alpha}_l = \bar{y}_j - \bar{y}_l.$$

Since $\bar{y}_j$ and $\bar{y}_l$ are computed from different observations, those in group j and l respectively, they are independent. Moreover, they are both normally distributed, so Infobox 4.2 implies that $\bar{y}_j - \bar{y}_l$ is normally distributed with mean $\alpha_j - \alpha_l$ and variance

$$\mathrm{Var}(\bar{y}_j - \bar{y}_l) = \mathrm{Var}(\bar{y}_j) + \mathrm{Var}(\bar{y}_l) = \frac{1}{n_j}\sigma^2 + \frac{1}{n_l}\sigma^2 = \left(\frac{1}{n_j} + \frac{1}{n_l}\right)\sigma^2$$

and finally we get the corresponding standard error

$$SE(\hat{\alpha}_j - \hat{\alpha}_l) = s\sqrt{\frac{1}{n_j} + \frac{1}{n_l}}. \tag{5.15}$$

The formulas above of course also apply for two samples ($k = 2$). We go through the computations for the salmon data in Example 5.9 (p. 125).

Example 5.4. Effect of antibiotics on dung decomposition (continued from p. 51). Table 5.1 lists the results from the antibiotics example. The third and fourth columns list the group means and their standard errors. The fifth and sixth columns list the estimated expected differences between the control group and each of the other groups as well as the corresponding standard errors.

Table 5.1: Estimates for group means and comparison to the control group for the antibiotics data

Antibiotics	n_j	$\hat{\alpha}_j$	$SE(\hat{\alpha}_j)$	$\hat{\alpha}_j - \hat{\alpha}_{control}$	$SE(\hat{\alpha}_j - \hat{\alpha}_{control})$
Control	6	2.603	0.0497	—	—
α-Cypermethrin	6	2.895	0.0497	0.2917	0.0703
Enrofloxacin	6	2.710	0.0497	0.1067	0.0703
Fenbendazole	6	2.833	0.0497	0.2300	0.0703
Ivermectin	6	3.002	0.0497	0.3983	0.0703
Spiramycin	4	2.855	0.0609	0.2517	0.0786

The standard errors in the table were computed as follows. We already computed s in Example 3.4,

$$s^2 = \frac{SS_e}{df_e} = \frac{0.4150}{34 - 6} = 0.01482; \quad s = \sqrt{0.01482} = 0.1217,$$

so the standard error for the control group, say, is computed by

$$SE(\alpha_{control}) = \frac{s}{\sqrt{6}} = \frac{0.1217}{\sqrt{6}} = 0.497$$

whereas the standard error for the comparison between the control group and the spiramycin group, say, is computed by

$$SE(\alpha_{spiramycin} - \alpha_{control}) = s\sqrt{\frac{1}{4} + \frac{1}{6}} = 0.1217 \cdot \sqrt{\frac{1}{4} + \frac{1}{6}} = 0.0786.$$

Notice that the standard errors for estimates involving the spiramycin group are larger than for the other group because there are fewer observations in that group. □

5.2.6   One sample

For completeness we repeat the formulas for a single sample case with the new terminology. Let $y_1, \ldots, y_n$ be independent and $N(\mu, \sigma^2)$ distributed. The estimates are

$$\hat{\mu} = \bar{y} \sim N(\mu, \sigma^2/n), \quad \hat{\sigma}^2 = s^2 = \frac{1}{n-1} \sum_{i=1}^{n} (y_i - \bar{y})^2,$$

and hence the standard error of $\hat{\mu}$ is

$$\mathrm{SE}(\hat{\mu}) = s \sqrt{\frac{1}{n}} = \frac{s}{\sqrt{n}}.$$

Example 5.5. Crab weights (continued from p. 68). For the crab weight data we get

$$\hat{\mu} = \bar{y} = 12.76, \quad s = 2.25, \quad \mathrm{SE}(\hat{\mu}) = 2.25 \cdot \sqrt{1/162} = 0.177.$$

In particular, the average weight of crabs is estimated to 12.76.   □

5.2.7   Bias and precision. Maximum likelihood estimation

Recall that the least squares estimates are the parameter values that make the best model fit in the sense that the observed and expected values are as close as possible, measured through the sum of squared deviations. In principle, this is just one definition of "best fit" and we could use other estimation methods or estimation principles.

In Section 5.2.3 we emphasized that the least squares estimates are unbiased in linear models: on average the estimates yield the correct values. Moreover, we computed the standard errors of the least squares estimates in order to quantify the variability of the estimate. The smaller the standard error, the more precise the estimate.

Bias and *precision* are illustrated in Figure 5.2. Assume that four different estimation methods are available for estimation of a two-dimensional parameter. One of the methods could be least squares estimation. Each "target board" corresponds to one of the estimation methods, and the centers of the circles represent the true value of the parameter. Imagine that 60 datasets are available and that we have used the four estimation methods for each dataset and plotted the estimates as points in the figure.

The estimates in the two left circles are scattered randomly around the center of the circle; that is, on average the estimates are close to the true value of the parameter. We say that there is low bias, perhaps even no bias. In the right-hand part of the figure the estimates are systematically positioned "north-east" of the circle centers, meaning that the estimates on average yield

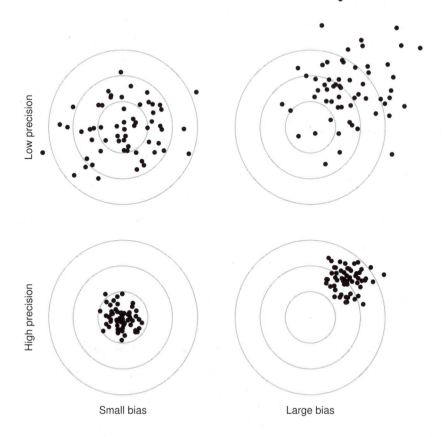

Figure 5.2: Bias and precision for four imaginary estimation methods. The centers of the circles represent the true value of the parameter whereas the points correspond to estimates based on 60 different datasets.

something different from the true value, and we say that the estimates are biased. We may also distinguish different estimation methods by the variability of the estimates. The estimates illustrated in the two top plots exhibit large variation — they are less precise — compared to the estimates illustrated in the bottom.

We generally prefer estimates with small bias and low variability (high precision). This corresponds to the lower left circle. For the linear models we already know that the least squares estimates are unbiased. Moreover, it turns out that they have the largest precision among all estimators that are unbiased, and in that sense the method of least squares is the best possible for linear models. For other data types and model types the situation may be less clear. It may not be possible to obtain unbiased estimates, and there is thus a trade-off between bias and precision. Is an estimate with a small bias but a large precision preferable to an estimate with a large bias but a small precision, or vice versa?

For smaller samples and non-standard models this can be a genuine dilemma. For a large class of models, however, the so-called *maximum likelihood* method is preferable to other methods — at least for large samples or, more precisely, as the sample size approaches infinity. This is due to three asymptotic properties:

- Maximum likelihood estimates are asymptotically unbiased; that is, they approach the correct values on average as the sample size increases.

- Maximum likelihood estimates exhibit the smallest variability, as the sample size increases, among estimates that are asymptotically unbiased.

- The distribution of the maximum likelihood estimates approaches the normal distribution as the sample size increases.

The idea behind the maximum likelihood principle is the following. For some values of the parameters the data that we observe are quite likely to occur — in linear regression this happens if the parameters correspond to a line close to the observations. For other parameter values it is very unlikely that such data should occur — for example, a slope parameter equal to 0 for data where there is obviously a non-zero slope. In short, the maximum likelihood estimates are the values of the parameters that make the observed values the most likely.

For linear models, least squares estimation and maximum likelihood estimation coincide, so in fact we already performed maximum likelihood estimation. The maximum likelihood estimation principle applies much more generally — it can (in principle at least) be used for any type of model — and likelihood-based analysis represents a cornerstone in statistical theory. Interested readers may consult Blæsild and Granfeldt (2003) for a short introduction or Azzalini (1996) for a comprehensive presentation on likelihood theory. We will use maximum likelihood estimation for logistic regression models in Chapter 12.

5.3   Confidence intervals

Consider a parameter θ in a linear model. The least squares estimate, $\hat{\theta}$, gives us the value of θ that makes the best model fit in the least squares sense.

It is also called a *point estimate* since it provides a single value. In itself, $\hat{\theta}$ is not that useful, since it gives no information about its precision: how much can we trust it? Such information is provided by the standard error, and hence the pair consisting of both $\hat{\theta}$ and $\mathrm{SE}(\hat{\theta})$ is much more informative.

Alternatively, we may use *confidence intervals*, also called *interval estimates*. A confidence interval for a parameter is an interval that includes the parameter values that are — in a certain sense and with a certain degree of reliance — in accordance with the data. We will be more specific about this interpretation of confidence intervals in Section 5.3.3.

Consider first the one sample case where $y_1, \ldots, y_n$ are independent and each y_i is $N(\mu, \sigma^2)$ distributed, and recall from Infobox 4.3 that $\bar{y}$ is $N(\mu, \sigma^2/n)$ distributed. It follows from Section 4.1.4 that $\bar{y}$ is in the interval $\mu \pm 1.96 \cdot \sigma / \sqrt{n}$ with probability 95%,

$$P\left(\mu - 1.96\frac{\sigma}{\sqrt{n}} < \bar{y} < \mu + 1.96\frac{\sigma}{\sqrt{n}}\right) = 0.95.$$

If we isolate μ in both inequalities we get

$$P\left(\bar{y} - 1.96\frac{\sigma}{\sqrt{n}} < \mu < \bar{y} + 1.96\frac{\sigma}{\sqrt{n}}\right) = 0.95, \qquad (5.16)$$

expressing that the interval $\bar{y} \pm 1.96\sigma / \sqrt{n}$ — which depends on $y_1, \ldots, y_n$ through $\bar{y}$ — includes the population mean μ for 95% of all samples. If σ was a known value, this would be a 95% confidence interval for μ. Unfortunately, we do not know the value of σ, only its estimate s. If we replace σ by s we have to make the interval somewhat wider than the above in order to retain the same probability, due to the uncertainty in the estimate. We need the t distribution for this purpose.

5.3.1   The t distribution

In order to define confidence intervals we need the t *distribution*, which is most easily introduced via the one sample case. Remember that if $y_1, \ldots, y_n$ are independent and $N(\mu, \sigma^2)$ distributed, then the sample mean $\bar{y}$ is normally distributed, too: $\bar{y} \sim N(\mu, \sigma^2/n)$. By standardization as in Infobox 4.2(c) we get

$$Z = \frac{\sqrt{n}(\bar{y} - \mu)}{\sigma} \sim N(0, 1).$$

Hence, for example,

$$P\left(-1.96 < \frac{\sqrt{n}(\bar{y} - \mu)}{\sigma} < 1.96\right) = 0.95 \qquad (5.17)$$

since the 97.5% quantile of the standard normal distribution is 1.96.

However, the value of σ is unknown, so we cannot compute $(\bar{y} - \mu)/\sigma$ for our data. If we replace σ with its estimate s and consider instead

$$T = \frac{\sqrt{n}(\bar{y} - \mu)}{s},$$

then extra uncertainty is introduced through the estimate of σ, and the distribution is changed. In particular, (5.17) is not true when σ is replaced by s. Intuitively we would expect the following properties to hold for the distribution of T:

Symmetry. The distribution of T is symmetric around zero, so positive and negative values are equally likely.

Dispersion. Values far from zero are more likely for T than for Z due to the extra uncertainty. This implies that the interval $(-1.96, 1.96)$ in (5.17) should be wider when σ is replaced by s in order for the probability to be retained at 0.95.

Large samples. When the sample size increases then s is a more precise estimate of σ, and the distribution of T more closely resembles the standard normal distribution. In particular, the distribution of T should approach $N(0,1)$ as n approaches infinity.

It can be proven that these properties are indeed true. The distribution of T is called the *t distribution with $n - 1$ degrees of freedom* and is denoted $t(n-1)$ or t_{n-1}, so we write

$$T = \frac{\sqrt{n}(\bar{y} - \mu)}{s} \sim t_{n-1}.$$

Note that the degrees of freedom for the t distribution coincide with the degrees of freedom associated with s.

The density of the t distribution depends on the degrees of freedom. In the left panel of Figure 5.3 the density for the t_r distribution is plotted for various degrees of freedom r together with the density of the standard normal distribution, $N(0,1)$. The interpretation of the density is the same as in Section 4.1.1: the probability that T falls within an interval (a, b) is the area under the density curve between a and b; cf. the right panel of Figure 5.3. In particular, it is most likely for T to attain values in areas where the density is large. From the figure we see that our intuition above was indeed correct (check the properties yourself).

In order to make probability statements similar to (5.17) for T — when we replace σ by s — we need quantiles for the t distribution. Table 5.2 shows the 95% and 97.5% quantiles for the t distribution for a few selected degrees of freedom for illustration. The quantiles are denoted $t_{0.95,r}$ and $t_{0.975,r}$ and are also illustrated for $r = 4$ in Figure 5.4. For data analyses where other degrees of freedom are in order, you should look up the relevant quantiles in a statistical table (Appendix C.3) or use a computer (Section 5.5.4).

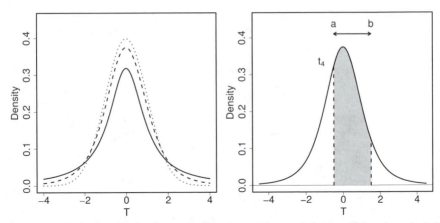

Figure 5.3: Left: density for the t_r distribution with $r = 1$ degree of freedom (solid) and $r = 4$ degrees of freedom (dashed) as well as for $N(0,1)$ (dotted). Right: the probability of an interval is the area under the density curve, illustrated by the t_4 distribution.

Table 5.2: 95% and 97.5% quantiles for selected t distributions

Quantile	t_1	t_2	t_5	t_{10}	t_{25}	t_{50}	t_{100}	$N(0,1)$
95%	6.314	2.920	2.015	1.812	1.725	1.676	1.660	1.645
97.5%	12.706	4.303	2.571	2.228	2.086	2.009	1.984	1.960

Finally, a curious historical remark: The t distribution is often called *Student's t distribution* because the distribution result was first published in 1908 under the pseudonym "Student". The author was a chemist, William S. Gosset, employed at the Guinness brewery in Dublin. Gosset worked with what we would today call quality control of the brewing process. Due to time constraints, small samples of 4 or 6 were used, and Gosset realized that the normal distribution was not the proper one to use. The Guinness brewery did not want their rivals to know that they were controlling their brew with such sophisticated methods, so they let Gosset publish his results under pseudonym only.

5.3.2 Confidence interval for the mean for one sample

We just saw that

$$T = \frac{\sqrt{n}(\bar{y} - \mu)}{s} \sim t_{n-1}$$

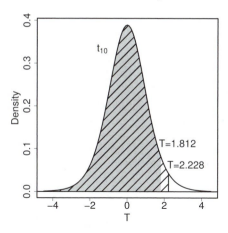

Figure 5.4: Density for the t_{10} distribution. The 95% quantile is 1.812, as illustrated by the gray region which has area 0.95. The 97.5% quantile is 2.228, illustrated by the dashed region with area 0.975.

if $y_1, \ldots, y_n$ are independent and all $N(\mu, \sigma^2)$ distributed. If we denote the 97.5% quantile in the t_{n-1} distribution by $t_{0.975, n-1}$, then

$$P\left(-t_{0.975,n-1} < \frac{\sqrt{n}(\bar{y} - \mu)}{s} < t_{0.975,n-1}\right) = 0.95. \qquad (5.18)$$

If we move around terms in order to isolate μ, we get

$$
\begin{aligned}
0.95 &= P\left(-t_{0.975,n-1} \cdot \frac{s}{\sqrt{n}} < \bar{y} - \mu < t_{0.975,n-1} \cdot \frac{s}{\sqrt{n}}\right) \\
&= P\left(\bar{y} - t_{0.975,n-1} \cdot \frac{s}{\sqrt{n}} < \mu < \bar{y} + t_{0.975,n-1} \cdot \frac{s}{\sqrt{n}}\right).
\end{aligned}
$$

For a random sample of size n from the $N(\mu, \sigma^2)$ distribution, the interval

$$\left(\bar{y} - t_{0.975,n-1} \cdot \frac{s}{\sqrt{n}}, \ \bar{y} + t_{0.975,n-1} \cdot \frac{s}{\sqrt{n}}\right)$$

therefore includes the true parameter value μ with a probability of 95%. The interval is called a 95% *confidence interval* for μ — or we say that the interval has 95% *coverage* for μ — and it is sometimes written as

$$\bar{y} \pm t_{0.975,n-1} \cdot \frac{s}{\sqrt{n}}. \qquad (5.19)$$

Notice that the estimate of μ and its corresponding standard error are $\hat{\mu} = \bar{y}$ and $\mathrm{SE}(\hat{\mu}) = s/\sqrt{n}$, so the 95% confidence interval for μ may also be written as

$$\hat{\mu} \pm t_{0.975,n-1} \cdot \mathrm{SE}(\hat{\mu}).$$

That is, the confidence interval has the form

$$\text{estimate} \pm t\text{-quantile} \cdot \text{SE(estimate)} \qquad (5.20)$$

This is important because, as we shall see, confidence intervals for any parameter in a linear model can be constructed in this way.

Example 5.6. Crab weights (continued from p. 68). The interesting parameter for the crabs data is the population mean, μ. The 97.5% quantile in the $t(161)$ distribution is $t_{0.975,161} = 1.975$; so from the estimate and standard error (Example 5.5, p. 113) we compute the 95% confidence interval for μ to

$$12.76 \pm 1.975 \cdot 0.177 = 12.76 \pm 0.35 = (12.41, 13.11).$$

$\square$

Notice the close resemblance between (5.19) and the confidence interval $\bar{y} \pm 1.96\sigma/\sqrt{n}$, which would be correct if σ was known: σ has been replaced by its estimate s and the normal quantile has been replaced by a t quantile. If n is large then the t quantile will be close to 1.96 and it does not matter much which of the quantiles is used. Moreover, s will be a precise estimate of the unknown σ when n is large, so it will be almost as if σ was known. For small samples there is a difference, though, and the t quantile is more correct to use.

5.3.3   Interpretation of the confidence interval

It is easy the get the interpretation of confidence intervals wrong, so we need to be careful: What does a confidence level or a coverage of 95% mean? It is most easily understood in terms of replications of the experiment. If we repeated the experiment or data collection procedure many times and computed the interval $\bar{y} \pm t_{0.975,n-1} \cdot s/\sqrt{n}$ for each of the samples, then 95% of those intervals would include the true value of μ. This is illustrated in the left panel of Figure 5.5. We have drawn 50 samples of size 10 from $N(0,1)$, and for each of these 50 samples we have computed and plotted the confidence interval. The vertical line corresponds to the true value, which is zero. Zero is included in the confidence interval in all but three cases (corresponding to 94%).

If we are interested in another confidence level than 95%, the only thing to be changed is the t quantile. For example, the 75% confidence interval for μ is given by

$$\hat{\mu} \pm t_{0.875,n-1} \cdot \text{SE}(\hat{\mu}).$$

We use the 87.5% quantile because the central 75% corresponds to all but the 12.5% most extreme values on both ends. Similarly, we should use the 95% quantile for computation of a 90% confidence interval.

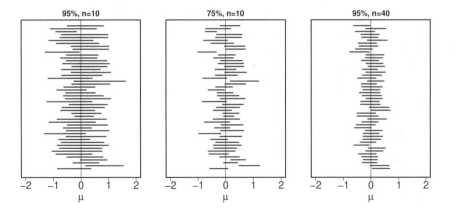

Figure 5.5: Confidence intervals for 50 simulated data generated from $N(0,1)$. Each sample consists of $n = 10$ observations.

The 75% confidence intervals are plotted in the middle panel of Figure 5.5. The 50 samples are the same as those used for the left panel, so for each sample the 95% and the 75% confidence intervals are centered around the same point ($\hat{\mu} = \bar{y}$). Compared to the left panel, we see that the 75% confidence intervals are more narrow such that the true value is excluded more often, with a probability of 25% rather than 5%. This reflects that our confidence in the 75% confidence interval is smaller compared to the 95% confidence interval.

When the sample size increases then $s/\sqrt{n}$ and the t quantile decrease, so the confidence intervals become more narrow. This reflects the larger amount of information in the data. The right panel of Figure 5.5 shows 95% confidence intervals for 50 samples of size 40. Compared to the left panel, the confidence intervals are more narrow but still include the true value for all but a few cases — with a probability of 95%.

As argued above, the interpretation of confidence intervals is based on replication of the experiment or data collection. From a practical point of view this might not seem very useful as there is only one dataset and thus one confidence interval available. The true value is either inside the interval or it is not, but we will never know. We can, however, interpret the values in the confidence interval as those parameter values for which it is reasonable to believe that they could have generated the data. If we use 95% confidence intervals, and if the true parameter value is μ_0, then

- the probability of observing data for which the corresponding confidence interval includes μ_0 is 95%

- the probability of observing data for which the corresponding confidence interval does not include μ_0 is 5%

In other words, if the true value is μ_0, then it is quite unlikely to observe data for which the confidence interval does not include μ_0.

As a standard phrase we may say that *the 95% confidence interval includes those values that are in agreement with the data on the 95% confidence level*. For the crab weight data we computed the 95% confidence interval (12.41, 13.11) for the mean μ in Example 5.6 (p. 120). We conclude that an average crab weight (in the population) between 12.41 and 13.11 is in accordance with the observed data on the 95% confidence level.

5.3.4   Confidence intervals for linear models

Above we computed confidence intervals in the one sample case. It turns out that we can use the exact same machinery for all linear models.

Consider the linear model

$$y_i = \mu_i + e_i, \quad i = 1, \ldots, n$$

from Section 5.1, where $e_1, \ldots, e_n$ are iid. $N(0, \sigma^2)$ and μ_i is described by explanatory variables and parameters $\theta_1, \ldots, \theta_p$. From Section 5.2 we know that the least squares estimator $\hat{\theta}_j$ of the jth parameter θ_j satisfies

$$\hat{\theta}_j \sim N(\theta_j, k_j \sigma^2) \quad \text{so} \quad \frac{\hat{\theta}_j - \theta_j}{\sigma \sqrt{k_j}} \sim N(0, 1),$$

where k_j is some constant that does not depend on the data. Replacing σ by its estimate

$$s = \sqrt{s^2} = \sqrt{\frac{1}{n-p} \sum_{i=1}^{n} (y_i - \hat{\mu}_i)^2}$$

makes the denominator equal to $SE(\hat{\theta}_j)$ and changes the distribution to a t distribution,

$$T = \frac{\hat{\theta}_j - \theta_j}{s \sqrt{k_j}} = \frac{\hat{\theta}_j - \theta_j}{SE(\hat{\theta}_j)} \sim t_{n-p}. \tag{5.21}$$

Note that the degrees of freedom, $n - p$, for the t distribution is equal to the denominator in the formula for s^2. Repeating the computations from the one sample case, we get a 95% confidence interval

$$\hat{\theta}_j \pm t_{0.975, n-p} \cdot s \cdot \sqrt{k_j}$$

or in terms of the standard error,

$$\hat{\theta}_j \pm t_{0.975, n-p} \cdot SE(\hat{\theta}_j). \tag{5.22}$$

We recognize that the interval has the same structure as (5.20).

The confidence interval (5.22) corresponds to a *confidence level* or a *coverage* of 0.95. It differs from situation to situation what is a relevant confidence level, but we often use 0.95 or 0.90. In general we denote the confidence level $1 - \alpha$, such that 95% and 90% confidence intervals corresponds to $\alpha = 0.05$ and $\alpha = 0.10$, respectively. The relevant t quantile is $1 - \alpha/2$, assigning probability $\alpha/2$ to the left as well as to the right of $\hat{\theta}_j$. For example, use the 97.5% quantile in order to compute a 95% confidence interval, or the 95% quantile for the 90% confidence interval. This α-notation may seem peculiar at the moment, but it is closely related to the significance level of a hypothesis test which will be introduced in Chapter 6.

The following box summarizes the theory on confidence intervals. In particular, it is worth emphasizing the interpretation.

Infobox 5.3: Confidence intervals for parameters in linear models

Construction A $1 - \alpha$ confidence interval for a parameter θ is of the form

$$\hat{\theta}_j \pm t_{r,1-\alpha/2} \cdot \text{SE}(\hat{\theta}_j),$$

where the degrees of freedom is equal to residual degrees of freedom, $r = \text{df}_e$.

Interpretation *The $1 - \alpha$ confidence interval includes the values of θ for which it is reasonable, at confidence degree $1 - \alpha$, to believe that they could have generated the data.* If we repeated the experiment many times then a fraction $1 - \alpha$ of the corresponding confidence intervals would include the true value θ.

What happens when $1 - \alpha$ changes? *The larger the $1 - \alpha$ (the smaller the α), the wider the confidence interval,* as large confidence in the interval requires it to be wide.

What happens when n changes? *The larger the sample size the more narrow the confidence interval:* More observations contain more information, so a smaller interval is sufficient in order to retain the same confidence level.

What happens when σ changes? *The larger the standard deviation the wider the confidence interval,* as a large standard deviation corresponds to large variation (uncertainty) in the data.

Now, we compute some confidence intervals in the data examples illustrating linear regression, one-way ANOVA, and the setup with two samples. This is quite easy now since we have already computed the estimates and the corresponding standard errors.

Example 5.7. Stearic acid and digestibility of fat (continued from p. 110). Consider first the slope parameter β, which describes the association between the level of stearic acid and digestibility. Estimates and standard errors were computed in Example 5.3 (p. 110), and since there are $n = 9$ observations and $p = 2$ mean parameters, we need quantiles from the t distribution with 7 degrees of freedom.

For illustration we compute both the 90% and the 95% confidence interval. Since $t_{0.95,7} = 1.895$ and $t_{0.975,7} = 2.365$, we get

$$90\% \text{ CI: } -0.9337 \pm 1.895 \cdot 0.0926 = -0.9337 \pm 0.1754 = (-1.11, -0.76)$$
$$95\% \text{ CI: } -0.9337 \pm 2.365 \cdot 0.0926 = -0.9337 \pm 0.2190 = (-1.15, -0.71).$$

Hence, decrements between 0.76 and 1.11 percentage points of the digestibility per unit increment of stearic acid level are in agreement with the data on the 90% confidence level.

Consider then the expected digestibility for a stearic acid level of 20%, $\mu_0 = \alpha + \beta \cdot 20$. The 95% confidence interval for μ_0 is given by

$$77.859 \pm 2.36 \cdot 1.1096 = 77.859 \pm 2.624 = (75.235, 80.483),$$

where again the estimate and the standard error are taken from Example 5.3 (p. 110). In conclusion, expected values of digestibility percentage corresponding to a stearic acid level of 20% between 75.2 and 80.5 are in accordance with the data on the 90% confidence level.

Similarly we can calculate the confidence interval for the expected digestibility percentage for other values of the stearic acid level. The lower and upper limits are shown in Figure 5.6. The width of the confidence band is smallest close to $\bar{x}$ and becomes larger as x_0 moves away from $\bar{x}$. This reflects that the data contain the most information about the area close to $\bar{x}$, as is also clear from the expression (5.14) for $SE(\hat{\mu}_0)$. $\qquad\square$

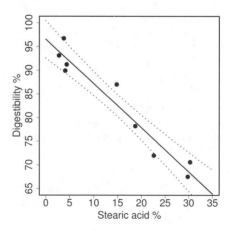

Figure 5.6: Pointwise 95% confidence intervals for the regression line.

Example 5.8. Effect of antibiotics on dung decomposition (continued from p. 112). Estimates and standard errors for the antibiotics data were listed in Table 5.1 (p. 112). Let us compute the 95% confidence level for the difference in organic material between the control group and the spiramycin group. There are $p = 6$ mean parameters in the one-way ANOVA model (one per group) and $n = 34$ observations, so we need $t_{0.975,28}$. It turns out to be 2.048, so we get the following 95% confidence interval for $\alpha_{\text{spiramycin}} - \alpha_{\text{control}}$:

$$0.2517 \pm 2.048 \cdot 0.0786 = 0.2517 \pm 0.1610 = (0.0907, 0.4127).$$

That is, values between 0.0907 and 0.4127 of the expected difference between the control group and the spiramycin group are in accordance with the data on the 95% confidence level. In particular, we see that the value zero is not included in the interval, indicating that zero is an unlikely value of the difference — or more precisely that the observed data are unlikely if there is no difference between the two groups. We will be more specific about the importance of this in the following chapter on hypothesis tests.   □

Example 5.9. Parasite counts for salmons (continued from p. 49). The statistical model for the salmon data is given by

$$y_i = \alpha_{g(i)} + e_i, \quad i = 1, \dots, 26,$$

where $g(i)$ is either "Ätran" or "Conon" and $e_1, \dots, e_{26}$ are iid. $N(0, \sigma^2)$. In other words, all observations are independent, Ätran observations are $N(\alpha_{\text{Ätran}}, \sigma^2)$ distributed, and Conon observations are $N(\alpha_{\text{Conon}}, \sigma^2)$ distributed.

We already computed the group means and group standard deviations in Example 3.1 (p. 49). With the notation from that example,

$$\hat{\alpha}_{\text{Ätran}} = \bar{y}_1 = 32.23, \quad \hat{\alpha}_{\text{Conon}} = \bar{y}_2 = 21.54$$

$$\hat{\sigma}^2 = s^2 = \frac{1}{24}(12 \cdot s_1^2 + 12 \cdot s_2^2) = \frac{1}{2}(7.28^2 + 5.81^2) = 43.40, \quad s = 6.59.$$

In particular, the difference in parasite counts is estimated to

$$\hat{\alpha}_{\text{Ätran}} - \hat{\alpha}_{\text{Conon}} = 32.23 - 21.54 = 10.69$$

with a standard error of

$$\text{SE}(\hat{\alpha}_{\text{Ätran}} - \hat{\alpha}_{\text{Conon}}) = s\sqrt{\frac{2}{13}} = 2.58.$$

The 95% confidence interval for the difference $\alpha_{\text{Ätran}} - \hat{\alpha}_{\text{Conon}}$ is thus

$$10.69 \pm 2.064 \cdot 2.58 = (5.36, 16.02)$$

Table 5.3: Mixture proportions and optical densities for 10 dilutions of a standard dissolution with Ubiquitin antibody

Mixture proportion, m	100	200	400	800	1600
Optical density, y	1.04	0.71	0.35	0.19	0.09
	1.11	0.72	0.38	0.26	0.11

since the 97.5% quantile in t_{24} is 2.064. In particular, we see that zero is not included in the confidence interval, so the data is not in accordance with a difference of zero between the stock means. In other words, the data suggests that Ätran salmons are more susceptible than Conon salmons to parasites during an infection. □

Example 5.10. ELISA experiment. As part of a so-called ELISA experiment, the optical density was measured for various dilutions of a standard dissolution with ubiquitin antibody. For each dilution, the mixture proportion describes how many times the original ubiquitin dissolution has been thinned.

The data are listed in Table 5.3 and plotted in Figure 5.7. The left plot clearly shows that the optical density does not depend linearly on the mixture proportion. Rather, as is shown in the right plot, the optical density is a linear function of the logarithmic mixture ratio. Hence, if m_i and y_i denote the mixture proportion and the optical density, respectively, for measurement i, then we consider the linear regression model

$$y_i = \alpha + \beta \cdot \log_2(m_i) + e_i$$

with the usual assumptions on the remainder terms. Notice that we have used the logarithm with base 2. This means that doubling of the mixture proportion corresponds to an increase of one in its logarithmic counterpart.

The parameter estimates for the intercept and the slope turn out to be (with the standard error in parentheses):

$$\hat{\alpha} = 2.605 \quad (\text{SE } 0.191), \qquad \hat{\beta} = -0.244 \quad (\text{SE } 0.022).$$

We conclude that the estimated effect of a doubling of the mixture proportion is a decrease of 0.244 in optical density. The 95% confidence interval for β is

$$\hat{\beta} \pm 2.306 \cdot \text{SE}(\hat{\beta}) = -0.244 \pm 2.306 \cdot 0.022 = (-0.295, -0.193),$$

where 2.306 is the 97.5% quantile in the t_8 distribution. This means that decreases between 19.3 and 29.5 when the mixture proportion is doubled are in agreement with the data on the 95% confidence level.

The regression model can be used to estimate (or predict) the optical density for a new mixture proportion, say 600. The expected optical density for such a mixture proportion is

$$\hat{\alpha} + \hat{\beta} \cdot \log_2(600) = 2.605 - 0.244 \cdot \log_2(600) = 0.353$$

and the corresponding confidence interval turns out to be $(0.276, 0.430)$. □

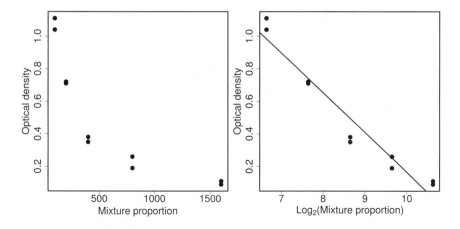

Figure 5.7: Scatter plot of the optical density against the mixture proportions (left) and against the logarithmic mixture proportions with the fitted regression line (right).

5.4 Unpaired samples with different standard deviations

Throughout this chapter we have assumed that the standard deviation is the same for all observations, and the situation with two independent samples with the same standard deviation is a special case of the one-way ANOVA setup. The assumption of variance homogeneity is essential, in particular for the computation of standard errors and confidence intervals. However, in the situation with two unpaired samples it is possible to handle the situation with different standard deviations in the two groups as well.

Assume that the observations $y_1, \ldots, y_n$ are independent and come from two different groups, group 1 and group 2. Both the mean and standard deviation are allowed to vary between groups, so observations from group 1 are assumed to be $N(\mu_1, \sigma_1^2)$ distributed and observations from group 2 are assumed to be $N(\mu_2, \sigma_2^2)$ distributed.

The estimates of the means are unchanged and thus equal to the group sample means,

$$\hat{\mu}_1 = \bar{y}_1, \quad \hat{\mu}_1 = \bar{y}_2.$$

The variance of their difference is

$$\mathrm{Var}(\hat{\mu}_2 - \hat{\mu}_1) = \mathrm{Var}(\hat{\mu}_2) + \mathrm{Var}(\hat{\mu}_1) = \frac{\sigma_1^2}{n_1} + \frac{\sigma_2^2}{n_2},$$

where we have used Infobox 4.2. Replacing the true (population) variances, σ_1^2 and σ_2^2, with their sample estimates, s_1^2 and s_2^2, yields the estimated variance and hence the standard error,

$$SE(\hat{\mu}_2 - \hat{\mu}_1) = \sqrt{\frac{s_1^2}{n_1} + \frac{s_2^2}{n_2}}. \tag{5.23}$$

In order to construct a confidence interval from (5.20), all we need is a t quantile, but since we do not have a pooled standard deviation it is not obvious how many degrees of freedom to use. It turns out that it is appropriate to use

$$r = \frac{\left(SE_1^2 + SE_2^2\right)^2}{\dfrac{SE_1^4}{n_1 - 1} + \dfrac{SE_2^4}{n_2 - 1}} \tag{5.24}$$

degrees of freedom, where $SE_1 = s_1/\sqrt{n_1}$ and $SE_2 = s_2/\sqrt{n_2}$.

The number r is not necessarily an integer. The corresponding $1 - \alpha$ confidence interval is

$$\hat{\mu}_2 - \hat{\mu}_1 \pm t_{1-\alpha/2,r} \cdot SE(\hat{\mu}_2 - \hat{\mu}_1).$$

Note that this confidence interval is only approximate, meaning that the coverage is only approximately (not exactly) 95%.

Example 5.11. Parasite counts for salmons (continued from p. 125). We already computed a 95% confidence interval for the difference between expected parasite counts for Ätran and Conon salmons under the assumption that the standard deviation is the same in both groups (Example 5.9, p. 125). If we are not willing to make this assumption, then we could compute the confidence interval based on (5.23) and (5.24) instead.

We get the standard error

$$SE(\hat{\mu}_2 - \hat{\mu}_1) = \sqrt{\frac{7.28^2}{13} + \frac{5.81^2}{13}} = 2.58,$$

exactly as in Example 5.9 because there are 13 fish in both samples. For the degrees of freedom we get $SE_1 = 7.28/\sqrt{13} = 2.019$, $SE_2 = 5.81/\sqrt{13} = 1.161$, and $r = 22.9$. The 97.5% quantile in t_{23} is 2.069, so the 95% confidence interval for $\alpha_{\text{Ätran}} - \alpha_{\text{Conon}}$ becomes

$$10.69 \pm 2.069 \cdot 2.58 = (5.35, 16.04),$$

almost the same as in Example 5.9, where the standard deviation was assumed to be the same for the two stocks. □

For the salmon data, the two confidence intervals (assuming equal standard deviations or not) were almost identical. This is so because there are the same number of observations in the two groups and because the sample standard deviations computed from the samples separately were close to each other. In other cases there may be a substantial difference between the two confidence intervals.

If the group standard deviations are close, we usually prefer to use the confidence interval based on the assumption of equal standard deviations, mainly because the estimate of the standard deviation is more precise, as it is based on all observations. The results are quite robust as long as the samples are roughly of the same size and not too small (Zar, 1999). Larger differences between the group standard deviations indicate that the assumption of equal standard deviations is not reasonable; hence we would rather use the confidence interval from the present section. In Case 3, Part II (p. 357) we will see an extreme example of such data.

Example 5.12. Vitamin A intake and BMR (continued from p. 82). Figure 4.9 showed histograms for men and women of the BMR variable, related to the basal metabolic rate. We concluded that the normal distribution was adequate to describe the data for each of the samples. The distribution for the sample of men seems to be slightly wider than for the sample of women, so if we want to estimate the difference in BMR between men and women we may want to allow for different standard deviations.

It turns out that the means of the BMR variable are 7.386 for men and 5.747 for women, respectively, whereas the standard deviations are 0.723 and 0.498. Hence the difference in expected BMR is estimated to

$$\hat{\alpha}_{men} - \hat{\alpha}_{women} = 7.386 - 5.747 = 1.639$$

and the corresponding standard error is

$$\text{SE}(\hat{\alpha}_{men} - \hat{\alpha}_{women}) = \sqrt{\frac{0.723^2}{1079} + \frac{0.498^2}{1145}} = 0.0265$$

because 1079 men and 1145 women participated in the study. The relevant degrees of freedom is 1900 and the relevant quantile is 1.961. This is of course close to the standard normal quantile because of the large number of observations. The 95% confidence interval becomes $(1.589, 1.692)$, so deviations in expected BMR between men and women in this interval are in accordance with the data. Notice that the confidence interval is very narrow. Again, this is due to the large samples, which imply that expected values are estimated with a large precision. □

5.5 R

As already mentioned in Sections 2.5 and 3.7.1, the lm() function is used to fit statistical models based on the normal distribution. Consider the situation with a response variable y and a single explanatory variable x which may be either quantitative (linear regression) or a factor (one-way ANOVA). In both cases the model is fitted with the command

```
model <- lm(y ~ x)
```

but recall the importance of x being coded correctly as either a factor or a quantitative variable (*cf.* Section 3.7.2 and the comments below).

The above call produces an object, which we have named model, from which we can extract important summaries from the model fit. The function summary() yields estimates with corresponding standard errors, and the function confint() computes confidence intervals:

```
> summary(model)
> confint(model)
```

We go through the output from these calls below. In principle, the output is completely comparable for regression models and ANOVA models, but due to parameterization issues in the ANOVA setup, we will discuss the output for the two model types separately. The output from summary() includes information about certain hypothesis tests — we discuss this part of the output in Section 6.6.

5.5.1   Linear regression

Let us first study the output corresponding to a linear regression model and run an analysis of the digestibility data. Recall the 90% confidence interval for the slope parameter from Example 5.7 (p. 123):

$$\beta: \quad -0.9337 \pm 1.895 \cdot 0.0926 = (-1.11, -0.76).$$

In particular, $\hat{\beta} = -0.9337$ and $SE(\hat{\beta}) = 0.0926$, values that are obtained from R as follows:

```
> modelDigest <- lm(digest~stearic.acid)  # Fit linear regression
> summary(modelDigest)                     # Estimates, SE's etc.

Call:
lm(formula = digest ~ stearic.acid)

Residuals:
    Min      1Q  Median      3Q     Max
-3.4308 -1.2249 -0.8724  2.3590  4.2860

Coefficients:
             Estimate Std. Error t value Pr(>|t|)
(Intercept)  96.53336    1.67518   57.63 1.24e-10 ***
stearic.acid -0.93374    0.09262  -10.08 2.03e-05 ***
---
Signif. codes:  0 '***' 0.001 '**' 0.01 '*' 0.05 '.' 0.1 ' ' 1
```

```
Residual standard error: 2.97 on 7 degrees of freedom
Multiple R-squared: 0.9356,Adjusted R-squared: 0.9264
F-statistic: 101.6 on 1 and 7 DF,  p-value: 2.028e-05
```

First the linear regression model is fitted and stored in the object modelDigest. The most important part of the output from summary() is the Coefficients part (roughly in the middle). It has the following structure:

- A list with one line per mean parameter (fixed effect parameter) θ_j. In the linear regression setup, there are two parameters: the intercept and the slope. Since stearic.acid is the explanatory variable, the slope parameter is named as such in the output.

- Four values are listed for each parameter: (i) The estimate, recognizing the value -0.93373 for the slope parameter; (ii) the standard error for the estimate, recognizing the value 0.09262 for the slope parameter; (iii) a t-test statistic for the hypothesis that the corresponding parameter is equal to zero; and (iv) the corresponding p-value. Hypothesis tests and p-values will be discussed in Chapter 6.

The output from summary() includes various other information. First, the call to lm() is repeated and then a few summaries on the non-standardized residuals are listed, which may be used for a rough check of symmetry of the residuals. After the Coefficients part we find the residual standard error, here $s = 2.97$, the residual degrees of freedom, here $df_e = 7$, and a few things that are not of interest at the moment.

By using the estimates and standard errors we can compute confidence intervals 'manually' from (5.20); see Section 5.5.4 for how the qt() function can be used to compute t quantiles. Alternatively — and more easily — we can make R do the computations for us and use the confint() function,

```
> confint(modelDigest, level=0.90)  # 90% confidence intervals
                  5 %       95 %
(Intercept)   93.359609 99.7071185
stearic.acid -1.109219 -0.7582645
> confint(modelDigest)               # Default is 95%
                  2.5 %      97.5 %
(Intercept)   92.572199 100.4945282
stearic.acid -1.152755  -0.7147285
```

Again, the output has one line per mean parameter, with the same names as in the output from summary, and we recognize the 90% confidence interval $(-1.109, -0.758)$ for the slope. Notice how we may specify the level of the confidence interval and that the 95% interval is computed if no level is specified.

Above we have concentrated on the values for the slope, β. For the intercept, α, we read the estimate 96.533 with a corresponding standard error

of 1.675 and a 95% confidence interval from 92.57 to 100.49. Estimation of
expected values of the response for fixed values of the explanatory variable
is very much related to prediction, and we will postpone the discussion of
computations of estimates and confidence intervals to Section 7.3.2.

Sometimes it is of interest to fit the model where the regression line is
forced to go through $(0,0)$. We say that we fit the model "without intercept"
(see also Section 2.5):

```
> modelDigest1 <- lm(digest~acid-1)  # Regr.line through (0,0)
```

5.5.2   One-way ANOVA

We use the antibiotics data (Example 3.2, p. 51) for illustration and repeat
Table 5.1 in order to easily compare with the output from R.

Table 5.4: Estimates for group means and comparison to the control group for the
antibiotics data (identical to Table 5.1)

Antibiotics	n_j	$\hat{\alpha}_j$	$SE(\hat{\alpha}_j)$	$\hat{\alpha}_j - \hat{\alpha}_{control}$	$SE(\hat{\alpha}_j - \hat{\alpha}_{control})$
Control	6	2.603	0.0497	—	—
α-Cypermethrin	6	2.895	0.0497	0.2917	0.0703
Enrofloxacin	6	2.710	0.0497	0.1067	0.0703
Fenbendazole	6	2.833	0.0497	0.2300	0.0703
Ivermectin	6	3.002	0.0497	0.3983	0.0703
Spiramycin	4	2.855	0.0609	0.2517	0.0786

Since the vector type contains text values, R automatically uses it as a
factor. Had the values been numerical, we should convert the vector to a
factor; see Section 3.7.2. The lm() and summary() calls are as follows:

```
> modelAntibio <- lm(org~type)     # Fit one-way ANOVA
> summary(modelAntibio)            # Estimates etc. with alphacyp
                                   # as reference

Call:
lm(formula = org ~ type)

Residuals:
     Min       1Q   Median       3Q      Max
-0.29000 -0.06000  0.01833  0.07250  0.18667

Coefficients:
            Estimate Std. Error t value Pr(>|t|)
(Intercept)  2.89500    0.04970  58.248  < 2e-16 ***
typeControl -0.29167    0.07029  -4.150 0.000281 ***
```

```
typeEnroflox -0.18500     0.07029  -2.632 0.013653 *
typeFenbenda -0.06167     0.07029  -0.877 0.387770
typeIvermect  0.10667     0.07029   1.518 0.140338
typeSpiramyc -0.04000     0.07858  -0.509 0.614738
---
Signif. codes:  0 '***' 0.001 '**' 0.01 '*' 0.05 '.' 0.1 ' ' 1

Residual standard error: 0.1217 on 28 degrees of freedom
Multiple R-squared: 0.5874,Adjusted R-squared: 0.5137
F-statistic: 7.973 on 5 and 28 DF,  p-value: 8.953e-05
```

The output before and after the `Coefficients` part is similar to the linear regression case; in particular, we find the residual standard error, $s = \mathrm{MS}_e = 0.1217$ on $\mathrm{df}_e = 28$ degrees of freedom (*cf.* Example 5.4 (p. 112)).

Let us turn to the `Coefficients` part of the output. As for linear regression, there is a line for each parameter in the model, but the interpretation of the parameters may appear strange. There is a parameter denoted `Intercept` followed by parameters `typeControl` through `typeSpiramyc` — but no `typeAlfacyp` (the word type of course refers to the name of the factor). The explanation is that R *uses one of the groups as a reference group and compares the other groups to the reference group.* More specifically:

- The α-cypermethrin group is used as the reference group — because it comes first in alphabetical order — and the parameter reported as `Intercept` is the mean parameter for this group; that is, $\alpha_{\mathrm{alphacyp}}$. Hence, we read $\hat{\alpha}_{\mathrm{alphacyp}} = 2.895$, which we recognize from Table 5.4.

- For the other groups, the parameter is the difference in mean for the group in question and the reference group. For example, the values in the `typeFenbenda` line relate to the parameter $\alpha_{\mathrm{fenbend}} - \alpha_{\mathrm{alphacyp}}$, so the estimate of this difference is

$$\hat{\alpha}_{\mathrm{fenbend}} - \hat{\alpha}_{\mathrm{alphacyp}} = -0.06167,$$

which we could also compute from Table 5.4 as $2.833 - 2.895$.

- The standard errors (and t-tests) reported in the output are for the differences as well, not for the original α's. For example, the difference $\alpha_{\mathrm{fenbend}} - \alpha_{\mathrm{alphacyp}}$ is estimated with a standard error of 0.07029. This is identical to the standard error for comparisons to the control group, which is listed in Table 5.4.

The final bullet represents an important "rule" regarding output from `summary()`: *the four values for a parameter always concern the same parameter.*

The reason why differences are reported is of course that those are typically of main interest. Rather than the average level of organic dung matter for each of the groups, we are interested in the effect of the different antibiotic types; that is, in differences between the α's. If we are interested in estimates for the α's themselves, they are easily recovered from the differences; for example:

$$\hat{\alpha}_{\text{fenbend}} = \hat{\alpha}_{\text{alphacyp}} + \hat{\alpha}_{\text{fenbend}} - \hat{\alpha}_{\text{alphacyp}} = 2.89500 - 0.06167 = 2.8333.$$

Alternatively — and if we want R to compute the corresponding standard errors — we can fit the same model, but without the intercept term. This is done by adding -1 to the model formula as follows (parts of the output have been deleted):

```
> modelAntibio1 <- lm(org~type-1)  # One-way ANOVA
                                   # without intercept
> summary(modelAntibio1)


Coefficients:
             Estimate Std. Error t value Pr(>|t|)
typeAlfacyp   2.89500    0.04970   58.25  <2e-16 ***
typeControl   2.60333    0.04970   52.38  <2e-16 ***
typeEnroflox  2.71000    0.04970   54.53  <2e-16 ***
typeFenbenda  2.83333    0.04970   57.01  <2e-16 ***
typeIvermect  3.00167    0.04970   60.39  <2e-16 ***
typeSpiramyc  2.85500    0.06087   46.90  <2e-16 ***

Residual standard error: 0.1217 on 28 degrees of freedom
```

We say that we fit the model without intercept. Compare to Table 5.4 and recognize estimates as well as standard errors.

It is important to realize that the two model fits, modelAntibio and modelAntibio1, specify *the same model*. Check for example that the residual standard errors are the same. The difference is the specification of the parameters — we say that they correspond to different *parameterizations* of the model. One parameterization is not better than another, but it may very well be that different parameterizations are useful for different purposes and it is therefore often useful to be able to "play around" with different parameterizations.

In the present example, yet another parameterization is perhaps the more natural: one of the treatments is a control and essentially we are interested in comparing the other groups to the control group. In other words, it would be useful to use the control group as the reference group rather than the α-cypermethrin group. We can change the reference level with the relevel() function as follows:

```
> mytype <- relevel(type, ref="Control") # New reference group
> modelAntibio2 <- lm(org~mytype)        # One-way ANOVA with
                                         # Control as reference
> summary(modelAntibio2)
```

```
Coefficients:
                Estimate Std. Error t value Pr(>|t|)
(Intercept)      2.60333    0.04970  52.379  < 2e-16 ***
mytypeAlfacyp    0.29167    0.07029   4.150 0.000281 ***
mytypeEnroflox   0.10667    0.07029   1.518 0.140338
mytypeFenbenda   0.23000    0.07029   3.272 0.002834 **
mytypeIvermect   0.39833    0.07029   5.667  4.5e-06 ***
mytypeSpiramyc   0.25167    0.07858   3.202 0.003384 **
---
Signif. codes:  0 '***' 0.001 '**' 0.01 '*' 0.05 '.' 0.1 ' ' 1

Residual standard error: 0.1217 on 28 degrees of freedom
```

We recognize the values from Table 5.4. The corresponding confidence intervals are easily obtained with confint():

```
> confint(modelAntibio2)      # Confidence intervals with
                              # Control as reference
                      2.5 %      97.5 %
(Intercept)     2.50152445 2.7051422
mytypeAlfacyp   0.14768716 0.4356462
mytypeEnroflox -0.03731284 0.2506462
mytypeFenbenda  0.08602049 0.3739795
mytypeIvermect  0.25435382 0.5423128
mytypeSpiramyc  0.09069268 0.4126407
```

A final comment on reparameterizations. The model fits modelAntibio, modelAntibio1, and modelAntibio2 corresponded to different parameterizations of the same model. In particular, the same model is fitted no matter if the intercept is included or not.

5.5.3   One sample and two samples

The one sample case can be handled with the lm() function, too, acknowledging that there are no explanatory variables, so the model should include only an intercept. For the crab weight data (Example 4.1, p. 68) the output reads as follows:

```
> modelCrab <- lm(wgt~1)  # Model fit for a single sample
> summary(modelCrab)

Coefficients:
            Estimate Std. Error t value Pr(>|t|)
(Intercept)   12.758      0.177    72.1   <2e-16 ***
---
```

```
Signif. codes:  0 '***' 0.001 '**' 0.01 '*' 0.05 '.' 0.1 ' ' 1
```

```
Residual standard error: 2.252 on 161 degrees of freedom
```

```
> confint(modelCrab)         # Conf. int. for a single sample
              2.5 %  97.5 %
(Intercept) 12.40887 13.1078
```

Recognize the estimate, standard error, and confidence interval from Example 5.6 (p. 120). Alternatively, and in some cases more easily, we can use the t.test() function:

```
> t.test(wgt)   # Analysis of a single sample
```

```
One Sample t-test
```

```
data:  wgt
t = 72.0963, df = 161, p-value < 2.2e-16
alternative hypothesis: true mean is not equal to 0
95 percent confidence interval:
 12.40887 13.10780
sample estimates:
mean of x
 12.75833
```

A lot of the output is concerned with a certain *t*-test and will be explained in Section 6.6.4, but the estimate and the confidence interval are also produced. The confidence level can be specified with the option conf.level, like in

```
> t.test(wgt, conf.level=0.90) # Change confidence level
```

The t.test() function is also useful for the analysis of two samples. The default call t.test(x,y) corresponds to the analysis of two independent samples with different standard deviations. If the standard deviations are assumed to be identical, then use the option var.equal=T, and if the samples are paired then use the option paired=T.

In Examples 5.9 and 5.11 (p. 125 and p. 128) we computed the confidence interval for the salmon data with and without the assumption of equal standard deviations. The confidence intervals could be computed by the following t.test() commands:

```
> atran <- c(31,31,32,22,41,31,29,40,41,39,36,17,29)
> conon <- c(18,26,16,20,14,28,18,27,17,32,19,17,28)
```

```
> t.test(conon, atran, var.equal=T) # Two independent samples,
                                     # equal standard deviations
```

```
Two Sample t-test

data:  conon and atran
t = -4.1381, df = 24, p-value = 0.0003715
alternative hypothesis: true difference in means is not
                        equal to 0
95 percent confidence interval:
 -16.025211  -5.359404
sample estimates:
mean of x mean of y
 21.53846  32.23077


> t.test(conon, atran) # Two independent samples,
                       # different standard deviations

Welch Two Sample t-test

data:  conon and atran
t = -4.1381, df = 22.874, p-value = 0.0004028
alternative hypothesis: true difference in means is not
                        equal to 0
95 percent confidence interval:
 -16.039136  -5.345479
sample estimates:
mean of x mean of y
 21.53846  32.23077
```

The lameness data from Example 3.5 (p. 57) are an example of paired data. We may either use the option paired=T to t.test() or directly run a one sample analysis on the differences. The relevant code and output is the following:

```
> healthy <- c(-0.9914,1.4710,1.2459,0.4024,0.0325,
+ -0.6396,0.7246,0.0604)
> lame <- c(4.3541,4.7865,6.1945,10.7383,3.3007,4.8678,
+ 7.8965,3.9338)
> t.test(healthy, lame, paired=T) # Two paired samples
> t.test(lame-healthy)            # Analysis of differences
```

It is left as an exercise (Exercise 5.4) to run the commands and interpret the output.

5.5.4   Probabilities and quantiles in the *t* distribution

Probabilities and quantiles in the *t* distributions are computed with the pt() and qt() functions. For example, the 95% quantile of the *t* distribution with 5 degrees of freedom is 2.015, according to Table 5.2. In other words, $P(T \leq 2.015) = 0.95$ if $T \sim t_5$. The cumulative distribution function mapping 2.015 to 0.95 is denoted pt, whereas the quantile function mapping 0.95 to 2.015 is denoted qt.

```
> pt(2.015, df=5) # P(T <= 2.015) if T~t_5
[1] 0.949997
> qt(0.95, df=5)  # 95% quantile in t_5
[1] 2.015048
```

Notice that it is of course necessary to specify the degrees of freedom so that R knows which t distribution to use.

5.6   Exercises

The exercises for this chapter involve estimation and confidence intervals, but some of them could easily be extended to involve hypothesis tests, too (Chapter 6). On the other hand, many questions in the exercises for Chapter 6 could be answered at this point, based on Chapter 5.

5.1 Points in a swarm. The following experiment was carried out in order to investigate how well people are able to estimate the number of points in a swarm of points (Rudemo, 1979). Seven persons were asked to look at a number of pictures with points. Each picture was shown four times to each person, in a random order and five seconds each time, and for each showing they were asked to estimate/guess the number of points in the picture. For a picture with 161 points, the average over the four estimates for the seven persons were as follows:

Person	1	2	3	4	5	6	7
Average guess	146	182	152.5	165	139.5	132	155

You may use (some of) the following R-output for your answers:

```
> y <- c(146, 182, 152.5, 165, 139.5, 132, 155)
> mean(y)
[1] 153.1429
> sd(y)
[1] 16.64010
> qt(0.975, df=5)
[1] 2.570582
> qt(0.975, df=6)
```

```
[1] 2.446912
> qt(0.975, df=7)
[1] 2.364624
```

1. Specify a statistical model; that is, make assumptions on the distribution of $y_1, \ldots, y_7$.

2. What are the parameters of the model and what is their interpretation? Estimate the parameters.

3. Compute a 95% confidence interval for the expected guess from a random person from the population.

4. Is there a systematic tendency to overestimate or underestimate the number of points?

   [Hint: Which value of the mean parameter corresponds to neither over- nor underestimation? What is the interpretation of the confidence interval?]

5. The experimenter would like the 95% confidence interval to have a length of at most 10. Calculate, just roughly, how many persons should be included in the experiment in order to obtain this.

   [Hint: Use (5.19) to compute the relevant n. You can use the approximation that the 97.5% quantile in the t distribution is around 2 (when the degrees of freedom is not very small).]

Assume now that a guess from a random person is normally distributed with mean 161 and standard deviation 16.

6. Calculate an interval where you would expect to find the guesses for 95% of the people in the population. Explain the difference in interpretation between this interval and the interval you computed in question 3.

5.2 ® **Phosphorous concentration.** In a plant physiological experiment the amount of water-soluble phosphorous (among others) was measured in the plants, as a percentage of dry matter (Skovgaard, 2004). The phosphorous concentration was measured nine weeks during the growth season, and the averages over the plants in the experiments are given in the table below:

Week	1	2	3	4	5	6	7	8	9
Phosphorous	0.51	0.48	0.44	0.44	0.39	0.35	0.28	0.24	0.19

1. Make a suitable plot of the data.

2. Specify a statistical model where the expected phosphorous concentration is a linear function of time. Estimate the parameters of the model.

3. Compute 95% confidence intervals for the parameters that describe the linear relationship between time and phosphorous.

4. Compute an estimate and a 95% confidence interval for the expected decrease over a period of three weeks.

   [Hint: What is the interpretation of the slope parameter? What is then the expected change over three weeks?]

5. Compute an estimate and a 90% confidence interval for the expected phosphorous concentration after 10 weeks.

5.3 **Malaria parasites.** This exercise is about the same data as those used in Case 2, Part I (p. 354). A medical researcher took blood samples from 31 children who were infected with malaria and determined for each child the number of malaria parasites in 1 ml of blood (Williams, 1964; Samuels and Witmer, 2003).

In Case 2 you are asked to define y as the natural logarithmic counts. The conclusion is that while the distribution of the original counts is highly skewed, the distribution of y is well approximated by a normal distribution.

1. Specify a statistical model for the logarithmic counts, $y_1, \ldots, y_{31}$.

2. Compute an estimate and a 95% confidence interval for the expected number of parasites for a random child infected with malaria.

   [Hint: For which variable (on which scale) is it appropriate to use the "usual" formulas for confidence intervals.]

3. Compute a 95% confidence interval for the expected number of parasites for a random child infected with malaria.

You can use the following R-output:

```
> mean(parasites)
[1] 12889.61
> sd(parasites)
[1] 27435.40
> mean(y)
[1] 8.02549
> sd(y)
[1] 1.85699
> qt(0.975, df=30)
[1] 2.042272
```

```
> qt(0.95, df=30)
[1] 1.697261
> qt(0.90, df=30)
[1] 1.310415
```

5.4 ℝ **Equine lameness.** Recall the data on equine lameness from Example 3.5 (p. 57). The data is available in the external file lameness.txt with variables lame and healthy.

1. Specify a statistical model for the data and explain how you would check if the model assumptions are reasonable.

   [Hint: With very few observations it is always difficult to validate the model, but this is not so much the point here. You should rather consider the following questions: Which variable is assumed to be described by a normal distribution? And how would you check it (if there were data from more than eight horses)?]

2. Read the data into R and use the command t.test(lame, healthy, paired=T) to compute an estimate and a confidence interval for the expected difference in the symmetry score for a random horse (difference between lame and healthy condition); see Section 5.5.3.

3. Use the command t.test(lame-healthy) and compare the output to that from question 2.

5.5 ℝ **Parasite counts for salmons.** In Section 5.5.3 we computed the estimate and the confidence interval for the difference between the expected number of parasites for Ätran and Conon salmons with the t.test() function. In this exercise you should use lm() for the same computations.

1. How should the data be organized in order for you to run the analysis with lm()?

   [Hint: How many variables should you have? Of which length and with which values?]

2. Fit the model with lm().

3. Compute the estimate and confidence interval by using summary() and confint() functions. In this analysis, is it assumed that the standard deviations are the same for Ätran and Conon salmons or not?

4. Compare the results to those from Example 5.9 (p. 125).

5.6 Heart weight for cats. A classical dataset contains data on body weight and heart weight for 97 male cats and 47 female cats (Fisher,

1947). In this exercise we shall consider the ratio between the heart weight and the body weight, so we define

$$y_i = \frac{\text{Heart weight in grams for cat } i}{\text{Body weight in kilograms for cat } i}.$$

The sample means and sample standard deviations of this ratio are listed for male and female cats:

Sex	n	$\bar{y}$	s
Male	97	3.895	0.535
Female	47	3.915	0.513

1. Specify a statistical model for the data that makes it possible to compare the ratio between heart weight and body weight for female and male cats.

2. Compute an estimate for the expected difference between the ratio for female and male cats. Compute also the pooled sample standard deviation.

3. Compute a 95% confidence interval for the expected difference between the ratio for female and male cats. Does the data support a belief that the ratio between heart weight and body weight differs between female and male cats?

5.7 ⓡ **Heart weight for cats.** This is a continuation of Exercise 5.6 where you can read the description of the data. The data are available in the dataset cats in the MASS package. The dataset contains the variables Sex (sex of the cat), Bwt (body weight in kilograms) and Hwt (heart weight in grams).

1. Use the following commands to make the variables directly available in R:

```
> library(MASS)
> data(cats)
> attach(cats)
```

2. If you did not solve Exercise 5.6 already, then answer question 1 from that exercise.

3. Use the t.test() function to compute the estimate and the 95% confidence interval for the expected difference between female and male cats of the ratio between heart weight and body weight.

[Hint: First, make a variable y with the ratios. Next, make variables yF and yM that contain the ratios for females and males, respectively. Finally, use the t.test() function.]

4. If you solved Exercise 5.6, then compare your results from this exercise to those from questions 2 and 3 in Exercise 5.6.

5.8 ℝ **Sorption of organic solvents.** In an experiment the sorption was measured for a variety of hazardous organic solvents (Ortego et al., 1995). The solvents were classified into three types (esters, aromatics, and chloroalkanes), and the purpose of this exercise is to examine differences between the three types. The data are listed in the table:

Solvent	Sorption				
Esters	0.29	0.06	0.44	0.61	0.55
	0.43	0.51	0.10	0.34	0.53
	0.06	0.09	0.17	0.60	0.17
Aromatics	1.06	0.79	0.82	0.89	1.05
	0.95	0.65	1.15	1.12	
Chloroalkanes	1.28	1.35	0.57	1.16	1.12
	0.91	0.83	0.43		

The external file `hazard.txt` contains the variables `type` and `sorption`.

1. Specify a statistical model that makes it possible to compare the sorption for the three materials.

2. Read the data into R and fit the model with `lm()`.

3. Compute the estimate for the expected sorption for each of the three types of solvents.

   [Hint: Use `summary()` combined with the `lm()` command from question 2. Which group does R use as the reference group? What is the interpretation of the estimates corresponding to the other types?]

4. Compute the estimates for the pairwise differences between the expected values for the three types. Use formula (5.15) to compute the corresponding standard errors. Can you find any of the standard errors in the output from the `summary()` command from question 3?

   [Hint: You need the pooled standard deviation s. You find it near the bottom of the output from `summary()`].

5. Compute 95% confidence intervals for the pairwise differences between the expected values for the types. What is the conclusion regarding differences between the three solvent types?

5.9 Digestibility coefficients. In an experiment with six horses the digestibility coefficient was measured twice for each horse: once after the horse had been fed straw treated with NaOH and once after the horse had been treated ordinary straw (Skovgaard, 2004). The results were as follows:

Horse	Ordinary	NaOH treated	Difference
1	40.70	55.10	14.40
2	48.10	60.30	12.20
3	47.00	63.10	16.10
4	35.40	45.00	9.60
5	49.60	57.50	7.90
6	41.30	58.80	17.50
$\bar{y}$	43.68	56.63	12.95
s	5.45	6.30	3.74

1. Specify a statistical model that makes it possible to compare the digestibility coefficient corresponding to NaOH-treated straw and ordinary straw.

2. Compute an estimate and a confidence interval for the expected difference between the digestibility coefficient for NaOH-treated straw and ordinary straw.

5.10 ℝ **Farm prices.** In February 2010, 12 production farms were for sale in a municipality on Fuen island in Denmark. The soil area in thousands of square meters and the price in thousands of DKK are listed in the table:

Soil area (1000 m^2)	Price (1000 DKK)	Soil area (1000 m^2)	Price (1000 DKK)
108	2495	246	9985
221	3975	454	13500
191	3995	511	14500
238	6850	946	23000
319	7600	1257	28200
354	9950	1110	34000

The external file `farmprice.txt` contains two variables, `area` and `price`, with the numbers from the table.

1. Read the data into R and make a plot of price against area.

2. Fit a linear regression model to the data and add the estimated regression line to the plot from question 1. What is the interpretation of the slope parameter?

3. Consider a random farm from the same area with a soil area of 100,000 square meters. How much would you expect it to cost? Compute the 95% confidence interval for the expected price.

Chapter 6

Hypothesis tests

Statistical tests are used to investigate if the observed data contradict or support specific biological assumptions. In short, a statistical test evaluates how likely the observed data is if the biological assumptions under investigation are true. If the data is very unlikely to occur given the assumptions, then we do not believe in the assumptions. Hypothesis testing forms the core of statistical inference, together with parameter estimation and confidence intervals, and involves important new concepts like *null hypotheses*, *test statistics*, and *p-values*. We will start by illustrating the ideas with an example.

Example 6.1. Hormone concentration in cattle. As part of a larger cattle study, the effect of a particular type of feed on the concentration of a certain hormone was investigated. Nine cows were given the feed for a period, and the hormone concentration was measured initially and at the end of the period. The data are given in the table below:

Cow	1	2	3	4	5	6	7	8	9
Initial (μg/ml)	207	196	217	210	202	201	214	223	190
Final (μg/ml)	216	199	256	234	203	214	225	255	182
Difference (μg/ml)	9	3	39	24	1	13	11	32	-8

The purpose of the experiment was to examine if the feed changes the hormone concentration. Initially, we may examine if the concentrations are generally increasing or decreasing from the start to the end of the experiment. The hormone concentration increases for eight of the nine cows (not cow 9) — for some cows quite substantially (for example, cow no. 3), for some cows only a little bit (for example, cow no. 5). In other words, there is certainly a tendency, but is it strong enough for us to conclude that the feed affects the concentration — or might the observed differences as well be due to pure chance? Is it likely that we would get similar results if we repeated the experiment, or would the results be quite different?

These data constitute an example of two paired samples, so we know from Sections 3.5 and 5.1.5 which model is appropriate: We consider the differences, denoted by $d_1, \ldots, d_9$, during the period and assume that they are independent and $N(\mu, \sigma^2)$ distributed. Then μ is the expected change in hormone concentration for a random cow, or the average change in the population, and $\mu = 0$ corresponds to no effect of the feed on the hormone concentration. The mean of the nine differences is $\bar{d} = 13.78$, and we would like to know if this reflects a real effect or if it might be due to chance.

A confidence interval answers the question, at least to some extent. The computations in the one sample model give

$$\hat{\mu} = \bar{d} = 13.78, \quad s = 15.25, \quad \text{SE}(\hat{\mu}) = 5.08$$
$$95\%-\text{CI} : 13.78 \pm 2.31 \cdot 5.08 = (2.06, 25.49).$$

The value zero is not included in the confidence interval, so *if* the true mean (population average of differences) is zero then it is unlikely to observe data like those we actually observed. Therefore, we believe that there is indeed an effect of the feed on the hormone concentration.

Another option is to carry out a hypothesis test. We are interested in the *null hypothesis*

$$H_0 : \mu = 0$$

corresponding to no difference between start and end measurements in the population. Now, the estimate $\hat{\mu}$ is our best "guess" for μ, so it seems reasonable to believe in the null hypothesis if $\hat{\mu}$ is "close to zero" and not believe in the null hypothesis if $\hat{\mu}$ is "far away from zero". We then need to decide what is "close" and what is "far away", so we might ask: *If μ is really zero (if the hypothesis is true), then how likely is it to get an estimate $\hat{\mu}$ that is as far or even further away from zero than the 13.78 that we actually got?*

A *t-test* can answer that question. Let

$$T_{\text{obs}} = \frac{\hat{\mu} - 0}{\text{SE}(\hat{\mu})} = \frac{13.78}{5.08} = 2.71. \tag{6.1}$$

The numerator is just the difference between the estimate of μ and the value of μ if the hypothesis is true. This difference is in itself not worth much. Had we used the unit $\mu g/l$ — microgram per liter rather than per milliliter — and thus coded the observations as 0.207, 0.216, *etc.*, then we would have obtained a mean difference of 0.01378 $\mu g/l$. This is much smaller but expresses exactly the same difference from the start to end of the experiment. We need a standardization and divide by the standard error.

Recall from Section 5.3.1 that if $\mu = 0$ (the hypothesis is true), then T_{obs} is an observation from the t_8 distribution (the number of degrees of freedom is $n - 1$, here $9 - 1$). Hence, standardization with $\text{SE}(\hat{\mu})$ brings the difference between $\hat{\mu}$ and zero to a known scale. In particular, we can compute the probability of getting a T-value which is *at least as extreme* — at least as far away from zero — as 2.71:

$$P(|T| \geq |T_{\text{obs}}|) = P(|T| \geq 2.71) = 2 \cdot P(T \geq 2.71) = 2 \cdot 0.013 = 0.026.$$

Here T is a t_8 distributed variable and the second equality follows from the symmetry of the t distribution. The computation is illustrated in the left part of Figure 6.1.

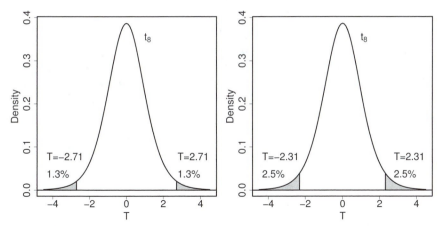

Figure 6.1: Illustration of the t-test for the cattle data. Left panel: the p-value is equal to the area of the gray regions; that is, p-value $= 2 \cdot 0.013 = 0.026$. Right panel: the critical values for the test are those outside the interval from -2.31 to 2.31, because they give a p-value smaller than 0.05.

This probability is called the *p-value* for the test and is the probability — if the null hypothesis is true — of getting a T-value that fits even worse with the hypothesis than the observed one, T_{obs}. If the p-value is small then the observed value T_{obs} is extreme, so we do not believe in the null hypothesis and reject it. If, on the other hand, the p-value is large then the observed value T_{obs} is quite likely, so we cannot reject the hypothesis. The p-value can be computed by many pocket calculators or with a statistical software program.

In this case the p-value is only 2.6%, so the T-value of 2.71 is quite unlikely if the true value of μ is zero (the null hypothesis is true). If we repeat the experiment many times (with nine cows each time) and there is indeed no effect of the feed, then in only 2.6% of the cases we would get a T-value larger than 2.71 or smaller than -2.71. Hence, we *reject* the null hypothesis and conclude that the data provides evidence that feed affects the hormone concentration (although not very strong evidence).

Usually we use a *significance level* of 5%; that is, we reject the hypothesis if the p-value is ≤ 0.05 and fail to reject it otherwise. This means that the null hypothesis is rejected if $|T_{\text{obs}}|$ is larger than or equal to the 97.5% quantile of the t_{n-1} distribution, which is in this case 2.31. Hence, an observed T-value outside $(-2.31, 2.31)$ leads to rejection of the null hypothesis on the 5% significance level. These values are called the *critical values* for the test; see the right panel of Figure 6.1 for illustration.

If we repeat the experiment many times and the hypothesis is true, then the p-value is the relative frequency of experiments for which the T-value is outside $(-2.71, 2.71)$, assuming that the hypothesis is true. This is illustrated in Figure 6.2: We have simulated 1000 samples of size nine from the $N(0, 15^2)$ distribution. In particular, the mean is zero so the hypothesis is true. Think of each of these datasets as the observed differences between start and end hormone concentrations in a replication of the experiment. For each dataset we have computed T_{obs}, and Figure 6.2 shows a histogram of these T-values together with the density of the t_8 distribution. The dashed lines correspond to 2.71 and -2.71. For 28 of the simulated datasets the T-value was numerically larger than 2.71, corresponding to a relative frequency of 0.028, very close to the p-value of 0.026.   □

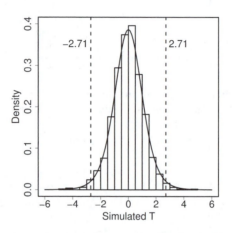

Figure 6.2: Histogram of the T-values from 1000 simulated datasets of size nine from the $N(0, 15^2)$ distribution. The density for the t_8 distribution is superimposed, and the dashed lines correspond to ± 2.71 (2.71 being the observed value from the cattle dataset).

We have introduced several concepts and topics that are worthy of notice at this point. Each of the topics will be discussed more thoroughly later in the chapter.

Null hypothesis. A *null hypothesis* is a simplification of the statistical model and is as such always related to the statistical model. Hence, no null hypothesis exists without a corresponding statistical model. A null hypothesis typically describes the situation of "no effect" or "no relationship", such that rejection of the null hypothesis corresponds to evidence of an effect or relationship.

Test statistic. A *test statistic* is a function of the data that measures the discrepancy between the data and the null hypothesis — with certain values contradicting the hypothesis and others supporting it. Values contradicting the hypothesis are called *critical* or *extreme*.

p-**value.** The test statistic is translated to a *p-value* — the probability of observing data which fit as bad or even worse with the null hypothesis than the observed data if the hypothesis is true. A small p-value indicates that the observed data are unusual if the null hypothesis is true, hence that the hypothesis is false.

Rejection. The hypothesis is rejected if the p-value is small; namely, below (or equal to) the significance level, which is often taken to be 0.05. With statistics we can at best reject the null hypothesis with strong certainty, but we can never confirm the hypothesis. If we fail to reject the null hypothesis, then the only valid conclusion is that the data *do not contradict the null hypothesis*. A large p-value shows that the data are in fine accordance with the null hypothesis, but *not that it is true*.

Type I and type II errors. The conclusion from a hypothesis test is drawn based on probabilities. This has the consequence that sometimes we draw the wrong conclusion. Rejection of a true hypothesis is called a *type I error*, whereas a *type II error* is committed when a false hypothesis is not rejected. The rate of type I errors is equal to the significance level, but reducing this rate comes at the expense of increasing the rate of type II errors.

Quantification of effects. Having established a significant effect by a hypothesis test, it is of great importance to *quantify the effect*. For example, how much larger is the expected hormone concentration after a period of treatment? Moreover, what is the precision of the estimates in terms of standard errors and/or confidence intervals?

6.1 Null hypotheses

Sometimes data are collected in order to answer specific scientific questions such as "does the treatment work?" or "is there any relationship between age and the concentration of a certain substance in the blood?" For the statistical analysis these questions are translated to assumptions or restrictions on the statistical model. Consider the statistical model

$$y_i = \mu_i + e_i \quad i = 1, \ldots, n,$$

where the means $\mu_1, \ldots, \mu_n$ are described through parameters $\theta_1, \ldots, \theta_p$ and explanatory variables as in (5.1), and $e_1, \ldots, e_n$ are iid. $N(0, \sigma^2)$. A *hypothesis* or a *null hypothesis* is a simplifying assertion — expressed by the parameters — about the model. This puts requirements on the statistical model: It should be possible to express the relevant hypotheses in terms of the parameters, so also for that reason it is important to think carefully about which statistical model to use for your data.

For a single sample we may be interested in a specific value μ_0 of the mean; for example, $\mu_0 = 0$, as for the cattle data in Example 6.1 (p. 145). If the hypothesis is true then each $y_i \sim N(\mu_0, \sigma^2)$, imposing the restriction that the distribution has mean μ_0.

In the linear regression of y on x,

$$y_i = \alpha + \beta \cdot x_i + e_i, \quad i = 1, \ldots, n,$$

the slope parameter β measures the (linear) relationship between x and y, and the assumption of *no association between x and y* is expressed by the hypothesis

$$H_0 : \beta = 0.$$

Under the null hypothesis (that is, if H_0 is true), all $y_i \sim N(\alpha, \sigma^2)$. This is a simplification of the original linear regression model.

As a third example, consider the one-way ANOVA model

$$y_i = \alpha_{g(i)} + e_i, \quad i = 1, \ldots, n,$$

where the n observations come from k different groups. The hypothesis that there is no difference among the group means is expressed as

$$H_0 : \alpha_1 = \alpha_2 = \cdots = \alpha_k. \tag{6.2}$$

If the hypothesis is true, then all $y_i \sim N(\alpha, \sigma^2)$, where α is the common value of the α_j's. This corresponds to the model for a single sample.

More generally, we say that the model under the null hypothesis — the original statistical model with the extra restriction(s) imposed by the hypothesis — is a *sub-model* of the original one. Note that the words "the hypothesis", "the null hypothesis", and even just "the null" are used synonymously. The model under the null hypothesis is often called the *null model*.

There is a corresponding *alternative hypothesis* to every null hypothesis. The alternative hypothesis describes what is true if the null hypothesis is false. Usually the alternative hypothesis is simply the complement of the null hypothesis; for example,

$$H_A : \beta \neq 0$$

in the linear regression model, but sometimes a one-sided or directional alternative,

$$H_A : \beta > 0$$

or

$$H_A : \beta < 0,$$

is appropriate. If the alternative is $H_A : \beta > 0$, say, then a negative estimate of β will be in favor of H_0 rather than in favor of the alternative. The alternative hypothesis should be decided *before* the data analysis, not after examination of the data. In particular, the sign of the estimate must not be used to decide the alternative hypothesis. In the one-way ANOVA the alternative to the hypothesis (6.2) is that not all k α's are the same: there are at least two groups, j and l say, for which $\alpha_j \neq \alpha_l$.

In many cases, the interest is in identifying certain effects; for example, that of a treatment or that of an association between two variables. This situation corresponds to the alternative hypothesis, whereas the null hypothesis corresponds to the situation of "no effect" or "no association". One reason for this is that "no effect" or "no association" can be precisely expressed in terms of the model.

This may all seem a little counterintuitive, but the machinery works like this: with a statistical test we *reject* a hypothesis if the data and the hypothesis are in contradiction; that is, if the null model fits poorly to the data. Hence, if we reject the null hypothesis then we believe in the alternative, which states that there is an effect. Having rejected the hypothesis of no effect, we use the estimates to identify and quantify the effects. In principle we never accept a hypothesis. If we fail to reject the hypothesis we say that the data does not provide evidence against it, or that the data supports it. This is not a proof that the null hypothesis is true, however; only an indication that the original model does not describe the data (significantly) better than the model under the null hypothesis.

Example 6.2. Stearic acid and digestibility of fat (continued from p. 26). Recall that a linear regression of digestibility on stearic acid is used for the data. As described above, the hypothesis of no relationship between stearic acid and digestibility is expressed by

$$H_0 : \beta = 0,$$

which is tested against the alternative hypothesis $H_A : \beta \neq 0$. Under H_0, all $y_i \sim N(\alpha, \sigma^2)$, so this is the null model.

Assume for a moment that there is theoretical evidence that the digestibility is 75% for a stearic acid level of 20%. The investigator wants to test if the collected data support or contradict this theory. Then the relevant hypothesis and its alternative are

$$H_0 : \alpha + \beta \cdot 20 = 75, \qquad H_A : \alpha + \beta \cdot 20 \neq 75,$$

involving both α and β. □

Example 6.3. Lifespan and length of gestation period. A horse breeder is convinced that there is a positive association between the length of the gestation period (period from conception to birth) and the lifespan (duration of life) such that horses that have developed longer in the protective embryonic stage also tend to live longer. In order to examine if this is true the breeder collected information about the gestation period and the lifespan for seven horses. The data are listed in Table 6.1 and plotted in Figure 6.3. Note that

Table 6.1: Lifespan and length of gestation period and age for seven horses

Horse	1	2	3	4	5	6	7
Gestation period (days)	370	325	331	334	350	366	320
Lifespan (years)	24.0	25.5	20.0	21.5	22.0	23.5	21.0

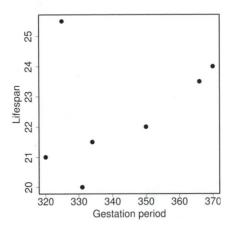

Figure 6.3: Lifespan and length of gestation period and age for seven horses.

the second observation seems to be very different from the others; we will briefly come back to that in Example 6.8 (p. 158).

Consider the linear regression model with lifespan as the response variable and length of gestation period as explanatory variable,

$$\text{lifespan}_i = \alpha + \beta \cdot \text{gestation}_i + e_i, \quad i = 1, \dots, 7,$$

where $e_1, \dots, e_n$ are iid. $N(0, \sigma^2)$. No association between the two variables is expressed by the hypothesis $\beta = 0$. If the breeder is convinced that a long gestation period corresponds to a long lifetime, then the alternative hypothesis should be $\beta > 0$,

$$H_0 : \beta = 0, \quad H_A : \beta > 0.$$

$\square$

Example 6.4. Effect of antibiotics on dung decomposition (continued from p. 51). The primary interest of the antibiotics study was to investigate if there are differences in the amount of organic material among the antibiotics groups. With the one-way ANOVA model

$$y_i = \alpha_{g(i)} + e_i, \quad i = 1, \dots, n,$$

the hypothesis of *no difference* between the group averages is

$$H_0 : \alpha_{\text{control}} = \alpha_{\text{alphacyp}} = \alpha_{\text{enroflox}} = \alpha_{\text{fenbend}} = \alpha_{\text{ivermect}} = \alpha_{\text{spiramycin}}.$$

This should be tested against the alternative hypothesis that at least two groups are different. Under the null hypothesis all y_i are $N(\alpha, \sigma^2)$ distributed, where α is the common mean shared by all groups, so this is the null model.

If it is of particular interest to compare the spiramycin group, say, to the control group, then the relevant hypothesis is

$$H_0 : \alpha_{\text{control}} = \alpha_{\text{spiramycin}}$$

or

$$H_0 : \alpha_{\text{control}} - \alpha_{\text{spiramycin}} = 0,$$

corresponding to no difference between the two groups. In this case there are no restrictions on the remaining parameters α_{alphacyp}, α_{enroflox}, α_{fenbend}, and α_{ivermect}. □

In the situation with comparison of two independent samples, the relevant hypothesis is $H_0 : \alpha_1 = \alpha_2$, where α_1 and α_2 are the expected values in the two groups; *cf.* Section 3.3.

Notice that there is not always a relevant hypothesis to test. For example, the crab weight data from Example 4.1 (p. 68) were collected in order to obtain knowledge about the distribution and variation of the weight of this particular type of crabs, but there does not seem to be one value that is of particular interest.

6.2 *t*-tests

We still consider the model

$$y_i = \mu_i + e_i \quad i = 1, \ldots, n,$$

where the means $\mu_1, \ldots, \mu_n$ are described through parameters $\theta_1, \ldots, \theta_p$ and explanatory variables as in (5.1), and $e_1, \ldots, e_n$ are iid. $N(0, \sigma^2)$. Denote the least squares estimates by $\hat{\theta}_1, \ldots, \hat{\theta}_p$. Consider a single parameter θ_j in the model and the hypothesis

$$H_0 : \theta_j = \theta_0 \tag{6.3}$$

for a fixed value θ_0.

As an example, the population average of the differences, μ, played the role of θ_j, and θ_0 was zero for the cattle data (Example 6.1, p. 145). The hypothesis $H_0 : \theta_j = \theta_0$ is more generally applicable than it appears to be at first sight because the θ_j appearing in the hypothesis may be a linear combination of the original parameters. In Example 6.2 (p. 151), for the digestibility data the slope parameter β or the expected value $\alpha + \beta \cdot 20$ played the role of θ_j, and θ_0 was equal to zero and 75, respectively. Moreover, the hypothesis in Example 6.4 (p. 152) of no difference between two antibiotic types was of this type, too, with $\theta_j = \alpha_{\text{control}} - \alpha_{\text{spiramycin}}$ and $\theta_0 = 0$. The general hypothesis in a one-way ANOVA model of no difference between the k groups is not of this type, however; see Section 6.3.1.

Data for which $\hat{\theta}_j$ is close to θ_0 support the hypothesis, whereas data for which $\hat{\theta}_j$ is far from θ_0 contradict the hypothesis; so it seems reasonable to consider the deviation $\hat{\theta}_j - \theta_0$. Recall from (5.21) that standardization of $\hat{\theta}_j - \theta_j$ with the standard error, $\text{SE}(\hat{\theta}_j)$, yields a t distribution; *i.e.*, $(\hat{\theta}_j - \theta_j)/\text{SE}(\hat{\theta}_j)$ follows the t_{n-p} distribution. Here θ_j is the true value. If the hypothesis is true we can replace θ_j by θ_0, so

$$T_{\text{obs}} = \frac{\hat{\theta}_j - \theta_0}{\text{SE}(\hat{\theta}_j)} \tag{6.4}$$

can be used as a test statistic. An extreme value of T_{obs} is an indication that the data are unusual under the null hypothesis, and the p-value measures how extreme T_{obs} is compared to the t_{n-p} distribution.

If the alternative is two-sided, $H_A : \theta_j \neq \theta_j$, then values of T_{obs} that are far from zero — both small and large values — are critical. Therefore, the p-value is

$$p\text{-value} = P(|T| \geq |T_{\text{obs}}|) = 2 \cdot P(T \geq |T_{\text{obs}}|),$$

where $T \sim t_{n-p}$. If the alternative is one-sided, $H_A : \theta_j > \theta_j$, then large values of T_{obs} are critical, whereas negative values of T_{obs} are considered in favor of the hypothesis rather than as evidence against it. Hence

$$p\text{-value} = P(T \geq T_{\text{obs}}).$$

Similarly, if the alternative is one-sided, $H_A : \theta_j < \theta_j$, then only small values of T_{obs} are critical, so the p-value is $P(T \leq T_{\text{obs}})$.

A limit for rejection/not rejection should be selected such that hypotheses are rejected if the p-value is smaller than (or equal to) this value. This *significance level* is usually denoted α. Notice that the significance level should be selected *before the analysis*. Tests are often carried out on the 5% level corresponding to $\alpha = 0.05$, but $\alpha = 0.01$ and $\alpha = 0.10$ are not unusual; see the discussion in Section 6.5.

For a hypothesis with a two-sided alternative, the hypothesis is thus rejected on the 5% significance level if T_{obs} is numerically larger than or equal to the 97.5% quantile in the t_{n-p} distribution; that is, if $|T_{\text{obs}}| \geq t_{0.975, n-p}$. Similarly, with a one-sided alternative, $H_A : \theta_j > \theta_j$, the hypothesis is rejected if $T_{\text{obs}} \geq t_{0.95, n-p}$. Otherwise, we fail to reject the hypothesis — the original model with a free θ_j does not describe the data significantly better than the null model with θ_j fixed at θ_0.

In order to evaluate if the null hypothesis should be rejected or not, it is thus enough to compare T_{obs} or $|T_{obs}|$ to a certain t quantile. Notice, however, that the p-value contains much more information than just the conclusion (rejection or not rejection) of the test: *the smaller the p-value, the more evidence against the hypothesis*, and the more affirmative we can be about the conclusion. A p-value of 0.04, say, indeed provides some evidence against the hypothesis, but not nearly as much as a p-value of, say, 0.0002. In the first case, the probability of observing data as extreme as those we actually did observe is 4% even if the hypothesis is true, whereas in the latter case it is highly unlikely to observe such extreme data. We recommend that the p-value is always reported.

The p-value, as well as the quantiles in the t distribution, can be computed by many pocket calculators and with statistical software programs. Moreover, quantiles of the t distributions are listed in statistical tables (Appendix C.3).

Let us summarize the results on the t-test:

Infobox 6.1: t-test

Let θ_0 be a fixed value. The hypothesis $H_0 : \theta_j = \theta_0$ is carried out on the test statistic

$$T_{obs} = \frac{\hat{\theta}_j - \theta_0}{SE(\hat{\theta}_j)},$$

which should be compared to the t_{n-p} distribution. For the two-sided alternative $H_A : \theta_j \neq \theta_0$, small and large values are critical,

$$p\text{-value} = P(|T| \geq |T_{obs}|) = 2 \cdot P(T \geq |T_{obs}|)$$

and H_0 is rejected on the 5% significance level if $|T_{obs}| \geq t_{0.975, n-p}$. For the one-sided alternative $H_A : \theta_j > \theta_0$, only large values are critical,

$$p\text{-value} = P(T \geq T_{obs})$$

and H_0 is rejected on the 5% significance level if $T_{obs} \geq t_{0.95, n-p}$.

The following example shows how to use the t-test for the situation with two independent samples with the same standard deviation.

Example 6.5. Parasite counts for salmons (continued from p. 49). Recall the salmon data with two samples corresponding to two different salmon stocks, Ätran or Conon. Assume that all observations are independent and furthermore that Ätran observations are $N(\alpha_{\text{Ätran}}, \sigma^2)$ distributed and that Conon observations are $N(\alpha_{\text{Conon}}, \sigma^2)$ distributed. If $\alpha_{\text{Ätran}} = \alpha_{\text{Conon}}$ then there is no difference between the stocks when it comes to parasites during infections. Hence, the hypothesis is $H_0 : \alpha_{\text{Ätran}} = \alpha_{\text{Conon}}$. If we define $\theta = \alpha_{\text{Ätran}} - \alpha_{\text{Conon}}$ then the hypothesis can be written as $H_0 : \theta = 0$.

We already computed the relevant values in Example 5.9 (p. 125):

$$\hat{\theta} = \hat{\alpha}_{\ddot{A}tran} - \hat{\alpha}_{Conon} = 32.23 - 21.54 = 10.69$$

$$SE(\hat{\theta}) = SE(\hat{\alpha}_{\ddot{A}tran} - \hat{\alpha}_{Conon}) = s\sqrt{\frac{2}{13}} = 2.58.$$

The t-test statistic is therefore

$$T_{obs} = \frac{\hat{\theta} - 0}{SE(\hat{\theta})} = \frac{10.69}{2.58} = 4.14$$

and the corresponding p-value is

$$p\text{-value} = 2 \cdot P(T \geq 4.14) = 0.00037, \quad T \sim t_{24}.$$

Hence, if there is no difference between the two salmon stocks then the observed value 4.14 of T_{obs} is very unlikely. We firmly reject the hypothesis and conclude that Ätran salmons are more susceptible to parasites than Conon salmons. Notice how we reached the same conclusion from the confidence interval in Example 5.9 (p. 125), since it did not include zero.

If we are not willing to assume variance homogeneity (same σ for both stocks), then we should compute $SE(\hat{\alpha}_1 - \hat{\alpha}_2)$ as in (5.23) and use the t distribution with r degrees of freedom; see (5.24).   □

As suggested by the salmon example above, there is a close connection between t-tests and confidence intervals. If the alternative is two-sided, then H_0 is rejected on the 5% level if $|T_{obs}|$ is outside the interval from $-t_{0.975,n-p}$ to $t_{0.975,n-p}$. Hence,

$$H_0 \text{ rejected} \quad \Leftrightarrow \quad T_{obs} \leq -t_{0.975,n-p} \quad \text{or} \quad T_{obs} \geq t_{0.975,n-p}$$

$$\Leftrightarrow \quad \frac{\hat{\theta}_j - \theta_0}{SE(\hat{\theta}_j)} \leq -t_{0.975,n-p} \quad \text{or} \quad \frac{\hat{\theta}_j - \theta_0}{SE(\hat{\theta}_j)} \geq t_{0.975,n-p}$$

$$\Leftrightarrow \quad \theta_0 \geq \hat{\theta}_j + t_{0.975,n-p} \cdot SE(\hat{\theta}_j) \text{ or } \theta_0 \leq \hat{\theta}_j - t_{0.975,n-p} \cdot SE(\hat{\theta}_j),$$

which is equivalent to θ_0 *not* being included in the 95% confidence interval for θ_j, which is $\hat{\theta}_j \pm t_{0.975,n-p} \cdot SE(\hat{\theta}_j)$. This result generalizes to $1 - \alpha$ confidence intervals and two-sided t-tests on level α.

Infobox 6.2: Relationship between t-tests and confidence intervals

$H_0 : \theta_j = \theta_0$ is rejected on significance level α against the alternative $H_A : \theta_j \neq \theta_0$ if and only if θ_0 is not included in the $1 - \alpha$ confidence interval.

The result in the infobox explains the formulation that we used in Section 5.3 about confidence intervals; namely, that a confidence interval includes the values that are in accordance with the data. This now has a precise meaning in terms of hypothesis tests. If the only aim of the analysis is to conclude whether a hypothesis should be rejected or not at a certain level α, then we get that information from either the t-test or the confidence interval. On the other hand, they provide extra information on slightly different matters. The t-test provides a p-value explaining how extreme the observed data are if the hypothesis is true, whereas the confidence interval gives us the values of θ that are in agreement with the data.

Now, let us illustrate how t-tests can be carried out in linear regression and for a single sample.

Example 6.6. Stearic acid and digestibility of fat (continued from p. 26). Recall the linear regression model for the digestibility data,

$$y_i = \alpha + \beta \cdot x_i + e_i, \quad i = 1, \ldots, n,$$

where $e_1, \ldots, e_n$ are iid. $N(0, \sigma^2)$. The hypothesis

$$H_0 : \beta = 0$$

that there is no relationship between the level of stearic acid and digestibility is tested by the test statistic

$$T_{obs} = \frac{\hat{\beta} - 0}{SE(\hat{\beta})} = \frac{-0.9337 - 0}{0.0926} = -10.08,$$

where the estimate $\hat{\beta} = -0.9337$ and its standard error $SE(\hat{\beta}) = 0.0926$ come from Example 5.3 (p. 110). The value of T_{obs} should be compared to the t_7 distribution. For the alternative hypothesis $H_A : \beta \neq 0$ we get

$$p\text{-value} = P(|T| \geq 10.08) = 2 \cdot P(T \geq 10.08) = 0.00002$$

and we conclude that there is strong evidence of an association between digestibility and the stearic acid level — the slope is significantly different from zero.

Assume that it is of interest to test if the expected digestibility percent differs from 75% for a stearic acid level of 20%. Hence the hypothesis and its alternative are

$$H_0 : \alpha + \beta \cdot 20 = 75, \qquad H_A : \alpha + \beta \cdot 20 \neq 75,$$

corresponding to $\theta = \alpha + \beta \cdot 20$ and $\theta_0 = 75$. The hypothesis is tested by

$$T_{obs} = \frac{\hat{\alpha} + \hat{\beta} \cdot 20 - 75}{SE(\hat{\alpha} + \hat{\beta} \cdot 20)} = \frac{77.859 - 75}{1.1096} = 2.58.$$

The estimate and standard error were computed in Example 5.3 (p. 110). The corresponding p-value is 0.036, which is computed in the t_7 distribution.

Hence, the data contradict the hypothesis (although not strongly) and indicate that the expected digestibility is not 75% for a stearic acid level of 20%. Note that this is in accordance with the 95% confidence interval (75.2, 80.5) from Example 5.7 (p. 123) for the expected value, as the interval does not include the value 75.   □

Example 6.7. Production control. A dairy company bought a new machine for filling milk into cartons and wants to make sure that the machine is calibrated correctly. The aim is an average weight of 1070 grams per carton (including the carton). A sample of 100 cartons with milk is chosen at random from the production line and each carton is weighed. It is assumed that the weights are independent and $N(\mu, \sigma^2)$ distributed. The relevant hypothesis is $H_0 : \mu = 1070$.

It turned out that

$$\hat{\mu} = \bar{y} = 1072.9 \text{ grams}, \quad s = 15.8 \text{ grams},$$

so $\text{SE}(\hat{\mu}) = s / \sqrt{100} = 15.8 / 10 = 1.58$ and the t-test statistic is

$$T_{obs} = \frac{\hat{\mu} - 1070}{\text{SE}(\hat{\mu})} = \frac{2.9}{1.58} = 1.83.$$

The corresponding p-value is

$$p\text{-value} = 2 \cdot P(T \geq 1.83) = 0.07, \quad T \sim t_{98}.$$

We fail to reject the hypothesis on the 5% significance level, but due to the low p-value we conclude nevertheless that there is a slight indication that the machine is calibrated incorrectly.   □

Example 6.8. Lifespan and length of gestation period (continued from p. 151). Recall the data on the association between the length of the gestation period and lifespan for horses. The null hypothesis and its alternative are given by

$$H_0 : \beta = 0, \quad H_A : \beta > 0,$$

where β is the slope in the linear regression with lifespan as response and gestation time as explanatory variable. We get

$$\hat{\beta} = 0.03262, \quad \text{SE}(\hat{\beta}) = 0.0401, \quad T_{obs} = \frac{\hat{\beta} - 0}{\text{SE}(\hat{\beta})} = 0.813$$

and under the null hypothesis T_{obs} is an observation from the t_5 distributed ($\text{df}_e = n - 2 = 5$ because there are seven observations). Since only positive values of β are included in the alternative hypothesis, only large values of $\hat{\beta}$ are in contradiction with the hypothesis. Negative values of $\hat{\beta}$ are considered as support to the hypothesis rather than as evidence against it, so

$$p\text{-value} = P(T \geq T_{obs}) = P(T \geq 0.813) = 0.23$$

and we fail to reject the hypothesis: the data do not provide evidence that the breeder is right about his theory.

Recall from Figure 6.3 that one horse was quite different from the others, and actually the conclusion changes if this observation is removed (Exercise 6.10). However, removing an observation without information that an error occurred during data collection is a quite dangerous strategy, particularly when there are so few observations. □

6.3 Tests in a one-way ANOVA

In the one-way ANOVA setup, the most important hypothesis is usually that of equal group means. If there are more than two groups the hypothesis imposes more than one restriction on the parameters in the model and is thus not covered by the t-test setup from the preceeding section. Instead, the hypothesis is tested with a so-called F-test.

6.3.1 The F-test for comparison of groups

Consider the one-way ANOVA model

$$y_i = \alpha_{g(i)} + e_i, \quad i = 1, \ldots, n,$$

where $g(i)$ is the group that observation i belongs to and $e_1, \ldots, e_n$ are independent and $N(0, \sigma^2)$ distributed. As usual, k denotes the number of groups. The null hypothesis that there is no difference between the groups is given by

$$H_0 : \alpha_1 = \cdots = \alpha_k$$

and the alternative is the opposite; namely, that at least two α's are different.

Now, if there are only two groups, the hypothesis is $H_0 : \alpha_1 = \alpha_2$ and the test could be carried out as a t-test by

$$T_{\text{obs}} = \frac{\hat{\alpha}_1 - \hat{\alpha}_2}{\text{SE}(\hat{\alpha}_1 - \hat{\alpha}_2)}.$$

However, if there are three or more groups ($k \geq 3$), then this is not possible: which T-statistic should we use? Instead, we must find another test statistic.

Recall the distinction between *between-group variation* and *within-group variation* from Section 3.2. Let us formalize matters. First, recall from Sections 3.4 and 5.2.2 that the residual sum of squares in the ANOVA setup is given by

$$\text{SS}_e = \sum_{i=1}^{n} r_i^2 = \sum_{i=1}^{n} (y_i - \hat{y}_i)^2 = \sum_{i=1}^{n} (y_i - \bar{y}_{g(i)})^2.$$

SS_e describes the within-group variation since it measures squared deviations between the observations and the group means. The residual degrees of freedom is $df_e = n - k$, so the residual mean squares is

$$MS_e = s^2 = \frac{SS_e}{df_e} = \frac{SS_e}{n - k}.$$

Second, the part of the variation which is due to the difference between the groups is described: The group means are compared to the overall mean, $\bar{y} = \frac{1}{n} \sum_{i=1}^{n} y_i$, by

$$SS_{grp} = \sum_{j=1}^{k} n_j (\bar{y}_j - \bar{y})^2.$$

Note that there is a contribution for each observation due to the factor n_j for group j. When we examine the between-group variation, the k group means essentially act as our "observations"; hence there are $k - 1$ degrees of freedom, $df_{grp} = k - 1$, and the "average" squared difference per group is

$$MS_{grp} = \frac{SS_{grp}}{df_{grp}} = \frac{SS_{grp}}{k - 1}.$$

If there is no difference between any of the groups (H_0 is true), then the group averages $\bar{y}_j$ will be of similar size and be similar to the total mean $\bar{y}$. Hence, MS_{grp} will be "small". On the other hand, if groups 1 and 2, say, are different (H_0 is false), then $\bar{y}_1$ and $\bar{y}_2$ will be somewhat different and cannot both be similar to $\bar{y}$ — hence, MS_{grp} will be "large". "Small" and "large" should be measured relative to the within-group variation, and MS_{grp} is thus standardized with MS_e. We use

$$F_{obs} = \frac{MS_{grp}}{MS_e} \tag{6.5}$$

as the test statistic and note that large values of F_{obs} are critical; that is, not in agreement with the hypothesis.

If the null hypothesis is true, then F_{obs} comes from a so-called *F distribution* with $(k - 1, n - k)$ degrees of freedom. Notice that there is a pair of degrees of freedom (not just a single value) and that the relevant degrees of freedom are the same as those used for computation of MS_{grp} and MS_e. The density for the F distribution is shown for three different pairs of degrees of freedom in the left panel of Figure 6.4.

Since only large values of F are critical, we have

$$p\text{-value} = P(F \geq F_{obs}),$$

where F follows the $F(k - 1, n - k)$ distribution. The hypothesis is rejected if the p-value is 0.05 or smaller (if 0.05 is the significance level). This is equivalent to F_{obs} being larger than the 95% quantile in the $F(k - 1, n - k)$ distribution, denoted $F_{0.95,k-1,n-k}$. See the right panel of Figure 6.4 for an imaginary situation where the hypothesis is not rejected.

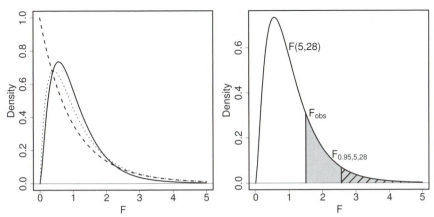

Figure 6.4: Left: Densities for the $F(5, 28)$ distribution (solid), the $F(2, 27)$ distribution (dashed), and the $F(4, 19)$ distribution (dotted). Right: The density for the $F(5, 28)$ distribution with the 95% quantile $F_{0.95, 5, 28} = 2.56$ and an imaginary value F_{obs}. The corresponding p-value is equal to the area of the gray region (including the dashed gray), whereas the dashed gray region has area 0.05.

The SS-values, the degrees of freedom, F_{obs}, and the corresponding p-value are often inserted in an analysis of variance table, as in Table 6.2. The quantiles of the F distributions are listed in statistical tables (see Appendix C.4), and the p-value can be computed by many pocket calculators or by a statistical software program. Let us summarize:

Infobox 6.3: F-test for comparison of groups

The hypothesis

$$H_0 : \alpha_1 = \cdots = \alpha_k$$

of no difference in group means in the one-way ANOVA model is tested with the test statistic

$$F_{obs} = \frac{MS_{grp}}{MS_e}$$

with large values critical and small values in favor of the hypothesis. The p-value is

$$p\text{-value} = P(F \geq F_{obs}),$$

where $F \sim F(k - 1, n - k)$. In particular, H_0 is rejected on the 5% significance level if $F_{obs} \geq F_{0.95, k-1, n-k}$.

Example 6.9. Effect of antibiotics on dung decomposition (continued from p. 51). We already computed the sum of squared residuals in Example 3.4 (p.

Table 6.2: Analysis of variance table

Variation	SS	df	MS	F_{obs}	p-value
Between groups	$\sum_{j=1}^{k} n_j(\bar{y}_j - \bar{y})^2$	$k-1$	$\frac{SS_{grp}}{df_{grp}}$	$\frac{MS_{grp}}{MS_e}$	$P(F \geq F_{obs})$
Residual	$\sum_{i=1}^{n}(y_i - \bar{y}_{g(i)})^2$	$n-k$	$\frac{SS_e}{df_e}$		
Total	$\sum_{i=1}^{n}(y_i - \bar{y})^2$	$n-1$			

56) and found $SS_e = 0.4150$. Moreover, we find $\bar{y} = 2.814$, so

$$
\begin{aligned}
SS_{grp} &= \sum_{j=1}^{6} n_j(\bar{y}_j - \bar{y})^2 \\
&= 6 \cdot (2.603 - 2.814)^2 + \cdots + 4 \cdot (2.855 - 2.814)^2 = 0.5908,
\end{aligned}
$$

where we have used the group means from Table 3.1 (p. 52). The corresponding degrees of freedom are

$$df_{grp} = k - 1 = 6 - 1 = 5, \quad df_e = n - k = 34 - 6 = 28.$$

Hence

$$F_{obs} = \frac{0.1182}{0.0148} = 7.97,$$

which should be compared to the $F(5, 28)$ distribution,

$$p\text{-value} = P(F \geq 7.97) = 9 \cdot 10^{-5} < 0.0001.$$

The values are listed in an ANOVA table as follows:

Variation	SS	df	MS	F_{obs}	p-value
Between types	0.5908	5	0.1182	7.97	<0.0001
Residual	0.4150	28	0.0148		
Total	1.0058	33			

Notice from the right panel of Figure 6.4 that the value 7.97 is very extreme — it is not even in the figure — corresponding to the very small p-value. We reject the hypothesis and conclude that there is strong evidence of group differences. Subsequently, we need to quantify the conclusion further: Which groups are different and how large are the differences? We will do so in the next section. $\qquad \square$

6.3.2 Pairwise comparisons and LSD-values

Sometimes interest is in particular groups from the experiment, and we want to compare group j and group l, say. Still, the analysis is carried out using *all* data since this makes the estimate of the standard deviation more precise. In a sense we "borrow information" from all observations when we estimate the standard deviation, even though we use only the data from the two groups in question to estimate the mean difference.

The relevant hypothesis is

$$H_0 : \alpha_j = \alpha_l \quad \text{or} \quad H_0 : \alpha_j - \alpha_l = 0,$$

and we consider the two-sided alternative $H_A : \alpha_j - \alpha_l \neq 0$. The test is carried out as a t-test in the usual way by

$$T_{\text{obs}} = \frac{\hat{\alpha}_j - \hat{\alpha}_l}{\text{SE}(\hat{\alpha}_j - \hat{\alpha}_l)} = \frac{\hat{\alpha}_j - \hat{\alpha}_l}{s\sqrt{\left(\frac{1}{n_j} + \frac{1}{n_l}\right)}};$$

cf. equation (5.15). The value should be compared to the t_{n-k} distribution, and the difference is significant on the 5% significance level if and only if $|T_{\text{obs}}| \geq t_{0.975,n-k}$. This is the case if and only if

$$|\hat{\alpha}_j - \hat{\alpha}_l| \geq t_{0.975,n-k} \cdot s\sqrt{\left(\frac{1}{n_j} + \frac{1}{n_l}\right)}.$$

The right-hand side of this equation is called the *least significant difference* — or the 95% *LSD-value* — for the difference between α_j and α_l. If the sample sizes in all k groups are the same, $n_1 = \cdots = n_k = n'$, then the LSD-value is

$$\text{LSD}_{0.95} = t_{0.975,n-k} \cdot s\sqrt{\frac{2}{n'}} \tag{6.6}$$

and it is the same for all pairs of α's. Hence, we can compare differences of $\hat{\alpha}_j$'s to the LSD-value and see if there are significant differences. This should not be done uncritically, though, due to the multiple testing problem; see Section 6.5.1.

Example 6.10. Effect of antibiotics on dung decomposition (continued from p. 51). The group means for the antibiotics data are (*cf.* Table 3.1, p. 52)

Antibiotics	n_j	$\bar{y}_j$
Control	6	2.603
α-Cypermethrin	6	2.895
Enrofloxacin	6	2.710
Fenbendazole	6	2.833
Ivermectin	6	3.002
Spiramycin	4	2.855

Table 6.3: Binding rates for three types of antibiotics

Antibiotic	Binding rate				Mean
Chloramphenicol	29.2	32.8	25.0	24.2	27.80
Erythromycin	21.6	17.4	18.3	19.0	19.08
Tetracycline	27.3	32.6	30.8	34.8	31.38

Recall also that $s = \sqrt{0.01482} = 0.1217$. The 95% LSD-value for comparison with spiramycin is

$$2.048 \cdot 0.1217 \sqrt{\left(\frac{1}{6} + \frac{1}{4} \right)} = 0.161,$$

whereas the 95% LSD-value for all other comparisons is

$$2.048 \cdot 0.1217 \sqrt{\left(\frac{1}{6} + \frac{1}{6} \right)} = 0.144.$$

For the spiramycin group, we find that $\hat{\alpha}_{\text{spiramycin}} - \hat{\alpha}_{\text{control}} = 0.252 > 0.161$, so the group is significantly different from the control group. On the other hand, there is no significant difference between the enrofloxacin group and the control group since $\hat{\alpha}_{\text{enroflox}} - \hat{\alpha}_{\text{control}} = 0.107 < 0.144$.

Using the same arguments for the remaining three antibiotic types, we conclude that the amount of organic material is significantly lower for the control groups than for all other groups, except the enrofloxacin group (compare to Figure 3.2 and Table 5.1). □

Example 6.11. Binding of antibiotics. When an antibiotic is injected into the bloodstream, a certain part of it will bind to serum protein. This binding reduces the medical effect. As part of a larger study, the binding rate was measured for 12 cows which were given one of three types of antibiotics: chloramphenicol, erythromycin, and tetracycline (Ziv and Sulman, 1972).

The binding rates (measured as a percentage) are listed in Table 6.3 and plotted in Figure 6.5. The figure indicates that the variation is roughly the same in all three groups, although perhaps slightly smaller in the erythromycin group.

The one-way analysis of variance model has three mean parameters: α_{chloro}, α_{eryth}, and α_{tetra}. They are interpreted as expected or average binding rates for the three types of medicine. First we test if the three α's are equal,

$$H_0 : \alpha_{\text{chloro}} = \alpha_{\text{eryth}} = \alpha_{\text{tetra}}$$

and get

$$F_{\text{obs}} = 16.4, \quad p\text{-value} = 0.0099.$$

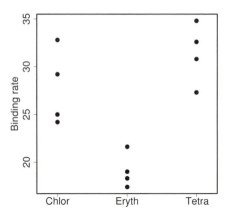

Figure 6.5: Binding rates for three types of antibiotics.

We conclude that the three antibiotics do not have the same average binding rate.

There are four observations in each group, so there is one common LSD-value for all three pairwise comparisons. The pooled standard deviation turns out to be $s = 3.122$ and the 97.5% quantile in the t_9 distribution is 2.262. Thus from (6.6) we get

$$\text{LSD}_{0.95} = t_{0.975, n-k} \cdot s \sqrt{\frac{2}{n'}} = 2.262 \cdot 3.122 \cdot \sqrt{\frac{2}{4}} = 4.99.$$

Comparing the group means from Table 6.3, we see that erythromycin has a significantly lower binding rate than both chloramphenicol and tetracycline, but that there is no evidence of a difference between chloramphenicol and tetracycline.

The estimates for the pairwise differences in binding rates and the corresponding 95% confidence intervals are given by

$$\hat{\alpha}_{\text{chloro}} - \hat{\alpha}_{\text{eryth}} = 27.80 - 19.08 = 8.72 \quad (3.73, 13.72)$$
$$\hat{\alpha}_{\text{tetra}} - \hat{\alpha}_{\text{eryth}} = 31.38 - 19.08 = 12.30 \quad (7.31, 17.29)$$
$$\hat{\alpha}_{\text{tetra}} - \hat{\alpha}_{\text{chloro}} = 31.38 - 27.80 = 3.58 \quad (-1.42, 8.57)$$

As we already knew from the comparison with the LSD-value, zero is included in the confidence interval for $\alpha_{\text{tetra}} - \alpha_{\text{eryth}}$. $\qquad\square$

6.4 Hypothesis tests as comparison of nested models

In Section 6.3.1 the F-test statistic F_{obs} for the group effect in a one-way ANOVA was constructed and interpreted as a comparison of the between-group variation and the within-group variation:

$$F_{obs} = \frac{MS_{grp}}{MS_e} = \frac{SS_{grp}/(k-1)}{SS_e/(n-k)}. \tag{6.7}$$

We will now interpret the F-statistic from another point of view.

Recall the residual sum of squares,

$$SS_e = \sum_{i=1}^{n} r_i^2 = \sum_{i=1}^{n} (y_i - \hat{y}_i)^2.$$

A good model fit corresponds, all things considered, to numerically small residuals and thus to a small SS_e; so we can think of SS_e for the one-way ANOVA model as a measure of the model fit when the group means are allowed to differ. Under the null hypothesis, the model is $y_i = \alpha + e_i$, where α is the common mean. This model is referred to as the null model. The residual sum of squares corresponding to this model is

$$SS_{total} = \sum_{i=1}^{n} (y_i - \bar{y})^2,$$

which is always larger than SS_e and is a measure of the model fit under the restriction that all group means are the same.

Hence, we may think of the difference $SS_{total} - SS_e$ as a measure of the difference between the models in their ability to describe the data. If there is a large difference, then the one-way ANOVA model is better; if the difference is only small, then the two models are almost equally good. In other words, small values of $SS_{total} - SS_e$ are in agreement with the hypothesis whereas large values contradict the hypothesis. The difference should be measured relative to the starting point, SS_e, and if it is standardized with the degrees of freedom we get

$$\frac{(SS_{total} - SS_e)/(k-1)}{SS_e/(n-k)}. \tag{6.8}$$

It turns out that the $SS_{total} = SS_{grp} + SS_e$ — meaning that "the total variation" can be split into the between-group variation and the within-group variation. Therefore, (6.8) is equal to F_{obs}:

$$F_{obs} = \frac{SS_{grp}/(k-1)}{SS_e/(n-k)} = \frac{(SS_{total} - SS_e)/(k-1)}{SS_e/(n-k)}.$$

In other words, we may think of the F-test statistic for comparison of groups as a standardized difference between the residual sums of squares under the null hypothesis and under the one-way ANOVA model.

This idea can be generalized to test (linear) hypotheses in all linear models. Consider a linear model as defined in Section 5.1. We will refer to this as the *full model*. A null hypothesis imposes restrictions on the statistical model and thus defines a new statistical model, the *model under the null* or the *null model*; cf. Section 6.1. The hypothesis test should therefore compare the ability of these two models to describe the data. Operationally, we can think about the test in the way that we fit both models — with and without the restrictions imposed by the hypothesis — and compare the two fits. Notice that we say that two statistical models are *nested* if one of them is a specialization (or a sub-model) of the other. Hence, the null model and the full model are nested.

Let SS_{full} and df_{full} be the residual sum of squares and the degrees of freedom corresponding to the full model, and let SS_0 and df_0 be the residual sum of squares and degrees of freedom corresponding to the model under the null hypothesis. Furthermore, define the *F-test statistic*

$$F_{obs} = \frac{(SS_0 - SS_{full})/(df_0 - df_{full})}{SS_{full}/df_{full}}. \tag{6.9}$$

Since the full model is more flexible, it has smaller residual sum of squares so $SS_0 \leq SS_{full}$, and $SS_{full} - SS_0$ describes the reduction in model fit imposed by the null hypothesis. This should be measured relative to the original model fit; hence, we divide by SS_{full}, or rather by SS_{full}/df_{full}. Furthermore, it should be measured relative to the "amount of restrictions" that the hypothesis imposes. If the full model is much more flexible than the null model, then we would expect $SS_0 - SS_{full}$ to be larger than if the full model is only slightly more flexible than the null model. The degrees of freedom for a model is

$$df = n - p,$$

where p is the number of parameters used to describe the mean structure of the model. Many parameters mean much flexibility, so $df_0 - df_{full}$ is a quantification of the difference in flexibility or complexity between the two models. It simply counts how many more parameters are used in the full model compared to the null model. Hence, we divide by $df_0 - df_{full}$. If the null hypothesis is true then F_{obs} from (6.9) is an observation from the F distribution with $(df_0 - df_{full}, df_{full})$ degrees of freedom, so

$$p\text{-value} = P(F \geq F_{obs})$$

is computed in the F distribution with these degrees of freedom. Recognize that the degrees of freedom are the same as those used in the definition of F_{obs}.

Example 6.12. Stearic acid and digestibility of fat (continued from p. 26). Consider the linear regression model

$$y_i = \alpha + \beta x_i + e_i$$

for the digestibility data. We know from Example 5.3 (p. 110) that $SS_e = 61.76$ and df $= n - 2 = 7$. Under the hypothesis that $\beta = 0$, the model degenerates to the one sample model,

$$y_i = \alpha + e_i,$$

which has $n - 1 = 8$ degrees of freedom and a residual sum of squares that amounts to 958.53. Hence, the F-test statistic is

$$F_{obs} = \frac{(958.53 - 61.76)/(8 - 7)}{61.76/7} = 101.63$$

and the p-value is

$$p\text{-value} = P(F \geq 101.63) = 0.00002, \quad F \sim F_{1,7}.$$

In Example 6.6 (p. 157) we tested the same hypothesis with a t-test and got $T_{obs} = -10.08$ and p-value $= 0.00002$. The p-value is the same for the t-test and the F-test. This is of course no coincidence, but is true because $F_{obs} = T_{obs}^2$, here $101.63 = (-10.08)^2$, such that the F_{obs} is "large" exactly if the T_{obs} is "far from zero". $\qquad\Box$

As illustrated by the digestibility example, hypotheses of the type $H_0 : \theta_j = \theta_0$ with a restriction on a single parameter can be tested with a t-test or an F-test. The F-test statistic is equal to the t-test statistic squared, and the p-values are the same. In other words, the tests are equivalent and we yield the same conclusion from both tests.

6.5   Type I and type II errors

Conclusions from a hypothesis test are drawn on the basis of probabilities, and we never know if we draw the correct conclusion. We can, however, control the probabilities of making errors.

Four scenarios are possible as we carry out a hypothesis test: the null hypothesis is either true or false, and it is either rejected or not rejected. The conclusion is correct whenever we reject a false hypothesis or do not reject a true hypothesis. Rejection of a true hypothesis is called a *type I error*, whereas a *type II error* refers to not rejecting a false hypothesis; see the chart below.

	Reject	Not reject
H_0 correct	type I	OK
H_0 false	OK	type II

Now, recall that the p-value is the probability of observing a more extreme value of the test statistic than the one we actually observed — if the hypothesis is true. And recall that we reject the hypothesis if $p \leq 0.05$ if we use a 5% significance level. This means that *if the hypothesis is true, then we will reject it with a probability of 5%.* In other words: *The probability of committing a type I error is 5%.*

For a general significance level α, the probability of committing a type I error is α. Hence, by adjusting the significance level we can change the probability of rejecting a true hypothesis. This is not for free, however. If we decrease α we make it harder to reject a hypothesis — hence we will accept more false hypotheses, so the rate of type II errors will increase.

The probability that a false hypothesis is rejected is called the *power* of the test. We would like the test to have large power and at the same time a small significance level, but these two goals contradict each other so there is a trade-off. As mentioned already, $\alpha = 0.05$ is the typical choice. Sometimes, however, the scientist wants to "make sure" that false hypotheses are really detected; then α can be increased to 0.10, say. On the other hand, it is sometimes more important to "make sure" that rejection expresses real effects; then α can be decreased to 0.01, say.

The situation is analogous to the situation of a medical test: Assume for example that the concentration of some substance in the blood is measured in order to detect cancer. If the concentration is larger than a certain threshold, then the "alarm goes off" and the patient is sent for further investigation. But how large should the threshold be? If it is large, then some patients will not be classified as sick although they are (type II error). On the other hand, if the threshold is low, then patients will be classified as sick although they are not (type I error).

6.5.1 Multiple testing. Bonferroni correction

Consider a one-way ANOVA with k groups. Sometimes it is of interest to compare the groups two-by-two. There is, however, a problem with comparing all groups pairwise — or at least with the interpretation, if focus is on rejection/not rejection. The problem arises because many tests are carried out. In each test there is a small risk of drawing the wrong conclusion (type I and type II errors), but the risk that at least one conclusion is wrong increases quite fast as the number of tests increases.

For example, consider the situation of six treatment groups; this gives 15 possible pairs of treatments. Assume that we carry out all 15 tests. If there is no difference between any of the treatments — all the hypotheses are true — then in each test there is a 5% risk of rejecting the hypothesis. If the 15 tests were independent (actually they are not, but the point is the same), then the chance of drawing only correct conclusions is $0.95^{15} = 0.46$ since each test is correct with probability 95%. In other words, there is a $1 - 0.46 = 0.54$ risk of drawing the wrong conclusion at least once!

In general, if we carry out m independent tests on the 5% significance level, then the risk of committing at least one type I error is

$$1 - 0.95^m.$$

The graph of the function is plotted in Figure 6.6, and we see for example that the probability of rejecting at least one true hypothesis is around 0.90 when 45 independent tests are carried out. This corresponds to pairwise comparison of 10 groups.

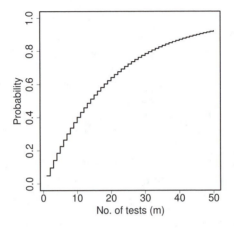

Figure 6.6: The risk of making at least one type I error out of m independent tests on the 5% significance level.

One possible solution to this problem is to adjust the requirement for rejecting a hypothesis. There are other such corrections for multiple tests, all of which make it harder for each hypothesis to be rejected.

The simplest correction is called the *Bonferroni correction*. It simply multiplies each p-value by the number of tests that is carried out. In the above example with six treatments leading to 15 comparisons, the p-values should then be multiplied by 15. Hence, if the usual p-value was computed to 0.0071, then the corrected p-value is

$$15 \cdot 0.0071 = 0.107.$$

In other words, if we use the Bonferroni corrected p-value to decide whether to reject a hypothesis or not, then the limit for rejection/not rejection on the original p-value scale is

$$0.05/15 = 0.0033.$$

However, the correction comes at a price: the risk of not classifying true differences as significant increases considerably, and one should be quite careful with the conclusions (as always). Some advice may be the following:

- If possible, carry out an overall test for all groups simultaneously. In the one-way ANOVA this corresponds to the hypothesis $H_0 : \alpha_1 = \alpha_2 = \cdots = \alpha_k$. This test investigates if there are any differences at all with a single test, thus without the problem of multiple testing. If the null hypothesis is rejected, then pairwise comparisons can be carried out to identify which groups differ from each other, but with care.

- Decide beforehand which pairwise comparisons you want to make. Then fewer tests are carried out, and the problem of multiple testing is diminished.

- Always state the p-values rather than just their status as significant or not significant. This provides information on the strength of the conclusions.

6.5.2 Summary of hypothesis testing

The procedure for hypothesis tests that we have described in this chapter is not restricted to the hypotheses in linear models we have considered so far. For later use, we therefore end this chapter by summarizing how hypothesis tests are carried out.

Infobox 6.4: General procedure for hypothesis tests

Significance level. Choose a significance level α; for example, 0.05.

Null hypothesis. Define the null hypothesis and the corresponding alternative such that it answers the scientific/biological question of interest.

Test statistic. Choose an adequate test statistic, decide which values are critical for the hypothesis, and find the distribution of the test statistic under the null hypothesis.

p-value. Compute the p-value and conclude if the hypothesis should be rejected (p-value $\leq \alpha$) or not (p-value $> \alpha$).

Conclusion. Translate the conclusion (rejection or not) from the test to a conclusion regarding the original biological question and quantify significant effects (if any).

6.6 R

Having specified the statistical model with lm(), as explained in Section 5.5, certain hypothesis tests are easily carried out with output from the summary() and anova() functions. As illustrated in Section 5.5.2, a model can have several different parameterizations, which may be useful for different hypotheses. Hence, as always, we need to be careful with the specification of the model.

6.6.1 *t*-tests

Consider the output from the one-way ANOVA model on the antibiotics data again (copied from Section 5.5.2):

```
> modelAntibio2 <- lm(org~mytype) # One-way ANOVA with control
                                  # as reference group
> summary(modelAntibio2)

Coefficients:
                Estimate Std. Error t value Pr(>|t|)
(Intercept)      2.60333    0.04970  52.379  < 2e-16 ***
mytypeAlfacyp    0.29167    0.07029   4.150 0.000281 ***
mytypeEnroflox   0.10667    0.07029   1.518 0.140338
mytypeFenbenda   0.23000    0.07029   3.272 0.002834 **
mytypeIvermect   0.39833    0.07029   5.667 4.5e-06 ***
mytypeSpiramyc   0.25167    0.07858   3.202 0.003384 **
---
Signif. codes:  0 '***' 0.001 '**' 0.01 '*' 0.05 '.' 0.1 ' ' 1

Residual standard error: 0.1217 on 28 degrees of freedom
```

Recall the structure of the output: one line for each parameter in the model, here the level for the control group and the differences between the control group and each of the other groups. Four values are listed for each parameter:

- The estimate. For example, the difference between the Fenbendazole group and the control group is estimated to 0.2300. The parameter corresponding to the Intercept line is the mean for the reference group, here the control group, so $\hat{\alpha}_{control} = 2.603$.

- The standard error of the estimate.

- The *t*-test statistic, T_{obs}, for the null hypothesis that the corresponding parameter is equal to zero. As we know, T_{obs} is simply the estimate divided by its standard error; *i.e.*, the *t*-test statistic for the hypothesis of no difference between the Fenbendazole group and the control group is $0.2300/0.07029 = 3.272$.

- The p-value for the t-test just mentioned. The p-value is supplied with a number of stars that state the significance status of the test. The translation of the stars is given below the list: two stars, for example, means that the p-value is between 0.001 and 0.01, whereas three stars mean that the p-value is smaller than 0.001. No star indicates that the p-value is larger than 0.1, hence we easily see that there are no indications of an effect of Enrofloxacin compared to the control group.

Notice how the careful choice of parameterization with the control group as reference pays off at this point: our primary interest is in comparing the treatment groups to the control, and the relevant estimates and tests are those reported in the output with this particular parameterization.

Also, notice how R automatically carries out t-tests for all parameters in the model. It is important to stress that not all these hypotheses are necessarily interesting or relevant, and hence should not automatically be reported by the user. For example, the hypothesis that α_{control} — the intercept in the current parameterization of the model — is zero is irrelevant but is nevertheless tested in the Intercept line. Biological arguments should decide which hypotheses to test and thus which parts of the output to use in the analysis.

6.6.2 The F-test for comparing groups

In the one-way ANOVA analysis, we want to test the hypothesis of an overall effect, $H_0 : \alpha_1 = \cdots = \alpha_k$. The test is obviously not one of those carried out in the summary() output, but the anova() function reports the relevant values.

```
> anova(modelAntibio2)  # Test for equal means in one-way ANOVA
Analysis of Variance Table

Response: org
          Df  Sum Sq Mean Sq F value    Pr(>F)
mytype     5 0.59082 0.11816  7.9726 8.953e-05 ***
Residuals 28 0.41500 0.01482
---
Signif. codes:  0 '***' 0.001 '**' 0.01 '*' 0.05 '.' 0.1 ' ' 1
```

The output contains one line per "source of variation", and for each source it lists the degrees of freedom, the SS-value, and the MS-value. Moreover, the F-test for the effect of mytype is carried out: the value of F_{obs} and the associated p-value are reported. Notice how the degrees of freedom for the test, here $(5, 28)$, also appear in the output.

6.6.3   Hypothesis tests as comparison of nested models

The general F-test is carried out by the anova() function if it is given two arguments rather than just one, as before. For the comparison of groups in the one-way ANOVA model, for example, we compare the full model, modelAntibio2 from above, to the model under the hypothesis. Under the hypothesis, the distribution is the same for all groups, corresponding to a single sample, so we fit the model with intercept only.

```
> modelAntibioHyp <- lm(org~1)            # No group effect
> anova(modelAntibioHyp, modelAntibio2) # Compare models
Analysis of Variance Table

Model 1: org ~ 1
Model 2: org ~ mytype
  Res.Df     RSS Df Sum of Sq      F    Pr(>F)
1     33 1.00582
2     28 0.41500  5   0.59082 7.9726 8.953e-05 ***
---
Signif. codes:  0 '***' 0.001 '**' 0.01 '*' 0.05 '.' 0.1 ' ' 1
```

The output has a line for each model with the residual degrees of freedom (Res.Df) and the sum of squared residuals (RSS). Moreover, the deviation between the sums of squared residuals (Sum of Sq), the F-test statistic, and the p-value are reported. Notice how this is the same test as the one carried out by anova(modelAnitbio2) above.

The same method applies to the digestibility data (Example 6.12, p. 168):

```
> modelDigest <- lm(digest~stearic.acid) # Linear regression
> modelDigestHyp <- lm(digest ~ 1)        # No covariate effect
> anova(modelDigestHyp, modelDigest)      # Compare models
Analysis of Variance Table

Model 1: digest ~ 1
Model 2: digest ~ stearic.acid
  Res.Df    RSS Df Sum of Sq      F    Pr(>F)
1     8 958.53
2     7  61.76  1   896.76 101.63 2.028e-05 ***
---
Signif. codes:  0 '***' 0.001 '**' 0.01 '*' 0.05 '.' 0.1 ' ' 1
```

First the ordinary linear regression model is fitted; next the model with zero slope (that is, with an intercept only). These are compared with anova() command. From the summary() output in Section 5.5.1 we read $T_{obs} = -10.08$ and $p = 0.00002$ for the same hypothesis. We see that the p-values are identical and that the t-test statistic squared is equal to the F-test statistic $(-10.08)^2 = 101.6$.

6.6.4 Tests for one and two samples

We already used the t.test() function for computation of estimates and confidence intervals for one and two samples in Section 5.5.3. As the name suggests, the function also carries out certain *t*-tests. For the cattle data (Example 6.1, p. 145) we get the following:

```
> dif <- c(9, 3, 39, 24, 1, 13, 11, 32, -8)  # Data
> t.test(dif)                                 # Test for mean = 0
```

```
        One Sample t-test

data:  dif
t = 2.7125, df = 8, p-value = 0.02655
alternative hypothesis: true mean is not equal to 0
95 percent confidence interval:
  2.064870 25.490686
sample estimates:
mean of x
 13.77778
```

We recognize the estimate, the confidence interval, the *t*-test statistic, and the *p*-value from the example.

Notice that the default hypothesis is that the mean is equal to zero. With the notation from (6.3), this corresponds to $\theta_0 = 0$. If we are interested in another value of θ_0, then we need to specify the value in the call to t.test(). In Example 6.7 (p. 158) on production control the relevant hypothesis was $H_0 : \mu = 1070$. If milkweight is a variable containing the observations then the following line of code would make the relevant *t*-test:

```
> t.test(milkweight, mu=1070)   # Test for mean = 1070
```

The t.test() function can also be used to carry out the *t*-test for comparison of two samples. In Exercises 6.4 and 6.5 you are asked to run the commands below and interpret the output.

For two independent samples, R as default allows for different standard deviations for the two samples. If the standard deviations are assumed to be equal, then the option var.equal=T should be used, as explained in Section 5.5.3. For the salmon data (Example 3.1, p. 49 and Example 6.5, p. 155) the relevant commands are

```
> t.test(conon, atran, var.equal=T)  # Same standers deviations
> t.test(conon, atran)               # Different standard dev.
```

Analysis of paired data (see Section 3.5) can be performed in two ways with t.test(). Consider the lameness data (Example 3.5, p. 57) and assume that the variables lame and healthy contain the data. The hypothesis of interest is $H_0 : \delta = 0$, where δ is the expected value for the difference between the symmetry score corresponding to lameness and the healthy condition. The relevant commands are

```
> t.test(lame, healthy, paired=T) # Paired samples
> t.test(lame-healthy)            # Analysis of differences
```

The first command is similar to the command for two independent samples except that the `paired=T` option specifies that the observations are paired. In particular, the two variables must have the same length. The second command performs a one sample analysis on the differences between the two variables. In both cases the hypothesis is tested against the two-sided alternative $H_A : \delta \neq 0$.

The symmetry score is constructed such that it takes small values if the gait pattern is symmetric and large values if the gait pattern is asymmetric. Hence, in this case it would be reasonable to consider the alternative $H_A : \delta > 0$. This alternative is easily specified to `t.test()` with the `alternative` option:

```
> t.test(lame-healthy, alternative="greater") # One-sided
                                              # alternative
```

For the alternative $H_A : \delta < 0$, the option should be `alternative="less"`.

6.6.5 Probabilities and quantiles in the F distribution

Probabilities and quantiles in the F distributions are computed with the `pf()` and `qf()` functions. Of course, the pair of degrees of freedom should be specified in the commands.

```
> pf(1.5, df1=5, df2=28)   # P(F <= 1.5) in F(1,28)-dist.
[1] 0.7783989
> pf(1.5, 5, 28)           # P(F <= 1.5) in F(1,28)-dist.
[1] 0.7783989
> qf(0.95, df1=5, df2=28)  # 95% quantile in F(1,28)-dist.
[1] 2.558128
> qf(0.95, 5, 28)          # 95% quantile in F(1,28)-dist.
[1] 2.558128
```

From the output we see that $P(F \leq 1.5) = 0.78$ for a $F(5,28)$-variable and that the 95% quantile in the $F(5,28)$ distribution is 2.56.

6.7 Exercises

6.1 Length of gestation period for horses. Recall the data from Exercise 4.5, where the length of the gestation period was measured for

13 horses. The mean and standard deviation were calculated to 341.08 and 3.07, respectively. You may also use the R-output

```
> pt(1.27, df=12)
[1] 0.8859207
> qt(0.975, df=12)
[1] 2.178813
```

1. Specify a statistical model for the data and estimate the parameters.

2. The average length of the gestation period for (the population of) horses is supposed to be 340 days. Carry out a hypothesis test to examine if the current data contradict or support this belief.

3. Compute a confidence interval for the expected length of the gestation period. Is 340 included in the confidence interval? How does this relate to the test in question 2?

6.2 **Stomach experiment.** Fifteen subjects participated in an experiment related to overweight and got a standardized meal (Skovgaard, 2004). The interest was, among others, to find relationships between the time it takes from a meal until the stomach is empty again and other variables. One such variable is the concentration of a certain hormone.

Person	Hormone conc.	Time to empty
1	0.33	17.1
2	0.29	17.1
3	0.29	14.3
4	0.32	16.1
5	0.31	17.4
6	0.25	13.5
7	0.30	17.1
8	0.17	15.0
9	0.31	16.8
10	0.38	17.3
11	0.21	15.3
12	0.30	12.5
13	0.20	13.1
14	0.18	12.7
15	0.26	13.0

You may use the R-output in the end of the exercise to answer the questions.

1. Specify a statistical model for the data and estimate the parameters.

2. Is there evidence of an association between the hormone concentration and the time it takes for the stomach to become empty?

3. Consider two persons' hormone concentrations, 0.20 and 0.30, respectively. Calculate an estimate and a 95% confidence interval for the expected difference in time it takes for their stomachs to become empty.

4. Assume that a physiological theory suggests that the time until the stomach becomes empty increases by 1.2 when the hormone concentration increases by 0.1. Do the current data contradict or support this theory?

In the following R-output the variables conc and empty contain the hormone concentrations and the time measurements:

```
> model <- lm(empty~conc)
> summary(model)

Coefficients:
            Estimate Std. Error t value Pr(>|t|)
(Intercept)    9.966      1.917   5.198 0.000172 ***
conc          19.221      6.861   2.802 0.014983 *

Residual standard error: 1.548 on 13 degrees of freedom
Multiple R-Squared: 0.3765,   Adjusted R-squared: 0.3285
F-statistic: 7.849 on 1 and 13 DF,  p-value: 0.01498

> qt(0.95, df=13)
[1] 1.770933
> pt(0.95, df=13)
[1] 0.8202798
> qt(0.975, df=13)
[1] 2.160369
> pt(1.052, df=13)
[1] 0.844019
```

6.3 ® **Fertility of lucerne.** Ten plants were used in an experiment of fertility of lucerne (Petersen, 1954). Two clusters of flowers were selected from each plant and pollinated. One cluster was bent down, whereas the other was exposed to wind and sun. At the end of the experiment, the average number of seeds per pod was counted for each cluster and the weight of 1000 seeds was registered for each cluster.

The external dataset lucerne.txt contains five variables

plant is the plant number

`seeds.exp` is the seed count for the exposed cluster

`seeds.bent` is the seed count for the bent cluster

`wgt.exp` is the weight of 1000 seeds from the exposed cluster

`wgt.bent` is the weight of 1000 seeds from the bent cluster

1. Read the data into R with `read.table()`, print the dataset on the screen, and make sure that you understand the structure of the dataset.

First, we will examine the seed count in order to see if there is evidence of a difference between the seed counts per pod for exposed and bent clusters.

2. Think of some relevant plots of the seed data and make them with R. What do they tell you?

3. Specify a statistical model that can be used to compare the seed counts for the bent and the exposed clusters. Specify the relevant hypothesis.

4. Use the `t.test()` function to carry out a test for the hypothesis. What is the conclusion?

   [Hint: You can do this in two ways, as explained in Section 6.6.4. Try both ways and make sure you get the same result.]

5. Compute an estimate and a confidence interval for the difference between the seed counts for bent and exposed clusters. What is the relation to the hypothesis test?

Finally, let us consider the weights of 1000 seeds; that is, the variables `wgt.exp` and `wgt.bent`.

6. Examine if the seed weights differ between the exposed and the bent clusters. Also, compute an estimate and a confidence interval of the difference.

6.4 ⓡ **Parasites counts for salmons.** Recall the data from Example 3.1 on parasites in two salmon stocks. In Section 5.5.3 we ran the commands

```
> t.test(conon, atran, var.equal=T)
> t.test(conon, atran)
```

and used the output to state estimates and confidence intervals. In this exercise we will use the output to test the hypothesis $H_0 : \alpha_{\text{Åtran}} = \alpha_{\text{Conon}}$ of no difference between the means for the two stocks.

1. Is the output from t.test(conon, atran, var.equal=T) in agreement with the results from Example 6.5 (p. 155)? What is the assumption on the standard deviations in the two groups?

2. Use the output from t.test(conon, atran) to carry out the hypothesis test without the assumption on identical standard deviations. Does the conclusion change?

For the remaining questions, the data should be organized as two vectors of length 26; namely, count, containing the parasite counts for all fish, and stock, with values atran or conon.

3. Construct the vectors count and stock. Use the command lm(count~stock) to fit the model (see also Exercise 5.5).

4. Use the summary() function to carry out the *t*-test for H_0.

5. Use the anova() function to carry out the *F*-test for H_0. Which of the above calls to t.test does this correspond to?

6. Fit the model with no effect of stock (lm(count~1)) and use the anova() function with two arguments to test the hypothesis.

6.5 ℝ **Equine lameness.** Recall the data on equine lameness from Example 3.5 (p. 57). The data are available in the external file lameness.txt with variables lame and healthy.

1. Solve Exercise 5.4 if you did not do so already.

2. The veterinarians want to know if the symmetry score changes when horses become lame on a forelimb. What is the hypothesis of interest? Use the output from the t.test() commands to test the hypothesis against a two-sided alternative. Make sure you get the same results with both commands. What is the conclusion?

3. Test the hypothesis again, now using the one-sided alternative that the expected value is larger for lame horses than for healthy horses. Explain the difference between this test and the test from question 2.

6.6 ℝ **Soybeans.** An experiment with 26 soybean plants was carried out as follows. The plants were pairwise genetically identical, so there were 13 pairs in total. For each pair, one of the plants was "stressed" by being shaken daily, whereas the other plant was not shaken. After a period the plants were harvested and the total leaf area was measured for each plant.

The external dataset soybean.txt contains the following variables:

pair is the number of the plant pair

`stress` is the total leaf area (cm^2) of the stressed plant

`nostress` is the total leaf area (cm^2) of the control plant

1. Why is it not reasonable to assume independence of all 26 measurements?

2. Specify a statistical model that makes it possible to examine if stress affects plant growth (leaf area).

3. Carry out a hypothesis test in order to examine if there is an effect of stress on the growth of the plants. Compute also a 95% confidence interval for the expected difference in leaf area between plants that are stressed and plants that are not stressed.

   [Hint: Use `t.test(stress,nostress, paired=T)`. Why should you write `paired=TRUE`?]

4. Try the command `t.test(stress,nostress, paired=F)`. How should the data have been collected in order for this command to be appropriate? What is the corresponding model? Do you get the same results (*p*-values and confidence intervals) if you run this? Explain why the results are different.

6.7 Ⓡ **Weight gain for chickens.** Twenty chickens were fed with four different feed types — five chickens for each type — and the weight gain was registered for each chicken after a period (Anonymous, 1949). The results are listed in the table below. The external file `chicken.txt` contains two variables, `feed` and `gain`.

Feed type	Weight gain				
1	55	49	42	21	52
2	61	112	30	89	63
3	42	97	81	95	92
4	169	137	169	85	154

1. Specify a statistical model for the data. Is there evidence that the chickens grow differently depending on which feed type they are fed?

   [Hint: Remember to write `factor(feed)` as the explanatory variable in R. Why?]

2. Assume that feed type 1 is a control feed. Is the weight gain significantly different from this control feed for all other feed types? And how about feed types 2 and 3 — do they yield different weight gains?

   [Hint: You could (for example) calculate and use the LSD-value.]

3. Compute a 95% confidence interval for the expected difference in weight gain between feed types 1 and 3.

6.8 Scale in linear regression. Consider the linear regression model of y on x,

$$y_i = \alpha + \beta \cdot x_i + e_i,$$

where $e_1, \ldots, e_n$ are independent and $N(0, \sigma^2)$ distributed. Assume that our observations (x_i, y_i) yield estimates $\hat{\alpha}$, $\hat{\beta}$, and s.

1. Assume that each x_i is replaced by $x_i' = 2 \cdot x_i$ and that we run the regression of y on x'. How does it influence the estimates? How does it influence the test for the hypothesis $\beta = 0$?

2. Assume instead that each y_i is replaced by $y_i' = 3 \cdot y_i$ and that we run the regression of y' on x. Answer the same questions as in 1.

6.9 ℝ **Vitamin A storage in the liver.** In an experiment on the utilization of vitamin A, 20 rats were given vitamin A over a period of three days (Bliss, 1967). Ten rats were fed vitamin A in corn oil and ten rats were fed vitamin A in castor oil. On the fourth day the rats were killed and the vitamin A concentration in the liver was determined.

The external dataset oilvit.txt contains the variables type with levels corn and am and the variable avit with the vitamin A concentrations. The data are also used in Case 3, Part I (p. 356). Let us first recap the results from Case 3:

1. Specify a statistical model for the data.

2. Construct variables x and y with the commands

```
x <- avit[type=="corn"]
y <- avit[type=="am"]
```

Use the command t.test(x,y, var.equal=T) and interpret the output.

Now, think of the data from a one-way ANOVA point of view.

3. What are n, k, and n_j?

4. Use lm() to fit the model. Use anova() to test the hypothesis that there is no difference between the two groups.

5. Compare the results from questions 2 and 4. Do you get the same conclusion? Do you get the same p-value? What is the relationship between the t-test statistic T_{obs} and the F-test statistic F_{obs}?

6. Use the summary() command to get an estimate of the expected difference between the two groups. Compute also the corresponding 95% confidence interval. Is it the same as R provided with the t.test() command in question 2?

6.10 ℞ **Lifespan and length of gestation period.** Recall the horse gestation data from Example 6.3 (p. 151) and the analysis in Example 6.8 (p. 158). One of the data points is very different from the others; see Figure 6.3.

1. Repeat the analysis without the extreme observation. Is the conclusion the same as when all the observations are included?

2. Discuss why it may be dangerous to remove strange-looking observations from a dataset.

6.11 Multiple testing in microarray experiments. A biotechnician intends to do a microarray experiment where the effect of 478 genes are examined simultaneously. We assume that the 478 genes are independent, so the experiment corresponds to making 478 independent tests to see if there is any effect of each of the genes. What significance level should we use for each of the 478 tests if we wish to use Bonferroni correction for multiple testing and control the overall error rate so that it does not exceed 0.05?

6.12 ℞ **Analysis of gene expression levels.** Two groups were compared in an experiment with six microarrays. Two conditions (the test group and the reference group) were examined on each array and the amount of protein synthesized by the gene (also called the gene expression) was compared for the two groups.

The expression levels for the six arrays are shown below:

Test group	Reference group
444	568
604	669
500	566
614	655
602	628
427	701

Test if the expression level is identical for the test and reference groups.

6.13 ℞ **Tartar for dogs.** Analyze the data from Exercise 3.2. This includes specification of a statistical model, test for an overall effect of treatment, identification and quantification of significant treatment differences, and conclusion.

Chapter 7

Model validation and prediction

The results from a statistical analysis are valid only if the statistical model describes the data in a satisfactory way. It is therefore essential to be able to validate the model assumptions, and this is the topic of the first part of this chapter. Moreover, it is often of interest — perhaps even the primary goal of the analysis — to make predictions from the model, and this will constitute the second part of this chapter.

7.1 Model validation

In Chapters 5 and 6 we discussed confidence intervals and hypothesis tests. For example, in linear regression we used the fact that

$$T = \frac{\hat{\beta} - \beta_0}{s/\sqrt{SS_x}}$$

comes from the t distribution with $n - 2$ degrees of freedom in order to compute confidence intervals and in order to test hypotheses about the slope parameter β. However, the t distribution — and hence the confidence interval and the test — is valid only if the underlying model is appropriate for the data; *i.e.*, if

$$y_i = \alpha + \beta \cdot x_i + e_i \quad i = 1, \dots, n,$$

where $e_1, \dots, e_n$ are iid. $N(0, \sigma^2)$.

In other words, *if* the model assumptions are appropriate, then all the results from Chapters 5 and 6 are valid. On the other hand, if the model assumptions are *not* appropriate, then we do not know the distribution of T. In principle, we could still compute the test statistic T_{obs} for a hypothesis $H_0 : \beta = \beta_0$, but we would not know whether an observed value of, say, 2.3 is significant or not. Similarly, we could still compute what should be the 95% confidence interval, but we would not know its coverage. It could be quite different from 0.95 if the model assumptions are wrong.

Similarly, for the one-way ANOVA with the hypothesis $H_0 : \alpha_1 = \cdots = \alpha_k$: If the model assumptions are not true then the usual test statistic F_{obs} does not necessarily come from an F distribution, and we would not know if the observed value is large enough to be significant or not — we cannot draw conclusions from it. The point is that we need to verify if the model assumptions are reasonable for our data — to carry out *model validation* — in order to trust the results from the analysis.

7.1.1   Residual analysis

Recall the statistical model from Section 5.1, where $y_1, \ldots, y_n$ are assumed to be independent and normally distributed with a common standard deviation σ and mean depending on an explanatory variable (or several explanatory variables), $\mu_i = f(x_i; \theta_1, \ldots, \theta_p)$. As emphasized in Infobox 5.2, we may also write

$$y_i = \mu_i + e_i, \quad i = 1, \ldots, n$$

and impose assumptions on the remainder terms $e_1, \ldots, e_n$. This is essential for the model validation, so let us write the assumptions in detail.

Infobox 7.1: Model assumptions

Consider the model

$$y_i = \mu_i + e_i, \quad i = 1, \ldots, n,$$

where μ_i is a function of parameters and explanatory variables (linear in the parameters). The remainder terms $e_1, \ldots, e_n$ are assumed to

- have mean zero

- have the same standard deviation

- be normally distributed

- be independent

The last assumption of independence is hard to check from the data. Rather, it should be thought about already in the planning phase of the experiment: make sure that the experimental units (animals, plants, patients, *etc.*) do not share information. See Example 4.5 for examples with independent data as well as data that are not independent.

Sometimes it is desirable to sample data that are not independent. For example, an animal could be used as "its own control" in the sense that several treatments are tested on each animal. This is useful because the between-animal variation can be distinguished from the between-treatment variation, if the dependence is specified properly in the statistical model. If there are two treatments, this is the setup corresponding to two paired samples (*cf.* Section 3.5), and a one sample analysis can be applied on the differences. If there are more than two treatments tested on each experimental unit, the experimental design is called a block design and a two-way ANOVA model with blocks is an appropriate model; see Section 8.2.

The first three assumptions in Infobox 7.1 can all be checked from the data by examination of the residuals. Recall the definition of *fitted values*, or *predicted values*, and *residuals* from (5.6),

$$\hat{\mu}_i = \hat{y}_i = f(x_i, \hat{\theta}_1, \ldots, \hat{\theta}_p), \quad r_i = y_i - \hat{\mu}_i, \quad i = 1, \ldots, n.$$

The fitted value $\hat{\mu}_i$ is the estimate or prediction of the mean μ_i. Hence we may interpret the residual r_i as the prediction of the remainder term e_i in the model, and it makes sense to check the assumptions on $e_1, \ldots, e_n$ by examination of $r_1, \ldots, r_n$.

For some models the residuals do not have the same standard deviation, even though the remainder terms e_i's do. This is the case for the linear regression model, where residuals corresponding to x-values close to $\bar{x}$ have smaller variation compared to residuals for observations with x-values far from $\bar{x}$; cf. (5.14). Therefore, the residuals are usually standardized with their standard error and the residual analysis is carried out on the *standardized residuals*:

$$\tilde{r}_i = \frac{r_i}{\text{SE}(r_i)}.$$

The standardized residuals are easily calculated by statistical software programs, so we do not go into detail about exactly how to calculate them here*. The important thing is that $\tilde{r}_1, \ldots, \tilde{r}_n$ are standardized such that they resemble the normal distribution with mean zero and standard deviation one — if the model assumptions hold.

Hence, we check these properties on $\tilde{r}_1, \ldots, \tilde{r}_n$. The normality assumption is usually checked with a QQ-plot or a histogram (if there are enough observations), as described in Section 4.3.1. Due to the standardization, the points should be scattered around the line with intercept zero and slope one.

The assumptions on the mean and standard deviation are usually validated with a *residual plot*, where the standardized residuals are plotted against the predicted values. In linear regression the standardized residuals may be plotted against the explanatory variable instead of the predicted values. Since the predicted values are computed as a linear function of values of the explanatory variable, this gives the same picture except that the scale is transformed linearly. Let us take a look at some examples in order to describe what we should look for in such a residual plot.

Example 7.1. Stearic acid and digestibility of fat (continued from p. 26). The residual analysis for the digestibility data is illustrated in Figure 7.1. The left panel shows the residual plot. We see that the points are *spread out without any clearly visible pattern*. This is how it should be. In particular,

---

*Some programs use all observations for computation of the standard error in the denominator for $\tilde{r}_i$; some programs omit the corresponding observation y_i in the computation.

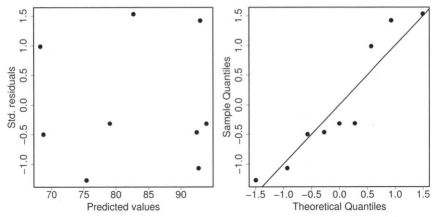

Figure 7.1: Residual analysis for the digestibility data: residual plot (left) and QQ-plot (right) of the standardized residuals. The straight line has intercept zero and slope one.

- there seem to be both positive and negative residuals in all parts of the plot (from left to right; for small, medium, as well as large predicted values). This indicates that the specification of the digestibility mean as a linear function of the stearic acid level is appropriate.

- there seems to be roughly the same vertical variation for small, medium, and large predicted values. This indicates that the standard deviation is the same for all observations.

- there are neither very small nor very large standardized residuals. This indicates that there are no outliers and that it is not unreasonable to use the normal distribution.

The last conclusion is supported by the QQ-plot in the right panel.   □

Example 7.2. Growth of duckweed (continued from p. 35). Recall the duckweed data, where the number of duckweed leaves (the response) was modeled as a function of time (days since the experiment started). The upper panel of Figure 2.8 clearly showed that the relationship between time and the number of duckweed leaves is non-linear, but let us fit the linear regression model for illustration, anyway. Hence, we consider the model

$$\text{Leaves}_i = \alpha + \beta \cdot \text{Days}_i + e_i, \quad i = 1, \dots, 14.$$

The left panel of Figure 7.2 shows the corresponding residual plot. There is a very clear pattern: for small and large predicted values the residuals are all positive, and in between the residuals are all negative. This tells us that the mean of the response has not been specified appropriately as a function of

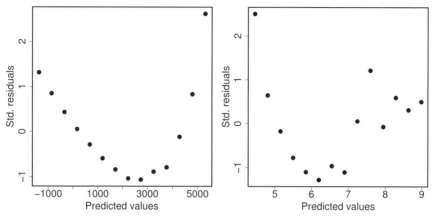

Figure 7.2: Residual plots for the duckweed data. Left panel: linear regression with the leaf counts as response. Right panel: linear regression with the logarithmic leaf counts as response.

the explanatory variable, so the model is not suitable (as we knew from the beginning).

In this example it was clear from the initial scatter plot of the data in Figure 2.8 that this would be case: the fitted line underestimates the number of leaves in the left and right part of the plot, yielding positive residuals, and overestimates in the middle, yielding negative residuals. In other examples, the picture might be less clear, and in general problematic patterns are more easily detected in the residual plot than in the plot of the original data.

The right panel of Figure 7.2 shows the residual plot for the linear regression of the natural logarithmic number of leaves on the days since the experiment started,

$$\log(\text{Leaves}_i) = \alpha + \beta \cdot \text{Days}_i + e_i.$$

This corresponds to an exponential relation between the original variables, as explained in Example 2.4. There is still a clear pattern in the residuals! This might be somewhat surprising: the fit is illustrated in the lower panels of Figure 2.8, and the agreement between the observations and the fitted values is striking. However, this merely reflects that there is very little variation (a small σ) — not that the specification of the remainder terms is appropriate. The logarithmic model is certainly preferable as the right-hand plot is better than the left-hand plot, but we should still be somewhat careful with the interpretation of confidence intervals, tests, *etc.*   □

Example 7.3. Chlorophyll concentration. An experiment with winter wheat was carried out at the Royal Veterinary and Agricultural University in Denmark in order to investigate if the concentration of nitrogen in the soil can be predicted from the concentration of chlorophyll in the plants. This could improve the adjustment of nitrogen supply. The chlorophyll concentration in the leaves as well as the nitrogen concentration in the soil were measured for 18 plants. The measurements are given in the table below:

Chlorophyll, C	391	498	597	648	606	630	364	546	648
Nitrogen, N	52	102	152	202	252	302	22	72	122
Chlorophyll, C	685	648	694	439	553	651	687	723	732
Nitrogen, N	172	222	272	28	78	128	178	228	278

The upper left panel of Figure 7.3 shows the data: the nitrogen concentration N is plotted against the chlorophyll concentration C. The other three panels show residual plots for three different models, the only difference being the choice of response:

Upper right : $N_i = \alpha + \beta \cdot C_i + e_i$

Lower left : $\log(N_i) = \alpha + \beta \cdot C_i + e_i$

Lower right : $\sqrt{N_i} = \alpha + \beta \cdot C_i + e_i$

For the regression with N as response — the upper right plot — there is an indication of a "trumpet shape": the variation seems to be larger for large predicted values (the right part of the plot) compared to small predicted values (the left part of the plot). This is quite often the case for biological data: large variation occurs for large predicted values, small variation for small predicted values. The problem can often be remedied by transformation. In particular, the logarithmic transformation has the property that it "squeezes" large values and in that way diminishes the variation for large values.

For this particular dataset, however, it seems like the log-transformation has been too powerful (see the lower left plot): there seems to be larger variation in the left part of the plot, for small predicted values. The square root transformation is sometimes a useful compromise between no transformation and the logarithmic transformation, and except for two large positive residuals, the variation seems to be constant across the different values of predicted values (lower right panel). We should be careful, however, not to over-interpret the findings as they are based on only 18 measurements.   □

As indicated by the examples, *the residual plot should not display any systematic patterns*. There should be roughly as many positive and negative residuals in all parts of the plot (left to right), and the vertical variation should be roughly the same in all parts of the plot (from left to right). Moreover, we can use the residual plot to detect *outliers* or extreme observations.

In Example 7.3 on the chlorophyll-nitrogen relation there were two standardized residuals that were somewhat larger than the others. *Very large* and *very small* residuals correspond to "unusual" observations and are potential outliers. The values in the example were not *very* large, though. Recall that 5% of the values in a sample from the standard normal distribution, $N(0,1)$,

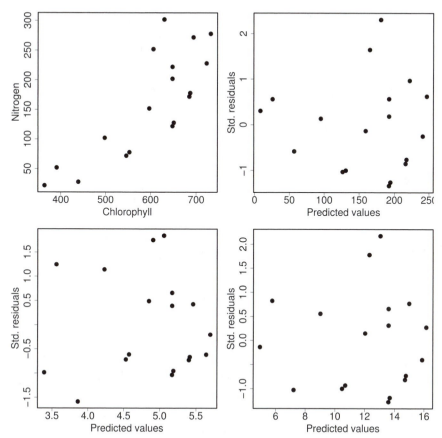

Figure 7.3: The chlorophyll data. Upper left panel: scatter plot of the data. Remaining panels: residual plots for the regression of N on C (upper right), for the regression of $\log(N)$ on C (lower left), and for the regression of $\sqrt{N}$ on C (lower right).

are numerically larger than 1.96. In other words, for a dataset with 100 observations, about 5 observations should be outside the interval from -2 to 2, and ideally 2–3 of 1000 observations should be outside the interval from -3 to 3.

Sometimes residuals are more extreme than in the example, and then the corresponding observation should be checked once again: Perhaps there was a mistake when the observation was registered. Perhaps it was taken on an animal that turned out to be sick. Perhaps it comes from a plant that did not get the water supply that it was supposed to. If so, the observations should be removed in the data analysis. If not, it is generally a dangerous strategy to remove "strange" observations from the dataset. Perhaps the extreme observation is due to a specific feature of the data which should be accounted for in the analysis. A better idea is to run the analysis both with and without the unusual observations and see if the main conclusions remain the same and are therefore valid in this sense.

Let us recap what we should look for in the residual analysis.

Infobox 7.2: Model validation based on residuals

Residual plot. Plot the standardized residuals against the predicted values. The points should be spread randomly in the vertical direction, without any systematic patterns. In particular,

- points should be roughly equally distributed between positive and negative values in all parts of the plot (from left to right).
- there should be roughly the same variation in the vertical direction in all parts of the plot (from left to right).
- there should be no too extreme points.

Systematic deviations correspond to problems with the mean structure, the variance homogeneity, or the normal distribution, respectively.

QQ-plot of standardized residuals. The points should be scattered around the straight line with intercept zero and slope one.

We should be careful, though, not to over-interpret tendencies in the residual analysis; in particular, when there are just a few observations. For example, if the sample size is 15 and the position of a single point in the residual plot makes you worried about variance homogeneity, then it is not really a pattern, and you most likely need not worry too much. Similarly, QQ-plots for 15, say, normally distributed variables may look quite different, so the deviation from a straight line should be quite substantial in order for you to worry about non-normality. For example, in Example 7.3 the tendencies were perhaps not that clear, after all.

The previous examples were all linear regression examples, but model validation is carried out in the same way, using the residuals, for one-way ANOVA models and other linear models.

Example 7.4. Effect of antibiotics on dung decomposition (continued from p. 51). Recall that a one-way ANOVA model with six groups is used for the antibiotics data. The residual analysis is illustrated in Figure 7.4. There are only six possible predicted values (one for each group), and the variation of the standardized residuals seems to be roughly the same for all groups. In Example 3.2 we noticed that the standard deviation was slightly smaller for the Spiramycin group compared to the others. We recognize this in the residual plot, too, but this could be due to the smaller number of observations in this

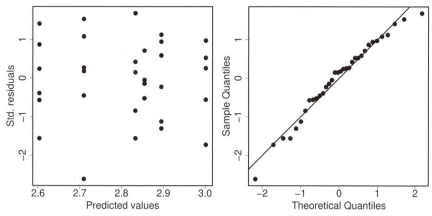

Figure 7.4: Residual analysis for the antibiotics data: residual plot (left) and QQ-plot (right) of the standardized residuals. The straight line has intercept zero and slope one.

group. The fewer the observations, the less likely it is to observe numerically large residuals. □

7.1.2 Conclusions from analysis of transformed data

In Example 7.2 (p. 188) on duckweed and Example 7.3 (p. 189) on chlorophyll we saw that the assumptions seemed to be more reasonable when a transformation of the data was used as the response. In that case we can trust the results from the model with the transformed response better compared to the results from an analysis where the model assumptions are clearly not true.

Confidence intervals should therefore be computed and tests should be carried out in the model with the transformed response. Then, however, the estimates and confidence intervals are also on the transformed scale; *e.g.*, the logarithmic scale. This is usually hard to interpret, so the estimates and confidence intervals should be *back-transformed* and reported on the original scale.

This was exactly what happened in Example 2.4 for the duckweed data, so let us take a look at that again.

Example 7.5. Growth of duckweed (continued from p. 35). The fitted model for the duckweed data with the logarithmic leaf counts as response was

$$\log(\text{Leaves}) = \hat{\alpha} + \hat{\beta} \cdot \text{Days} = 4.4555 + 0.3486 \cdot \text{Days},$$

which on the original scale corresponds to

$$\text{Leaves} = \exp(4.4555 + 0.3486 \cdot \text{Days})$$
$$= \exp(4.4555) \cdot \exp(0.3486 \cdot \text{Days})$$
$$= 86.099 \cdot \exp(0.3486 \cdot \text{Days}).$$

Hence, when the variable Days increases by one, then the fitted value of the variable Leaves increases by a factor of $\exp(0.3486) = 1.417$. This means that the number of leaves on average is increased by a factor of 1.417 per day of growth.

Moreover, the 95% confidence interval for the slope can be computed to

$$\beta: \quad 0.3486 \pm 0.0100 = (0.3386, 0.3586).$$

In order to get a 95% confidence interval for the factor, $\exp(\beta)$, by which the leaf count is increased from one day to the next, we take the exponential function at the endpoints of the confidence interval for β and get

$$\exp(\beta): \quad \left(\exp(0.3386), \exp(0.3586)\right) = (1.403, 1.431).$$

The interpretation is that growth factors between 1.403 and 1.431 per day are in accordance with the data on the 95% confidence level. (Recall, however, that the residual analysis was not completely satisfactory, so we should be a little careful with the interpretation.) □

Example 7.6. Growth prohibition. An experiment with weed control was carried out. A pesticide was applied to pots with weeds in different doses, and the growth of weeds over one week was measured for each pot. It was established that a relationship of the type

$$\text{growth} \approx c \cdot \left(\frac{1}{\text{dose}}\right)^b \tag{7.1}$$

was adequate to describe the association between dose and growth. Here b is a positive parameter corresponding to a decrease in growth for increasing dose. Taking logarithms on both sides transforms (7.1) to a linear relation: If we use the logarithm with base 2, $\log_2$, we get

$$\log_2(\text{growth}) \approx \log_2(c) - b \cdot \log_2(\text{dose}).$$

Hence the data is analyzed with a linear regression with response $y = \log_2(\text{growth})$ and explanatory variable $x = \log_2(\text{dose})$. In particular, an estimated slope $\hat{b}$ is computed with the interpretation that a doubling of the dose increases the weed growth by a factor of $(\frac{1}{2})^{\hat{b}}$. □

Sometimes the need for transformation is already detected in the initial graphical analysis of the data. For the duckweed data, for example, it was obvious from the very start that a linear regression of the number of leaves on the days variable would not be meaningful. Similarly, variance heterogeneity in a one-way ANOVA is often recognized from the parallel boxplots. In other situations (in particular, for more complicated models such as those we shall discuss in Chapter 8), the inappropriateness of a model is not detected until the residual analysis.

In any case the question arises: Which transformation should we use? As should be clear from the examples, the logarithmic transformation is extremely useful and works very well in many cases. It is theoretically justified because it corresponds to the standard deviation being proportional to the mean,

$$\frac{\text{sd}(y)}{\text{mean}(y)} = \text{constant.}$$

This is a property that is often satisfied for biological data. The constant is called the *coefficient of variation* and measures the variation per mean unit. Moreover, many parameters still have nice interpretations since differences on the logarithmic scale correspond to "incremental factors" on the original scale; *cf.* Example 7.5 on duckweed.

However, we also experienced that the logarithmic transformation does not always do the job, and we might try other transformations. The job is to *find a scale on which the model assumptions hold reasonably well*. There are various methods which can be used in order to find the "best" transformation within a class, but we will not go into details about that here. It may also be that it is not possible to find a useful transformation. In such cases different statistical models, for example non-parametric models, are in order.

7.2 Prediction

Prediction is the act of "forecasting" the value of new or future observations similar to those from the dataset. The statistical model is used for prediction and the statements are given in terms of probabilities. A 95% *prediction interval* is an interval that has a 95% probability of containing a new observation.

Example 7.7. Blood pressure. Assume that the blood pressure depends on age according to a linear regression. If a 35 year old woman has her blood pressure measured the physician will compare it to what is normal for the population — or rather to what is normal for 35 year old women. The prediction interval is an interval that includes the blood pressure for 95% of all (healthy) 35 year old women. The prediction interval is also called a "normal area". □

If the distribution of the observations is normal, $N(\mu, \sigma^2)$, with *known* parameters μ and σ, then we already know how to construct prediction intervals. With a probability of 95% a new observation will fall into the interval $\mu \pm 1.96\sigma$, which is then the 95% prediction interval.

Example 7.8. Beer content in cans. Assume that a machine that fills beer to 330 ml cans is such that the amount of beer filled into a can is normally distributed with mean μ and standard deviation σ. The predicted content of beer is then μ, and the 95% prediction interval is the interval that contains the beer content for 95% of the cans; that is, $\mu \pm 1.96\sigma$. For example, if $\mu = 335$ ml and $\sigma = 5$ ml, then 95% of the cans will have a content between 325.2 ml and 344.8 ml.

The standard deviation determines the precision of the machine and the brewer usually has no influence on it. Assume however that the mean μ can be adjusted in order to make the production satisfy requirements like "99% of the cans should contain at least 330 ml". Hence, we should choose μ such that

$$P(Y \geq 330) = 0.99,$$

where $Y \sim N(\mu, 25)$, still assuming that the standard deviation is 5 (so the variance is 25).

If we use property (c) from Infobox 4.2, we can rewrite the equation to a statement concerning a standard normal variable Z:

$$0.01 = P(Y \leq 330) = P\left(\frac{Y - \mu}{5} \leq \frac{330 - \mu}{5}\right) = P\left(Z \leq \frac{330 - \mu}{5}\right).$$

This equation holds if $(330 - \mu)/5$ is the 1% quantile of the standard normal distribution (which is equal to -2.326), so

$$\frac{330 - \mu}{5} = -2.326 \Leftrightarrow \mu = 330 + 5 \cdot 2.326 = 341.6.$$

In conclusion, the brewer should on average fill 341.6 ml into the cans in order to make sure that 99% of the cans contain at least 330 ml.   □

In the previous examples the distribution of the observations was implicitly assumed to be known — or at least estimated from a very large dataset, making the estimation error negligible. Most often this is not the case. Rather, the parameters of the distribution are unknown but estimated from a sample. In the beer example, the mean and standard deviation might not be known but estimated from a sample of, say, 1000 cans. From the values we compute estimates and their standard errors, but we do not get the true values. There is thus uncertainty about the parameter values, and this uncertainty should be taken into account in the computation of the prediction intervals.

7.2.1   Prediction in the linear regression model

Let us consider the linear regression model,

$$y_i = \alpha + \beta \cdot x_i + e_i, \quad i = 1, \ldots, n$$

and assume that we are interested in predicting the response corresponding to a value x_0 of the explanatory variable x. The value x_0 could be one of the values already used or measured in the experiment or it could be a new value.

Now, according to the model, the response is

$$y_0 = \alpha + \beta \cdot x_0 + e_0, \quad e_0 \sim N(0, \sigma^2)$$

and the prediction is of course obtained by inserting the estimates $\hat{\alpha}$ and $\hat{\beta}$,

$$\hat{y}_0 = \hat{\alpha} + \hat{\beta} \cdot x_0.$$

Notice that the predicted value $\hat{y}_0$ is identical to the estimated expected value $\hat{\mu}_0$; cf. Section 5.2.4. From the same section, formula (5.14), recall that $\hat{y}_0$ has standard error

$$\text{SE}(\hat{y}_0) = s \cdot \sqrt{\frac{1}{n} + \frac{(x_0 - \bar{x})^2}{SS_x}}.$$

The standard error takes into account the variation in the observed data $y_1, \ldots, y_n$: if we repeated the experiment we would get (slightly) different estimates. In other words, it takes into account the estimation error and thus gives rise to the *confidence interval*

$$95\% \text{ CI:} \quad \hat{\alpha} + \hat{\beta} \cdot x_0 \pm t_{0.975,n-2} \cdot s \cdot \sqrt{\frac{1}{n} + \frac{(x_0 - \bar{x})^2}{SS_x}} \qquad (7.2)$$

for the expected value $\mu_0 = \alpha + \beta x_0$.

However, y_0 is not exactly equal to $\alpha + \beta \cdot x_0$: it has mean $\alpha + \beta \cdot x_0$, but just as the original observations, it is subject to observation error. The observational error is here denoted e_0. It has standard deviation σ, and the prediction interval should take this source of variation into account, too. Intuitively, this corresponds to adding s^2 to the standard error — or adding a one under the square root — and one can show that this is the right thing to do. Hence, the 95% prediction interval is computed as follows:

$$95\% \text{ PI:} \quad \hat{\alpha} + \hat{\beta} \cdot x_0 \pm t_{0.975,n-2} \cdot s \cdot \sqrt{1 + \frac{1}{n} + \frac{(x_0 - \bar{x})^2}{SS_x}} \qquad (7.3)$$

The interpretation is that *a (new) random observation with $x = x_0$ will belong to this interval with probability 95%*. Of course, the probability 0.95 could be substituted by another by substituting the 97.5% quantile accordingly.

The prediction interval is valid for values of x_0 that are present in the original dataset as well as for new values — as long as the value is not too different for the x-values in the dataset. The point is that the model is only valid within the range of observations. Outside this range the data provide no information about the relation between x and y. We say that we are not allowed to extrapolate the results to values outside the range of observations.

7.2.2   Confidence intervals vs. prediction intervals

A prediction interval corresponds to a statement regarding the value of a *new observation*. In contrast, a confidence interval is a statement concerning the *expected value* for such an observation; that is, the average over all such observations. For example, the digestibility data gave rise to the confidence interval

$$95\% \text{ CI:}\quad 77.859 \pm 2.624 = (75.235, 80.483) \tag{7.4}$$

for the digestibility for a 20% stearic acid level; *cf.* Example 5.7 (p. 123). This interval is likely to include the average digestibility percent for a stearic acid level of 20%. A single observation may, however, very well be outside this interval due to the natural biological variation. The prediction interval is

$$95\% \text{ PI:}\quad 77.859 \pm 7.498 = (70.361, 85.357). \tag{7.5}$$

See Example 7.9. Hence, a new observation corresponding to 20% stearic acid level will belong to this interval with 95% probability.

The prediction interval is wider than the corresponding confidence interval, as the example as well as the interpretation and the formulas (7.2) and (7.3) show. Furthermore, notice that whereas we can make the confidence interval as narrow as we wish by increasing the sample size, this is not true for the prediction interval. No matter how precise we make the parameter estimates, the new observation will be subject to an observation error which we cannot reduce by increasing the sample size. For n very large, the estimates $\hat{\alpha}$, $\hat{\beta}$, and s approach the true population values and the prediction interval approaches the interval $y_0 \pm 1.96 \cdot \sigma$, corresponding to fixed values of the parameters.

The differences between confidence intervals and prediction intervals are summarized as follows:

Infobox 7.3: Confidence intervals and prediction intervals

Interpretation. The confidence interval includes the expected values that are in accordance with the data (with a certain degree of confidence), whereas a new observation will be within the prediction interval with a certain probability.

Interval widths. The prediction interval is wider than the corresponding confidence interval.

Dependence on sample size. The confidence interval can be made as narrow as we want by increasing the sample size. This is not the case for the prediction interval.

Example 7.9. Stearic acid and digestibility of fat (continued from p. 110). Consider the digestibility data and a new observation with stearic acid level

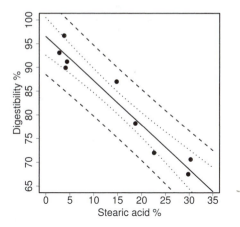

Figure 7.5: Predicted values (solid line), pointwise 95% prediction intervals (dashed lines), and pointwise 95% confidence intervals (dotted lines) for the digestibility data.

equal to $x_0 = 20\%$. Estimates and other useful values were computed in Example 5.3 (p. 110). In particular, the predicted digestibility percent was computed to

$$\hat{y}_0 = \hat{\alpha} + \hat{\beta} \cdot x_0 = 96.5334 - 0.9337 \cdot 20 = 77.859,$$

and the prediction interval becomes

$$95\% \text{ PI}: \quad 77.859 \pm t_{0.975,7} \cdot s \cdot \sqrt{1 + \frac{1}{9} + \frac{(20 - 14.5889)^2}{1028.549}}$$
$$= 77.859 \pm 2.365 \cdot 2.970 \cdot 1.0675$$
$$= 77.859 \pm 7.498$$
$$= (70.361, 85.357).$$

Hence, for a new observation with stearic acid level 20%, we would expect the digestibility percent to be between 70.4 and 85.4 with a probability of 95%.

We could make the same computations for other values of x_0 and plot the lower and upper limits of the prediction intervals. This is done in Figure 7.5 with dashed curves. The confidence limits are plotted as dotted curves. As we knew, the prediction intervals are wider than the confidence intervals. Also notice that the confidence bands and the prediction bands are not straight lines: the closer x_0 is to $\bar{x}$, the more precise the prediction — reflecting that there is more information close to $\bar{x}$. □

7.2.3   Prediction in the one sample case and in one-way ANOVA

Sometimes we are interested in predicting new observations from other types of models. Consider first the one sample model

$$y_i = \mu + e_i, \quad i = 1, \ldots, n,$$

where $e_1, \ldots, e_n$ are iid. $N(0, \sigma^2)$, and suppose that we want to predict the outcome of a new observation. A new observation may be written as $y_0 = \mu + e_0$, where $e_0 \sim N(0, \sigma^2)$. The prediction is $\hat{y}_0 = \hat{\mu}$ with 95% prediction interval

$$95\% \text{ PI}: \quad \hat{\mu} \pm t_{0.975, n-1} \cdot s \cdot \sqrt{1 + \frac{1}{n}}.$$

Comparing this to the confidence interval (5.19) for μ, we see that the confidence interval is adjusted by adding a one under the square root in order to also take into account the observation error e_0 of the new observation. This is the same as we did for linear regression.

Example 7.10. Beer content in cans (continued from p. 196). In the beer production example, assume that the mean and standard deviation are estimated to $\hat{\mu} = 332.5$ and $s = 6.2$ based on $n = 50$ observations. The 95% prediction interval is

$$95\% \text{ PI}: \quad 332.5 \pm 2.010 \cdot s \cdot \sqrt{1 + \frac{1}{50}} = 332.5 \pm 12.6 = (319.9, 345.1),$$

so 95% of the cans in the production will contain between 320 and 345 ml of beer.   □

Similarly, let us consider the one-way ANOVA model,

$$y_i = \alpha_{g(i)} + e_i, \quad i = 1, \ldots, n$$

with the usual assumptions, and suppose that we want to predict the outcome of a new observation in group j. This new observation can be written as $y_0 = \alpha_j + e_0$. The prediction is $\hat{y} = \hat{\alpha}_j$ and the corresponding 95% prediction interval is

$$95\% \text{ PI}: \quad \hat{\alpha}_j \pm t_{0.975, n-k} \cdot s \cdot \sqrt{1 + \frac{1}{n_j}},$$

where s is the pooled standard deviation.

More generally, let

$$y_i = \mu_i + e_i, \quad i = 1, \ldots, n,$$

where $\mu_i = f(x_i, \theta_1, \ldots, \theta_p)$ is a function of parameters $\theta_1, \ldots, \theta_p$ and the value of an explanatory variable x, and $e_1, \ldots, e_n$ are as usual. Consider a new observation for which the explanatory variable is x_0. Then the new observation is written as $y_0 = f(x_0, \theta_1, \ldots, \theta_p) + e_0$ and the prediction is

$$\hat{y} = f(x_0, \hat{\theta}_1, \ldots, \hat{\theta}_p).$$

Recall that the corresponding standard error has the form $s \cdot \sqrt{k}$, where k is a constant that does not depend on the data; *cf.* equation (5.12). Then the prediction interval is

$$95\% \text{ PI}: \quad \hat{y} \pm t_{0.975, n-p} \cdot s \cdot \sqrt{1+k}$$

and we see that the extra observational error e_0 is taken into account exactly as in the previous situations.

Example 7.11. Vitamin A intake and BMR (continued from p. 129). Recall the BMR data from Example 4.6 (p. 82). BMR is a variable related to the basal metabolic rate. In Example 5.12 (p. 129) we computed a confidence interval for the difference in expected value between men and women and concluded that it was very narrow due to the large samples (1079 men and 1145 women). The 95% confidence intervals for the expected values for each sex are narrow, too. For men we get

$$7.386 \pm 1.961 \cdot 0.723 \cdot \sqrt{\frac{1}{1079}} = 7.386 \pm 0.043 = (7.343, 7.429).$$

See Example 5.12 (p. 129) for values used in the formula. The corresponding prediction interval is

$$7.386 \pm 1.961 \cdot 0.723 \cdot \sqrt{1 + \frac{1}{1079}} = 7.386 \pm 1.418 = (5.968, 8.804),$$

which is much wider because the random variation for a single person is not affected by the large sample size. Notice how the prediction interval resembles the upper left panel of Figure 4.9 very well: the interval corresponds to the central 95% of the fitted distribution because the contribution from the estimation error is small due to the large sample size. □

7.3 R

7.3.1 Residual analysis

Predicted values and standardized residuals are extracted from a model object by the functions `fitted()` and `rstandard()`, respectively. For the digestibility data (Example 7.1, p. 187) we get the following predicted values and standardized residuals:

```
> modelDigest <- lm(digest~stearic.acid)   # Linear regression
> fitted(modelDigest)                        # Pred./fitted values
        1        2        3        4        5        6
 68.70786 68.24099 75.43080 79.07240 82.71399 92.70502
        7        8        9
 92.42490 93.91889 92.98515
> rstandard(modelDigest)                     # Std. residuals
         1          2          3          4          5
 -0.4990400  0.9858697 -1.2704454 -0.3144279  1.5304524
         6          7          8          9
 -1.0679066 -0.4645465 -0.3175299  1.4199368
```

These vectors can then be used to produce plots similar to those of Figure 7.1:

```
> plot(fitted(modelDigest), rstandard(modelDigest)) # Res. plot
> qqnorm(rstandard(modelDigest))          # QQ-plot of std. res.
> abline(0,1)                             # Compare to y=x line
```

The plot() command produces the residual plot; that is, a scatter plot of standardized residuals against predicted values. The qqnorm() command produces a QQ-plot that compares the empirical quantiles to the theoretical quantiles from the normal distribution. Finally, the abline() command adds the straight line with intercept zero and slope one to the QQ-plot.

Note that the predicted values may also be extracted with the predict() function and that the "raw" (not standardized) residuals r_i are extracted with the residuals() function.

7.3.2 Prediction

As described above, the fitted() function as well as the predict() function are applicable for computation of predicted values corresponding to the observations in a dataset. The function predict() is more generally applicable, however, and can also be used for prediction of new observations.

Consider again the digestibility data and assume that we wish to predict the digestibility percent for levels 10, 20, and 25 of stearic acid. The predicted values are calculated in the linear regression model as follows:

```
> new <- data.frame(stearic.acid=c(10,20,25))  # New values of
                                                # stearic acid
> new
  stearic.acid
1           10
2           20
3           25

> predict(modelDigest, new)  # Predictions of digestibility
```

```
       1        2        3
87.19595 77.85853 73.18982
```

First, a new data frame with the new values of the explanatory variables is constructed and then printed for clarity. It has three observations and a single variable, stearic.acid. It is important that the variable has the same name as the explanatory variable in the original dataset. The predict() command asks R to calculate the predicted values.

Most often we are not interested in the predicted values on their own, but also confidence or prediction intervals. The level of the intervals may be changed with the level option.

```
> predict(modelDigest, new, interval="prediction")  # 95% PI
       fit      lwr      upr
1 87.19595 79.72413 94.66776
2 77.85853 70.36037 85.35669
3 73.18982 65.44275 80.93689
> predict(modelDigest, new, interval="confidence")  # 95% CI
       fit      lwr      upr
1 87.19595 84.64803 89.74386
2 77.85853 75.23436 80.48270
3 73.18982 69.92165 76.45800
> predict(modelDigest, new, interval="prediction", level=0.90)
                                                    # 90% PI
       fit      lwr      upr
1 87.19595 81.20940 93.18250
2 77.85853 71.85087 83.86619
3 73.18982 66.98273 79.39691
```

Compare the output to the results from (7.4) and (7.5).

7.4   Exercises

7.1 ® **Age and body fat percentage.** In order to relate the body fat percentage to age, researchers selected nine healthy adults and determined their body fat percentage (Skovgaard, 2004). The table shows the results.

Age	23	28	38	44	50	53	57	59	60
Body fat %	19.2	16.6	32.5	29.1	32.8	42.0	32.0	34.6	40.5

1. Make an appropriate plot of the data that illustrates the relationship between age and fat percentage.

   2. Specify a statistical model and fit it with `lm()`.

   3. Make the residual plot; *i.e.*, a scatter plot of standardized residuals against predicted values. Does the model seem to be appropriate?

   4. Make a scatter plot of standardized residuals against age. Compare the plot to the plot from question 2 and explain the difference. Does it make a difference for the conclusion regarding model validation which of the plots we consider?

      [Hint: See the comment just above Example 7.1 (p. 187).]

   5. A 50 year old person has his fat percent measured to 42%. Is that an unusual value?

      [Hint: What kind of interval says something about the likely values for a new subject?]

7.2 ® **Tumor size.** An experiment involved 21 cancer tumors (Shin et al., 2004). For each tumor the weight was registered as well as the emitted radioactivity obtained with a special medical technique (scintigraphic images). Assume that we are interested in prediction of the tumor weight from the radioactivity. We want to use a linear regression model, and we use only data from 18 of the tumors since the linear relation seems to be unreasonable for large values of the radioactivity.

The external file `cancer2.txt` has three variables: `id` (tumor identification number), `tumorwgt` (tumor weight), and `radioact` (radioactivity).

   1. Which variable should be used as response and which should be used as explanatory variable when the purpose is prediction of the tumor weight? Make a scatter plot of the data and fit a linear regression model to the data.

   2. Make a residual plot for the model. Is it reasonable to assume variance homogeneity?

   3. Try to find a transformation of the tumor weight variable such that a linear regression model on the transformed data fits the data in a satisfactory way.

   4. Use the transformed model to predict the tumor weight for a tumor that has a radioactivity value of 8.

      [Hint: Make the prediction on the transformed scale and transform the prediction back to the original scale.]

7.3 **Hatching of cuckoo eggs.** Cuckoos place their eggs in other birds' nests for hatching and rearing. Several observations indicate that cuckoos choose the "adoptive parents" carefully, such that the cuckoo eggs are similar in size and appearance to the eggs of the adoptive species.

In order to examine this further, researchers investigated 154 cuckoo eggs and measured their size (Latter, 1905). The unit is half millimeters. The width of the eggs ranges from 30 half millimeters to 35 half millimeters. The eggs were adopted by three different species: wrens, redstarts, and whitethroats.

The distribution of eggs of different sizes is given for each adoptive species in the table. The data are stored in the R-dataset cuckoo, a part of which is shown in the beginning of the R-output appearing after the questions.

Adoptive	Width of egg (half mm)					
species	30	31	32	33	34	35
Wren	3	11	19	7	10	4
Redstart	0	5	13	11	6	1
Whitethroat	2	2	17	19	22	2

1. Make sure you understand the structure of the data — in the table as well as in the R-dataset. In particular, what are n, k, and the n_j's?

2. Analyze the data and draw conclusions. This includes: specification of the statistical model, hypotheses, and p-values; numerical results in the form of relevant estimates, confidence intervals, and interpretations and conclusions.

3. Compute a 95% prediction interval for the width of a random cuckoo egg found in a wren nest. Same question for redstart and whitethroat nests. What is the interpretation of the intervals? Why do the intervals not have the same length?

R-output:

```
> cuckoo                    # The dataset
       spec width
1      wren    30
2      wren    30
3      wren    30
.
.      [More lines here]
.
153 whitethroat    35
154 whitethroat    35
```

```
> model1 <- lm(width~spec)   # One-way ANOVA,
                             # redstart as reference group

> anova(model1)
Analysis of Variance Table

Response: width
          Df  Sum Sq Mean Sq F value  Pr(>F)
spec        2  10.268   5.134  3.7491 0.02576 *
Residuals 151 206.771   1.369

> summary(model1)

Coefficients:
                 Estimate Std. Error t value Pr(>|t|)
(Intercept)       32.5833     0.1950 167.067   <2e-16 ***
specwhitethroat    0.4010     0.2438   1.645    0.102
specwren          -0.1759     0.2518  -0.699    0.486

Residual standard error: 1.17 on 151 degrees of freedom
Multiple R-squared: 0.04731,Adjusted R-squared: 0.03469
F-statistic: 3.749 on 2 and 151 DF,  p-value: 0.02576

> model2 <- lm(width~spec-1)   # One-way ANOVA,
                               # no intercept term

> summary(model2)

Coefficients:
                 Estimate Std. Error t value Pr(>|t|)
specredstart      32.5833     0.1950   167.1   <2e-16 ***
specwhitethroat   32.9844     0.1463   225.5   <2e-16 ***
specwren          32.4074     0.1592   203.5   <2e-16 ***

Residual standard error: 1.17 on 151 degrees of freedom
Multiple R-squared: 0.9987, Adjusted R-squared: 0.9987
F-statistic: 4.006e+04 on 3 and 151 DF, p-value: <2.2e-16
```

7.4 ® **Pillbugs.** This exercise is about the same data as Case 2, Part II (p. 355). An experiment on the effect of different stimuli was carried out with 60 pillbugs (Samuels and Witmer, 2003). The bugs were split into three groups: 20 bugs were exposed to strong light, 20 bugs were exposed to moisture, and 20 bugs were used as controls. For each bug it was registered how many seconds it used to move six inches.

The external dataset `pillbug.txt` contains two variables, `time` and `group`.

1. Make a new variable, `logtime<-log(time)`, with the log-transformed values. Make two sets of parallel boxplots with `group` as explanatory variable: one with `time` as response and one with `logtime` as response. Which variable seems to be the more appropriate for a one-way ANOVA analysis? Why? (This examination is also part of Case 2.)

2. Fit a one-way ANOVA model with `time` as response variable and make the corresponding residual plot and QQ-plot of standardized residuals.

3. Fit a one-way ANOVA model with `logtime` as response variable and make the corresponding residual plot and QQ-plot of standardized residuals.

4. Based on the residual analysis, which of the models seem to be most appropriate? Compare to your answer in question 1.

5. Is there an effect of exposure? If so, is there a significant effect of both light and moisture?

   [Hint: Which model should you use to answer the questions, *cf.* question 4?]

6. Compute an estimate and a 95% confidence interval for the expected effect of the light treatment, measured on the logarithmic time scale. What does this imply for the effect of light, measured on the original time scale?

   [Hint for the second part: How should you "reverse" the results from the log-scale to the original time scale?]

7. A new pillbug is exposed to light, and it takes it 50 seconds to move the 6 inches. Is that unusual?

   [Hint: You should compute a prediction interval. Why? Which model should you use? How do you reverse the result to the original scale?]

7.5 ® **Seal populations.** The number of seals in a population were counted each year during a period of 11 years (Verzani, 2005). The counts are listed in the table below and the data are available on the external file `seal.txt` with variables `year` and `size`.

Year	1952	1953	1954	1955	1956	1957
Population	724	176	920	1392	1392	1448
Year	1958	1959	1960	1961	1962	
Population	1212	1672	2068	1980	2116	

1. Plot the size of the population against year and fit the corresponding linear regression model. What is the interpretation of the slope parameter? Has the size of the seal population changed significantly during the period?

2. What is the predicted size of the population for year 1963? Compute also the corresponding 95% confidence interval and the 95% prediction interval. What are the interpretations of the two intervals? Compare the intervals.

3. Compute confidence intervals and prediction intervals for each year in the dataset (1952–1962) and add them to the scatter plot from question 1. (See Figure 7.5 in the notes for a similar plot.)

4. Can you use the data to predict the size of the seal population for year 2010?

7.6 **Prediction of crab weight.** Recall the crab weight data from Example 4.1. Compute a 95% and a 90% prediction interval for the weight of a random crab from the same population.

Chapter 8

Linear normal models

In most experiments there are several variables that possibly affect the response and thus should be included in the statistical model. In the previous chapters we have seen several examples on models and data analyses but there has been only one explanatory variable. Linear regression dealt with the description of one quantitative variable as a linear function of another quantitative variable and one-way analysis of variance modeled the average effect of a quantitative variable for a number of pre-determined groups corresponding to a categorical explanatory variable.

While these two types of models may seem different at first, they are actually both special cases of a general class of models called the *linear models*. In particular, they have one thing in common: the models assume that the residuals are normally distributed. In this chapter we will first present extensions to both linear regression and one-way analysis of variance and then present the general framework for linear models.

8.1 Multiple linear regression

Linear regression models the relationship between the response variable, y, and a single explanatory variable, x. In multiple linear regression we assume that several continuous explanatory variables, or covariates, are measured for each observational unit. If we denote the d covariate measurements for unit i as $x_{ij}, j = 1, \ldots, d$ then the *multiple linear regression* model is defined by

$$y_i = \alpha + \beta_1 x_{i1} + \cdots + \beta_d x_{id} + e_i, \quad i = 1, \ldots, n, \quad (8.1)$$

where the residuals are assumed to be independent and normally distributed; *i.e.*,

$$e_i \sim N(0, \sigma^2).$$

The regression parameters $\beta_1, \ldots, \beta_d$ are interpreted as ordinary regression parameters: a unit change in x_k corresponds to an expected change in y of β_k if we assume that all other variables remain unchanged. Thus we can think of $\beta_1, \ldots, \beta_d$ as partial slopes and the intercept α as the expected value of an observation where $x_1 = \cdots = x_d = 0$.

The parameter estimates from (8.1) can be found using the least squares technique, as previously described in Sections 2.4.2 and 5.2.1, while the estimate of the variance can be obtained from the general formula (5.7), where p is the number of explanatory covariates plus 1. We need to add 1 since we also include the intercept, α, in the model (8.1), so in total there are $d + 1$ parameters: $\alpha, \beta_1, \ldots, \beta_d$.

Hypotheses for the multiple regression model typically consist of setting a single parameter to zero (*e.g.*, $H_0 : \beta_j = 0$), which corresponds to no effect or no influence of variable x_j on y given that the other explanatory variables remain in the model. This is a simple hypothesis, as described in Chapter 6, so we can test that with a t-test,

$$T_{\text{obs}} = \frac{\hat{\beta}_j - 0}{\text{SE}(\hat{\beta}_j)}, \tag{8.2}$$

which is t distributed with $n - d - 1$ degrees of freedom.

The standard error $\text{SE}(\hat{\beta}_j)$ in the denominator in (8.2) has the general form seen in (5.12) and the constant k_j only depends on the model and the data structure. It can be quite laborious to calculate this constant by hand for models more complicated than simple linear regression models because it involves matrix inversion of a $(d + 1) \times (d + 1)$ matrix. Suffice to say that it is possible and quite easy to do the calculations on a computer and that the general form resembles the approach in Section 5.2.3, where the squared deviance of the regression variable x_j is taken into account.

Example 8.1. Volume of cherry trees. The value of trees is determined by their volume, but it is difficult to measure the volume of a tree without cutting it down. The tree diameter and height are easy and cheap to measure without cutting down the tree, and the primary purpose of this experiment was to predict the tree volume from the diameter and height in order to be able to estimate the value of a group of trees without felling. The girth (diameter in inches $4\frac{1}{2}$ feet above ground), height (in feet), and volume (in cubic feet) were measured for each of 31 cherry trees. The data are shown in Table 8.1 and are from Ryan Jr. et al. (1985).

We start our initial exploration of the data by producing graphs that show the relationship between the variables. Figure 8.1 shows the relationship between height and volume and between diameter and volume. The left panel of Figure 8.2 shows the standardized residual plot obtained from the multiple linear regression model (8.1) with two explanatory variables:

$$v_i = \alpha + \beta_1 \cdot h_i + \beta_2 \cdot d_i + e_i, \qquad i = 1, \ldots, 31,$$

where $e_i \sim N(0, \sigma^2)$, and where v_i is the volume, h_i is the height, and d_i is the diameter.

The residual plot shown in the left panel of Figure 8.2 suggests that the residuals are not homogeneous: there are small standardized residuals for

Table 8.1: Dataset on diameter, volume, and height for 31 cherry trees

Tree	Diameter	Height	Volume	Tree	Diameter	Height	Volume
1	8.3	70	10.3	17	12.9	85	33.8
2	8.6	65	10.3	18	13.3	86	27.4
3	8.8	63	10.2	19	13.7	71	25.7
4	10.5	72	16.4	20	13.8	64	24.9
5	10.7	81	18.8	21	14.0	78	34.5
6	10.8	83	19.7	22	14.2	80	31.7
7	11.0	66	15.6	23	14.5	74	36.3
8	11.0	75	18.2	24	16.0	72	38.3
9	11.1	80	22.6	25	16.3	77	42.6
10	11.2	75	19.9	26	17.3	81	55.4
11	11.3	79	24.2	27	17.5	82	55.7
12	11.4	76	21.0	28	17.9	80	58.3
13	11.4	76	21.4	29	18.0	80	51.5
14	11.7	69	21.3	30	18.0	80	51.0
15	12.0	75	19.1	31	20.6	87	77.0
16	12.9	74	22.2				

small predicted values and slightly larger standardized residuals for large predicted values. Heterogeneity of the residuals is always a cause of concern, and in this case we might improve the fit if we can find a reasonable way to transform the response variable and/or the explanatory variables.

In this case we can use a well-known geometric shape to create an approximate model for the shape of a tree.

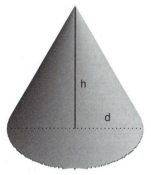

If we assume that the trunk of a tree can be viewed as a cone with diameter d and height h, we can use the result from geometry that gives the volume of a cone as

$$v = \frac{\pi}{12} \cdot h \cdot d^2. \qquad (8.3)$$

This model may work well for some tree species like conifer, which are generally all cone shaped, but it may not be entirely appropriate for other tree species.

Hence, we consider the following extension of (8.3):

$$v = c \cdot h^{\beta_1} \cdot d^{\beta_2}, \qquad (8.4)$$

where we place no restrictions on the exponents of d and h, and where the constant, c, can vary. If we take natural logarithms on both sides of the equa-

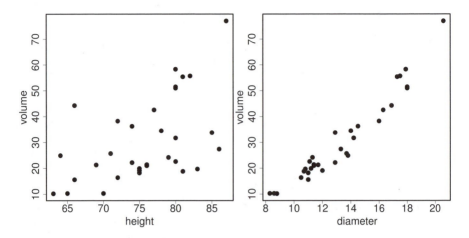

Figure 8.1: Scatter plot of volume against height (left panel) and volume against girth (right panel) for 31 cherry trees.

tion, we get the following model:

$$\log v_i = \alpha + \beta_1 \log h_i + \beta_2 \log d_i + e_i, \quad i = 1, \ldots, n, \qquad (8.5)$$

where $\alpha = \log c$ and where $e_1, \ldots, e_n$ are independent $N(0, \sigma^2)$. The model in (8.5) has the form of a multiple linear regression with two explanatory variables, $\log d_i$ and $\log h_i$, corresponding to x_{1i} and x_{2i} in (8.1).

The right panel of Figure 8.2 shows the standardized residual plot for the transformed data. If we compare this residual plot to the residual plot for the original model (left panel), it is clear that the fit has improved — the variance is more homogeneous and there is no apparent structure of the residuals. The estimates for the mean parameters from model (8.5) are calculated using least squares and are

$$\hat{\alpha} = -4.9907, \hat{\beta}_1 = 0.6490, \hat{\beta}_2 = 2.1381 \text{ and } \hat{\sigma} = 0.1151.$$

We use a computer to calculate the individual standard errors, $SE(\hat{\alpha})$, $SE(\hat{\beta}_1)$, and $SE(\hat{\beta}_2)$, and we can summarize the result in a table:

Parameter	Estimate	SE	T_{obs}	p-value
α	-6.6316	0.7998	-8.292	5.06e-05
β_1	1.1171	0.2044	5.464	7.81e-06
β_2	1.9827	0.0750	26.432	< 2e-16

The p-values are calculated from the t-test statistics by looking in a t distribution with $31 - 3 = 28$ degrees of freedom. Based on the summary table we conclude that both the height and the diameter are significant and therefore that if we want to model the tree volume, we get the best model when we include information on both diameter and height.

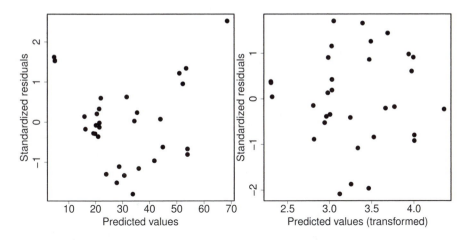

Figure 8.2: Residual plots for cherry tree data (left panel) and log-transformed cherry tree data (right panel).

The parameter estimates are based on the log-transformed values, and we can back-transform α using the exponential function in order to get the value to use with (8.4); *i.e.*, $\hat{c} = \exp(\hat{\alpha}) = 0.001318$. However, we can already now notice that both $\hat{\beta}_1$ and $\hat{\beta}_2$ are positive and rather close to the values from the mathematical representation (8.3), so the model predicts that the volume of trees increases with both diameter and height. Had we received negative values for any of these parameters we should certainly doubt the model given our knowledge about trees! □

The statistical model under the null hypothesis $H_0 : \beta_j = 0$ is a multiple linear regression model without the jth covariate, but with the other covariates remaining in the model. The test therefore examines if the jth covariate contributes to the explanation of variation in y *when the association between y and other covariates has been taken into account*. The last part is important for the conclusion of the test. Assume for a moment that the hypothesis for $\beta_2 = 0$ in the cherry tree example had not been rejected. The conclusion would *not* be that there is no association between height and volume, but rather that height does not provide extra information on the tree volume when the information from the diameter of the tree has been accounted for.

This discussion is closely related to the concept of *multicollinearity*, which has to do with dependence among covariates. Collinearity is a linear relationship between two explanatory variables and multicollinearity refers to the situation where two or more explanatory variables are highly correlated. For example, two covariates (*e.g.*, height and weight) may measure different aspects of the same thing (*e.g.*, size). More generally, covariates may almost be linear combinations of each other.

Multicollinearity may give rise to spurious results. For example, you may find estimates with the opposite sign compared to what you would expect, unrealistically high standard errors, and insignificant effects of covariates that you would expect to be significant. The problem is that it is hard to distinguish the effect of one of the covariates from the others. The model fits more or less equally well no matter if the effect is measured through one or the other variable. Multicollinearity results in large standard errors of the related explanatory variables but only affects the interpretation of the effect of the explanatory variables — not the ability of the model to predict values.

Example 8.2. Nutritional composition. Consider a study on low-fat milk products which is carried out in order to examine the effect of the nutritional composition on taste. A number of products have been scored by 20 subjects, and the average score (over the subjects) is the response. The explanatory variables consist of x_1, x_2, and x_3, where x_1 and x_2 denote the carbohydrates and protein content per 100 grams of the product and x_3 is the energy content in kJ (kiloJoule) per 100 grams. Then

$$x_3 \approx 17 \cdot x_1 + 17 \cdot x_2$$

since 1 gram of carbohydrate as well as 1 gram of protein corresponds to 17 kJ and since the products are low-fat products. An analysis may yield a non-significant effect of carbohydrate because the effect of this has already been taken into account through the other covariates.   □

Multiple linear regression models enable us to use the same machinery as previously and at the same time to include more than one explanatory quantitative variable. This allows for more complicated (and often more realistic) models where several factors influence the response variable. Moreover, it enables us to describe relationships between two variables that are not just straight lines but polynomials.

For example, we can model a quadratic relationship between two variables using a multiple regression model. The quadratic formula is

$$y = \alpha + \beta_1 \cdot x + \beta_2 \cdot x^2$$

so the *quadratic regression model* is given by

$$y_i = \alpha + \beta_1 x_i + \beta_2 x_i^2 + e_i, \quad i = 1, \ldots, n, \tag{8.6}$$

which is equivalent to (8.1) if we let the error terms be iid. $N(0, \sigma^2)$ and we set $x_{i1} = x_i$ and $x_{i2} = x_i^2$. This is a special case of the multiple regression model (8.1), so we can use the same approach as earlier. In particular, we can test if a quadratic model fits better than a straight line model if we test the hypothesis $H_0 : \beta_2 = 0$. If we reject the null hypothesis we must conclude that the quadratic model fits the data better than the simpler straight line model. If we fail to reject the null hypothesis we might as well use a straight line model to describe the relationship between the two variables.

Example 8.3. Tensile strength of Kraft paper. The following data come from Joglekar et al. (1989) and show the tensile strength in pound-force per square inch of Kraft paper (used in brown paper bags) for various amounts of hardwood contents in the paper pulp.

Hardwood	Strength	Hardwood	Strength	Hardwood	Strength
1.0	6.3	5.5	34.0	11.0	52.5
1.5	11.1	6.0	38.1	12.0	48.0
2.0	20.0	6.5	39.9	13.0	42.8
3.0	24.0	7.0	42.0	14.0	27.8
4.0	26.1	8.0	46.1	15.0	21.9
4.5	30.0	9.0	53.1		
5.0	33.8	10.0	52.0		

Figure 8.3 shows the paper strength plotted as a function of hardwood content and the residual plot from the quadratic regression model:

$$\text{strength}_i = \alpha + \beta_1 \cdot \text{hardwood}_i + \beta_2 \cdot \text{hardwood}_i^2 + e_i, \quad i = 1, \ldots, n,$$

with $e_i \sim N(0, \sigma^2)$. When we look at the observed strength as a function of hardwood content, it is clear that the strength of the paper starts to decline after the hardwood reaches 11% and that we need a model that is able to capture this change. The residual plot in Figure 8.3 shows a decent fit. There are maybe a little too few observations for the small predicted values to say anything definitive about variance homogeneity, but the general picture is fine.

The least squares estimates and their corresponding standard errors are

Parameter	Estimate	SE	t	p-value
α	-6.67419	3.39971	-1.963	0.0673
β_1	11.76401	1.00278	11.731	2.85e-09
β_2	-0.63455	0.06179	-10.270	1.89e-08

One thing worth noting here is that the parameter for the quadratic term, β_2, is highly significant. If we calculate the test statistic for the hypothesis $H_0 : \beta_2 = 0$, we get

$$T_{\text{obs}} = \frac{-0.63455}{0.06179} = -10.270,$$

which results in a p-value < 0.0001 if we compare the test statistic to a t distribution with $19 - 3 = 16$ degrees of freedom. This tells us that the quadratic regression model is significantly better than a simple linear regression model since we reject the hypothesis $H_0 : \beta_2 = 0$, which corresponds to a model reduction to the simple linear regression model. This is hardly surprising given the graph in Figure 8.3. $\qquad \square$

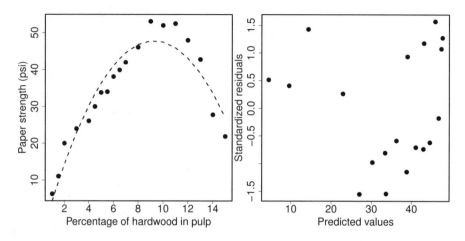

Figure 8.3: Left panel shows paper strength of Kraft paper as a function of hardwood contents in the pulp with the fitted quadratic regression line superimposed. Right panel is the residual plot for the quadratic regression model.

8.2   Additive two-way analysis of variance

In Chapters 3 and 5 we modeled the mean value of the observations for each category defined by one categorical explanatory variable. We can extend the one-way analysis of variance discussed in those chapters to a multi-way analysis of variance in the same way that we extended the simple linear regression model to account for multiple explanatory variables in Section 8.1. In the following we will look more closely at the two-way analysis of variance case, but the results apply to the more general multi-way analysis of variance model as well.

The *two-way analysis of variance* (also called the additive two-way analysis of variance model for reasons which will become clear in Section 8.4) is a special case of the multi-way analysis of variance, where there are two categorical explanatory variables. Let $g(i)$ and $h(i)$ denote the functions that define the groups of the two categorical variables for observation i, and consider the model

$$y_i = \alpha_{g(i)} + \beta_{h(i)} + e_i, \quad i = 1, \ldots, n. \tag{8.7}$$

Estimation of the mean parameters for a two-way analysis of variance is done using least squares, and the variance estimate is obtained from the general formula (5.7), where p is the effective number of parameters. The variance estimate for the additive two-way analysis of variance model is

$$s^2 = \frac{SS_e}{df_e} = \frac{1}{n - (k_1 + k_2 - 1)} \sum_i (y_i - \hat{y}_i)^2$$

$$= \frac{1}{n - (k_1 + k_2 - 1)} \sum_i (y_i - \hat{\alpha}_{g(i)} - \hat{\beta}_{h(i)})^2,$$

which has the exact same form we have seen in the previous chapters.

At first sight we might presume that there were $k_1 + k_2$ parameters in the model. However, one of these parameters is redundant and the effective number of parameters is $k_1 + k_2 - 1$, which can be seen as follows: We start by choosing one of the cells in the two-way table or design to be a reference cell and let the first parameter be the expected value in that cell. Then, for each of the remaining $k_1 - 1$ levels of variable x_1, we add a parameter for the contrast or difference between this level and the reference level. That yields $k_1 - 1$ extra parameters. Similarly, we add a parameter for each of the possible contrasts for variable x_2, which results in an extra $k_2 - 1$ parameters for a total of $1 + (k_1 - 1) + (k_2 - 1)$ parameters. Thus, the number of parameters in an additive two-way analysis of variance model is $k_1 + k_2 - 1$. For additive multi-way analysis of variance models the effective number of parameters is $1 + (k_1 - 1) + (k_2 - 1) + \cdots + (k_d - 1) = k_1 + k_2 + \cdots + k_d - (d - 1)$. We can see that this formula also fits with the results from Chapter 3, where we had a single parameter for each category. Section 8.3.2 below discusses the number of parameters for multi-way analysis of variance models and linear models in more detail.

Example 8.4. Cucumber disease. Consider the following dataset from de Neergaard et al. (1993) about a greenhouse experiment that was undertaken to examine how the spread of a disease in cucumbers depends on climate and amount of fertilizer. Two different climates were used: (A) change to day temperature 3 hours before sunrise and (B) normal change to day temperature. Fertilizer was applied in 3 different doses: 2.0, 3.5, and 4.0 units. The amount of infection on standardized plants was recorded after a number of days, and two plants were examined for each combination of climate and dose.

Infection	Climate	Dose	Infection	Climate	Dose
51.5573	A	2.0	48.8981	B	2.0
51.6001	A	2.0	60.1747	B	2.0
47.9937	A	3.5	48.2108	B	3.5
48.3387	A	3.5	51.0017	B	3.5
57.9171	A	4.0	55.4369	B	4.0
51.3147	A	4.0	51.1251	B	4.0

As we can see, we have 2 categorical variables: climate (with 2 possible categories) and dose (with 3 possible categories). If we index the observations from top to bottom, with climate A first followed by climate B, we get that

$$g(1) = g(2) = \cdots = g(6) = A, \ g(7) = \cdots = g(12) = B$$

and that

$$
\begin{aligned}
h(1) = h(2) = h(7) = h(8) &= 2.0 \\
h(3) = h(4) = h(9) = h(10) &= 3.5 \\
h(5) = h(6) = h(11) = h(12) &= 4.0.
\end{aligned}
$$

One way to think about the design of a two-way analysis of variance model is that we can place each of our observations in exactly one cell of the two-way table shown in Figure 8.2 and that the two-way analysis of variance model enables us to predict the average level of infection for every combination of dose and climate.

Table 8.2: Two-way table showing infection rate in cucumbers for different combinations of climate and fertilizer dose

		Dose	
Climate	2.0	3.5	4.0
A	51.5573	47.9937	57.9171
	51.6001	48.3387	51.3147
B	48.8981	48.2108	55.4369
	60.1747	51.0017	51.1251

There are $3 + 2 - 1 = 4$ parameters in the model, which might seem strange at first glance since there are six cells in Table 8.2, but that is because additive models only accommodate contrasts between different levels of a categorical variable. Because the model imposes restrictions to the expected values, we have that the average change from climate A to climate B for dose 2.0 has to be the same as the average change from A to B for doses 3.5 and 4.0.   □

Often we want to compare several parameters simultaneously, which corresponds to testing hypotheses like

$$H_0 : \alpha_1 = \cdots = \alpha_{k_1}$$

or

$$H_1 : \beta_1 = \cdots = \beta_{k_2}.$$

The alternative hypotheses are that at least two α's are unequal or that at least two of the β's are different, respectively. These two hypotheses are analogous to their one-way analysis of variance counterparts and state that there is no difference among the levels of the first and second explanatory variables, respectively. Just as in Section 6.3.1, the mean squared deviations are used to test the hypotheses. Define

$$MS_{x_1} = \frac{SS_{x_1}}{k_1 - 1} = \frac{1}{k_1 - 1} \sum_{j=1}^{k_1} n_{x_{1_j}} (\bar{y}_{x_{1_j}} - \bar{y})^2,$$

where x_1 is the first of the explanatory variables. The hypothesis H_0 is tested by comparing the test statistic

$$F_1 = \frac{MS_{x_1}}{MS_e}$$

to an F distribution with $(k_1 - 1, n - k_1 - k_2 + 1)$ degrees of freedom. Similarly, the hypothesis H_1 is based on the test statistic

$$F_2 = \frac{MS_{x_2}}{MS_e},$$

where x_2 is the second explanatory variable and where

$$MS_{x_2} = \frac{SS_{x_2}}{k_2 - 1} = \frac{1}{k_2 - 1} \sum_{j=1}^{k_2} n_{x_{2_j}} (\bar{y}_{x_{2_j}} - \bar{y})^2.$$

F_2 follows an F distribution with $(k_2 - 1, n - k_1 - k_2 + 1)$ degrees of freedom. Hypothesis tests in a multi-way analysis of variance are carried out completely analogously to the procedure described in Section 6.3.1. In both cases the F test statistic is the ratio between mean squares for the variable of interest and the residual mean squares. Notice that the residual mean squares is different for the one-way and two-way analysis of variance since in the two-way analysis of variance the effect of both grouping variables has been taken into account. All of the steps undertaken in the analysis can be summarized in an analysis of variance table, as shown in Table 8.3. Compared to the one-way analysis of variance in Table 6.2, we just add an extra line representing the contribution from the second explanatory variable.

Example 8.5. Cucumber disease (continued from p. 217). The two hypotheses of interest for the cucumber data are

$$H_0 : \alpha_A = \alpha_B$$

and

$$H_1 : \beta_{2.0} = \beta_{3.5} = \beta_{4.0}.$$

In order to test these hypotheses, we calculate the mean squared deviations

$$\begin{aligned} MS_{climate} &= \frac{SS_{climate}}{k_1 - 1} = \frac{1}{k_1 - 1} \sum_{j=1}^{k_1} n_{climate_j} (\bar{y}_{climate_j} - \bar{y})^2 \\ &= \frac{1}{1} \left[6 \cdot (\bar{y}_A - \bar{y})^2 + 6 \cdot (\bar{y}_B - \bar{y})^2 \right] \\ &= 6 \cdot (51.45 - 51.96)^2 + 6 \cdot (52.47 - 51.96)^2 = 3.127 \end{aligned}$$

Table 8.3: Analysis of variance table for the additive two-way model of the cucumber data

Variation	SS	df	MS	F_{obs}	p-value
Between climates	3.1270	1	3.1270	0.23	0.6434
Between doses	58.4264	2	29.2132	2.16	0.1776
Residual	108.1042	8	13.5130		

and

$$\begin{aligned}
\text{MS}_{\text{dose}} &= \frac{\text{SS}_{\text{dose}}}{k_2 - 1} = \frac{1}{k_2 - 1} \sum_{j=1}^{k_2} n_{\text{dose}_j} (\bar{y}_{\text{dose}_j} - \bar{y})^2 \\
&= \frac{1}{2} \left[4 \cdot (\bar{y}_{2.0} - \bar{y})^2 + 4 \cdot (\bar{y}_{3.5} - \bar{y})^2 + 4 \cdot (\bar{y}_{4.0} - \bar{y})^2 \right] \\
&= 2 \cdot \left[(53.06 - 51.96)^2 + (48.89 - 51.96)^2 + (53.95 - 51.96)^2 \right] \\
&= 29.213
\end{aligned}$$

and the test statistics are calculated as

$$F_{obs} = \frac{\text{MS}_{\text{climate}}}{\text{MS}_e} = \frac{3.1270}{13.5130} = 0.23$$

and

$$F_{obs} = \frac{\text{MS}_{\text{dose}}}{\text{MS}_e} = \frac{29.2132}{13.5130} = 2.16,$$

which are F distributed with $(1,8)$ and $(2,8)$ degrees of freedom, respectively. □

A consequence of the two-way additive model (8.7) is that the contrast between any two levels for one of the explanatory variables is the same for every category of the *other* explanatory variable. In order to see this, let Y_{1j} and Y_{2j} be two observations from climates A and B, respectively, but from the same dose. Then

$$Y_{1j} = \alpha_1 + \beta_j + e_1$$

and

$$Y_{2j} = \alpha_2 + \beta_j + e_2,$$

where e_1, e_2 are independent and $N(0, \sigma^2)$. The difference becomes

$$Y_{1j} - Y_{2j} = \alpha_1 + \beta_j + e_1 - (\alpha_2 + \beta_j + e_2) = \alpha_1 - \alpha_2 + e_1 - e_2,$$

which has mean $\alpha_1 - \alpha_2$ regardless of the dose category j. Thus the average difference between climates for dose 2.0 is the same as the difference between the climates for dose 3.5 and likewise for dose 4.0.

Table 8.4: Tristimulus brightness measurements of pork chops from 10 pigs at 1, 4, and 6 days after storage

Day	\multicolumn{10}{c}{Pig number}
	1	2	3	4	5	6	7	8	9	10
1	51.6	53.4	59.7	52.6	59.6	62.5	52.6	54.1	49.3	54.5
4	56.3	54.0	61.5	52.2	59.3	60.1	52.7	56.6	52.3	56.3
6	54.8	53.9	59.9	51.0	64.1	61.4	55.6	63.9	51.6	58.0

We can estimate the contrasts as in Section 5.2.5,

$$\widehat{\alpha_j - \alpha_l} = \bar{y}_j - \bar{y}_l,$$

where $\bar{y}_j$ and $\bar{y}_l$ are the marginal means of all the observations that belong to category j and l, respectively (for that explanatory variable). Likewise, the standard error is calculated as in formula (5.15):

$$\text{SE}(\hat{\alpha}_j - \hat{\alpha}_l) = s \sqrt{\left(\frac{1}{n_j} + \frac{1}{n_l} \right)}, \tag{8.8}$$

with n_j and n_l being the number of observations that belong to category j and l, respectively. The same calculations can be carried through for the β parameters.

Example 8.6. Pork color over time. Juncker et al. (1996) investigated the meat quality of pork by examining the color stability of pork chops. The investigators seek to examine if there is a systematic change in the brightness from a tristimulus color measurement. The color was measured from a pork chop from each of ten pigs at days 1, 4, and 6 after storage. Data are shown in Table 8.4.

Note that we here have two categorical explanatory variables: day and pig. The data are plotted in the left panel of Figure 8.4 with an *interaction plot*, where the change over days for each of the 10 pigs is plotted as a line. From Figure 8.4 there appears to be some small increase in brightness over time, but we can also see that the change is highly variable from pig to pig.

We include pig as an explanatory variable because we suspect that meat brightness might depend on which specific pig the pork was cut from: meat from some pigs has one natural color while meat from another pig might have another color. Hence, there could be an effect of "pig", and we seek to account for that by including pig in the model even though we are not particularly interested in being able to compare any pair of pigs like, say, pig 2 and pig 7. Excluding the explanatory variable pig from the analysis might blur the effects of day since the variation among pigs may be much larger than the variation between days.

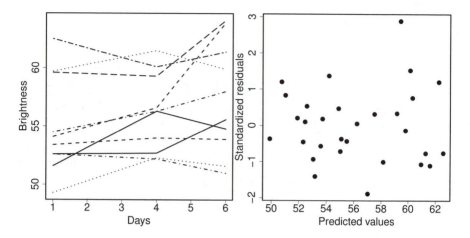

Figure 8.4: Interaction plot of the change in meat brightness for 10 pigs measured at days 1, 4, and 6 after storage. Right panel shows the residual plot for the two-way analysis of variance of the pork data.

Mathematically, we write the model as

$$y_i = \alpha_{p(i)} + \beta_{d(i)} + e_i, \quad i = 1, \ldots, N, \tag{8.9}$$

where $e_i \sim N(0, \sigma^2)$. The p-function defines which pig (levels $1, \ldots, 10$) each of the 30 observations corresponds to, while d is the similar function for days (levels 1, 4, and 6).

The right panel of Figure 8.4 shows the residual plot for the two-way analysis of variance. Apart from the single outlier with a standardized residual around 4, the residual plot gives us no reason to distrust the model. The outlier turns out to be a genuine observation, so we have no reason to discard it from the dataset, and we continue with the analysis.

The primary hypothesis of interest is

$$H_0 : \beta_1 = \beta_4 = \beta_6,$$

which corresponds to no change in brightness over time. The test statistic for the hypothesis H_0 is shown in the analysis of variance (Table 8.5), and we can conclude that there is a borderline significant effect of days. The test for pigs can also be seen in the analysis of variance table, and while this may be of little interest for the manufacturers producing pork, since that is nothing they can control, we can still see that it is highly significant. That means that there is a large variation among pigs and that we certainly would be wrong not to take that variation into account when we analyze the effect of days.

Now that we found a borderline significant effect of days, we should quantify this effect, by presenting the estimates of the contrasts so we can

Table 8.5: Analysis of variance table for the additive two-way model for the pig brightness data

Variation	SS	df	MS	F_{obs}	p-value
Between days	29.56	2	14.78	3.7174	0.04452
Between pigs	395.05	9	43.89	11.0395	<0.0001
Residual	71.57	18	3.98		

conclude which levels of days are different. The p-value from the test of H_0, 0.04452, only tells us that not all days have the same level. The contrasts give us more information about the different days and we can summarize the contrasts in a table:

Contrast	Estimate	SE	T_{obs}	p-value
$\widehat{\beta_4 - \beta_1}$	1.14	0.8918	1.278	0.2174
$\widehat{\beta_6 - \beta_1}$	2.43	0.8918	2.725	0.0139
$\widehat{\beta_6 - \beta_4}$	1.29	0.8918	1.447	0.1652

From the contrasts we see that the difference in days primarily stems from a difference between days 1 and 6. We can also see that the brightness scores decrease as time increases since the contrasts are all positive. The average difference in brightness between days 1 and 6 is 2.43. □

Example 8.6 is an example of a *block experiment* with pigs as "blocks". Sometimes observational units are grouped in such blocks and the observational units within a block are expected to be more similar than observations from different blocks. In Example 8.6 we expect observations taken on the same pig to be potentially more similar than observations taken on different pigs.

For block designs we need to take the blocking information into account in the statistical model, and this is exactly what we did in Example 8.6 by including an effect of pig in the model. We may care very little about identifying differences between two specific pigs, say, pig 3 and 7, but we need to make sure that the model at least allows for a pig effect *if* we believe that each animal could have its own influence on the response variable.

In agricultural/geographic research, "blocks" are often made up of observations taken from the same geographical area. For example, if we have multiple observations from a number of different fields or plots, then fields or plots take the role of blocks. In animal research we might have observations on more than one animal from a number of herds and then herds take the role of blocks; *i.e.*, it groups together observations that are potentially more identical than two random observations.

8.2.1   The additive multi-way analysis of variance

This section will briefly sketch how we can extend the two-way analysis of variance model to the more general additive multi-way analysis of variance model.

Let us assume that for each observational unit we have information on d different categorical variables, $x_1, \ldots, x_d$, besides the observed response y. The categorical variables need not have the same number of categories, so x_1 could have k_1 possible categories, x_2 could have k_2 possible categories, and so forth.

The general analysis of variance model is then given by

$$y_i = \alpha_{g(i)} + \beta_{h(i)} + \cdots + \gamma_{l(i)} + e_i, \quad i = 1, \ldots, n, \tag{8.10}$$

where the residuals are assumed to be independent and normally distributed, $e_i \sim N(0, \sigma^2)$. The grouping functions, $g, \ldots, l$ are used to identify to which combination of categories an observation belongs. The estimation techniques, tests, and contrasts discussed for the two-way analysis of variance model can be applied directly to the additive multi-way analysis of variance model.

8.2.2   Analysis of variance as linear regression

Recall that the general analysis of variance model (8.10) was given by

$$y_i = \alpha_{g(i)} + \beta_{h(i)} + \cdots + \gamma_{l(i)} + e_i, \quad i = 1, \ldots, n. \tag{8.11}$$

Thus for each explanatory variable we have a set of parameters that define the average level for the given category.

Each of the explanatory variables can be rewritten as a sum of *dummy variables* or *indicator variables*; i.e., variables that can only take the values 0 or 1. Consider for example the first explanatory variable and assume for now that it has k_1 categories such that we have k_1 parameters $\alpha_1, \ldots, \alpha_{k_1}$. We then define the dummy variables $x_{i1}^1, \ldots, x_{ik_1}^1$ in the following way:

$$x_{ij}^1 = \begin{cases} 1 & \text{if observation } i \text{ belongs to category } j \text{ of the first variable} \\ 0 & \text{otherwise} \end{cases}$$

which enables us to rewrite the term $\alpha_{g(i)}$ as $\alpha_1 \cdot x_{i1}^1 + \alpha_2 \cdot x_{i2}^1 + \cdots + \alpha_{k_1} \cdot x_{ik_1}^1$.

We can define similar dummy variables for each of the explanatory variables in the model. The analysis of variance model (8.11) is then equivalent to the model

$$y_i = \alpha_1 \cdot x_{i1}^1 + \alpha_2 \cdot x_{i2}^1 + \cdots + \alpha_{k_1} \cdot x_{ik_1}^1 + \beta_1 \cdot x_{i1}^2 + \cdots + \gamma_{k_d} \cdot x_{ik_d}^d + e_i, \tag{8.12}$$

where $k_1, k_2, \ldots, k_d$ are the number of categories for each of the d explanatory variables and where the superscript on the dummy variables, x, refers to which explanatory variable is used to define the dummy variable.

The formulation in (8.12) has the same representation as the multiple linear regression model (8.1), where the dummy variables take the place of the explanatory variables. Hence, by introducing dummy variables we can think of an analysis of variance model as a multiple linear regression model.

Example 8.7. Pork color over time (continued from p. 221). For the pork quality dataset we had two categorical explanatory variables: pig and day. The two-way analysis of variance model (8.9) is written as

$$y_i = \alpha_{p(i)} + \beta_{d(i)} + e_i, \quad i = 1, \ldots, n, \tag{8.13}$$

where the grouping functions p and d define the number of categories for each variable and therefore the number of parameters in the model. Since "pig" has 10 categories and "day" has 3 categories, we end up with parameters $\alpha_1, \ldots, \alpha_{10}, \beta_1, \beta_4$, and β_6. A given observation must belong to exactly one of the 10 pig categories and exactly one of the 3 day categories.

Now define a set of dummy variables, x_{ij}^{pig}, with $i = 1, \ldots, n$ and $j \in \{1, 2, \ldots, 10\}$ such that

$$x_{ij}^{pig} = \begin{cases} 1 & \text{if observation } i \text{ came from pig } j \\ 0 & \text{otherwise} \end{cases}$$

and another set of dummy variables, $x_{ij}^{day}, i = 1, \ldots, n, j \in \{1, 4, 6\}$, to replace the "day" variable

$$x_{ij}^{day} = \begin{cases} 1 & \text{if observation } i \text{ was taken on day } j \\ 0 & \text{otherwise} \end{cases}.$$

The table below illustrates the relationship between the original categorical variables and the dummy variables for this dataset:

		Dummy variables						
Pig	Day	x_1^{day}	x_4^{day}	x_6^{day}	x_1^{pig}	x_2^{pig}	$\cdots$	x_{10}^{pig}
1	1	1	0	0	1	0	$\cdots$	0
1	4	0	1	0	1	0	$\cdots$	0
1	6	0	0	1	1	0	$\cdots$	0
2	1	1	0	0	0	1	$\cdots$	0
2	4	0	1	0	0	1	$\cdots$	0
$\vdots$	$\vdots$	$\vdots$	$\vdots$	$\vdots$	$\vdots$	$\vdots$	$\ddots$	$\vdots$
10	6	0	0	1	0	0	$\cdots$	1

If we look at the original parameterization (8.9), then the first observation (taken on pig 1 at day 1) would be

$$y_1 = \alpha_1 + \beta_1 + e_1,$$

from which we can see that the mean value of y_1 is given by the sum of α_1 and β_1. If we instead write the same model using the dummy variables, we would get

$$
\begin{aligned}
y_i = {} & \alpha_1 \cdot x_{i1}^{\text{pig}} + \alpha_2 \cdot x_{i2}^{\text{pig}} + \cdots + \alpha_{10} \cdot x_{i10}^{\text{pig}} \\
& + \beta_1 \cdot x_{i1}^{\text{day}} + \beta_4 \cdot x_{i4}^{\text{day}} + \beta_6 \cdot x_{i6}^{\text{day}} + e_i.
\end{aligned}
\tag{8.14}
$$

From the definition of the dummy variables we see that for observation 1 there are exactly two dummy variables from the entire set $(x_{11}^{\text{pig}}, \ldots, x_{110}^{\text{pig}}, x_{11}^{\text{day}}, \ldots, x_{16}^{\text{day}})$ which are non-zero; namely, x_{11}^{pig} and x_{11}^{day}. Hence (8.14) reduces to

$$
y_1 = \alpha_1 \cdot 1 + \beta_1 \cdot 1 + e_1
$$

for observation 1, which is exactly identical to the parameterization we had above. The same result follows for each of the n observations.   □

The reparameterization of the analysis of variance model is quite important because it means that an analysis of variance model can be expressed in the multiple linear regression framework. It makes it possible to combine the analysis of variance models (*i.e.*, models with categorical explanatory variables) with multiple regression models into a single class of models: the linear model. So why is it called a *linear model*? Clearly the dummy variables cannot be considered linear as they attain only two different values. The reason for this term is that the parameters enter the statistical model in a linear fashion. Basically, we can write any linear regression model as $y_i = \sum_{j=1}^{p} \beta_j \cdot x_{ij}$, where p is the number of variables (regular continuous variables as well as dummy variables). It is worth emphasizing that we place no real restriction on the x's. The actual variables need not be continuous, and this very general class of models can be used to model complicated relationships as long as the model formula can be written on the form $y_I = \sum \beta_j f_j(x_{ij})$, where the functions f_j are known.

8.3   Linear models

The *linear model* or *linear normal model* for an observation y_i is given by

$$
y_i = \sum_{j=1}^{d} \beta_j \cdot x_{ij} + e_i, \ i = 1, \ldots, n,
\tag{8.15}
$$

where the error terms are assumed to be iid. $e_i \sim N(0, \sigma^2)$. Note that although the linear model (8.15) has the same form as a standard regression model, we learned from the previous section that categorical explanatory variables can be written in this form as well, so the linear model accommodates multiple categorical and quantitative explanatory variables simultaneously.

Specifying models in detail with formula (8.15) can be quite tedious if there are several explanatory variables — particularly if there are explanatory variables with several possible categories. One simplification is to write continuous explanatory variables of the form $\beta \cdot x$, to show that they have the interpretation of a proper regression slope, and write categorical explanatory variables as in (8.10) with a subscript indicating that the function assigns the category for each observation. A model that includes two categorical and two continuous explanatory variables may be written as

$$y_i = \alpha_{g(i)} + \gamma_{h(i)} + \beta_1 \cdot x_{i1} + \beta_2 \cdot x_{i2} + e_i. \tag{8.16}$$

This notation is simpler and easier to interpret than (8.15) but still requires a lot of definitions of parameters, grouping functions, *etc.* A much simpler standard for writing models is to use the statistical model formulas introduced in Section 5.1.2.

8.3.1 Model formulas

We already introduced model formulas for simple linear models in Section 5.1.2, but it is not until now, when several explanatory variables are in play, that they come to their proper right.

In Example 8.1 we investigated whether tree volume could be explained by girth and height, and in Example 8.6 we tried to explain pork color as a function of pig and storage time. For both situations we can write the proper mathematical model, (8.5) and (8.9), respectively, but we would like to introduce a more informal way of writing model formulae. The cherry tree model can be written as

```
volume = girth + height
```

while the pork color model would be

```
colour score = day + pig.
```

On the left-hand side of these model formulas is printed the response variable, while the explanatory variables are printed on the right-hand side. Note that we cannot differentiate between categorical variables (*e.g.*, day number and pig) and continuous variables (tree girth and height) from the model formulae alone. Information about the data types and the distribution of the residuals must be provided from the context or from elsewhere. Another thing that is hidden by the model formulas is the number of parameters that each explanatory variable introduces to the model. Tree girth and height were both continuous and therefore each resembles a single parameter (the partial slope effects of girth and height), while day and pig number correspond to 3 and 10 parameters, respectively.

The advantage of specifying statistical models using model formulas is that they are very condensed and directly specify the relationship that is modeled in a way that is easily understood. In the cherry tree example we might say "tree volume is modeled as a function of girth and height". In the following we shall use the model formulas as well as the model specifications (8.15) and (8.16) interchangeably.

Note that in general an intercept or grand mean is always understood to be included in the model formulas. Thus for the tree example

```
volume = girth + height
```

we have a standard multiple regression model that includes three parameters: intercept, partial slope of girth, and partial slope of height. Likewise, for models that contain a categorical variable, an intercept is always included and the intercept acts as a reference level in these situations*. If we want to specify a regression model formula with no intercept, we do so explicitly in the model formula by including a term "-1". Hence

```
volume = girth + height - 1
```

denotes a multiple regression formula which goes through the origin.

8.3.2   Estimation and parameterization

Mean parameters in the linear model can be estimated using least squares exactly as explained in Section 5.2.1. However, in some models with multiple categorical explanatory variables we end up with a model that is *overparameterized*; i.e., a model where we have more parameters than we are able to estimate.

We already discussed different parameterizations for one-way analysis of variance models in Sections 5.5.2 and 6.6. Let us return to that situation again and assume we wish to compare the means of three groups. We can parameterize this model in two ways: use a parameter to describe the mean for each group, or use one parameter to describe the mean of one of the groups (the reference group) and then use two parameters to describe the difference in group levels between the remaining two groups and the mean of the reference group. This is illustrated in Figure 8.5. The two parameterizations hold the same amount of information, and we can go from one parameterization to the other simply by adding or subtracting parameters.

---

*Different computer programs parameterize their models differently, and the interpretation of intercept can change accordingly although the models are the same. In both R and SAS, the default parameterization is to set one of the categories to a reference level and then parameterize the remaining categories as contrasts relative to that reference level.

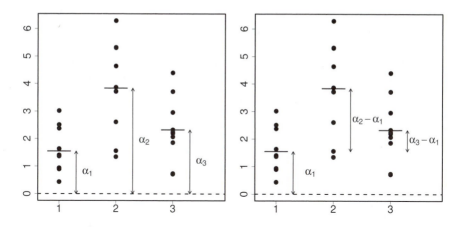

Figure 8.5: Different parameterizations for comparison of means in three groups. In the left panel we have three parameters — α_1, α_2, and α_3 — that each describe the average level in groups 1–3, respectively. In the right panel we have one parameter that describes the average level in group 1 and two parameters — the difference $\alpha_2 - \alpha_1$ and the difference $\alpha_3 - \alpha_1$ — that describe contrasts relative to the mean values of group 1.

Let us now consider the situation where we have two categorical variables, each with two categories. The additive model is given by

$$y_i = \alpha_{g(i)} + \beta_{h(i)} + e_i, \quad i = 1, \ldots, n. \tag{8.17}$$

We can think of this design as, for example, a feeding experiment where we wish to examine the average weight of some animal and where we have a reference feeding strategy and two substances we can add to the food. If we denote the average weight for the reference group by μ and the average increase in weight for substances 1 and 2 by α and β, respectively, then we can summarize the average values for the groups defined by different feeding strategies as shown in the design below:

	Substance 1	
Substance 2	Not added	Added
Not added	μ	$\mu + \alpha$
Added	$\mu + \beta$	$\mu + \alpha + \beta$

Here we have made one assumption: the effects of the two substances do not influence each other. This assumption is seen when we add both substance 1 and substance 2 to the feed since we expect the average weight to be $\mu + \alpha + \beta$. The design corresponds to the two-way additive setup from Section 8.2, and it is worth noting that even though we have four possible combinations of the two substances, we can describe the average value of all four combinations by only three parameters: μ, α, and β.

All statistical computer packages are able to handle overparameterization, and they automatically set one level of a categorical variable as a reference level and model the remaining levels as a set of contrasts relative to the reference level. When we look at the output from those programs we do not have to worry about overparameterization, but we should be very careful with the interpretation of the parameter estimates. Parameter estimates that belong to quantitative explanatory variables should be interpreted as partial regression slopes, while estimates for categorical variables should be read as contrasts relative to the reference category.

The variance estimate is estimated from the residuals as always,

$$s^2 = \frac{SS_e}{df_e} = \frac{1}{n-p} \sum_i (y_i - \hat{y}_i)^2, \tag{8.18}$$

where p is the effective number of parameters in the model. We have already seen how we should calculate p for multiple regression and for additive multi-way analysis of variance models, but if we use a statistical computer package it is automatically able to determine the effective number of parameters for us — even for complicated models[†].

8.3.3   Hypothesis testing in linear models

We use the same approach for hypothesis testing in linear models as we have used previously in the text. If we test a hypothesis that consists of a single restriction of a parameter,

$$H_0 : \beta_j = 0,$$

we use a t-test where we compare the test statistic

$$T_{obs} = \frac{\hat{\beta}_j - 0}{SE(\hat{\beta}_j)}$$

to a t distribution with $n - p$ degrees of freedom, where p is the effective number of parameters. If we have a compound hypothesis where we test restrictions on more that one parameter related to a variable X, *e.g.*,

$$H_1 : \beta_A = \cdots = \beta_D,$$

we use an F-test like (6.9) where we compare two models — the full model and the model under the hypothesis H_1:

$$F_{obs} = \frac{(SS_0 - SS_{full})/(df_0 - df_{full})}{SS_{full}/df_{full}}.$$

[†]We will not go into details about the technique used to determine the effective number of parameters but only provide a brief hint: the model can be written in matrix form $Y = X\beta + e$, where Y is the vector of responses, X is the design matrix that contains information from the covariates, β is a set of k parameters, and e is the vector of random normal errors. The effective number of parameters is the rank of the design matrix X.

The F-test statistic is F distributed with a pair of degrees of freedom that is easily computed by a statistical software package. This approach also works for more complicated hypotheses, and restrictions on parameters corresponding to several explanatory variables can also be tested with F tests. Note that the number of degrees of freedom can be quite counterintuitive for more complicated models but the algorithms in the statistical packages easily determine the number of degrees of freedom for both the numerator and the denominator.

One of the most important features about linear models is that they can accommodate multiple explanatory variables, so we can build complex models where several variables influence our response. However, when we have multiple explanatory variables there may not be an obvious sequence of hypotheses that we should test.

The additive two-way analysis of variance model (8.7) can be written as

$$y = A + B$$

where A and B are our two variables. Based on this model we can either test the hypothesis

$$H_0 : \alpha_1 = \cdots = \alpha_{k_1}$$

or the hypothesis

$$H_1 : \beta_1 = \cdots = \beta_{k_2},$$

where the α's and β's are the parameters belonging to A and B, respectively. If we test the hypothesis H_0 but fail to reject it, we essentially can use the model

$$y = B$$

where we have removed the insignificant terms, so only B remains in the model. Likewise, the hypothesis H_1 corresponds to the model

$$y = A$$

Both of these hypotheses may be equally interesting, but there is no way of knowing if we should test H_0 before H_1 or vice versa. If we are more interested in one hypothesis over another we should test the less interesting hypothesis first.

It is worth emphasizing that we need to *refit the model every single time we have reduced the complexity of the model*. Thus we cannot use the estimate for the residual standard error from the model $y = A + B$ if we fail to reject the hypothesis H_0. If we fail to reject H_0, then $y = B$ is the best model and we should determine estimates and contrasts for the β parameters in that model and not from the initial model $y = A + B$.

It should be clear why we need to refit the model if we look at the residual standard error. In the initial two-way analysis of variance model $y = A + B$ we estimate the residual standard error with $n - k_1 - k_2 + 1$ degrees of freedom. During our analysis we find out that we fail to reject

H_0, so we can safely use $y = B$ to describe our data. In this model we estimate the residual standard error with $n - k_1$ degrees of freedom, and since $n - k_1 > n - k_1 - k_2 + 1$, we expect this to be a slightly better estimate of the residual standard error since we use the data to determine fewer mean parameters.

Infobox 8.1: Model reduction steps

- Whenever we remove an explanatory variable from a model we should refit the new and reduced model before we use any estimates to draw conclusions.

- In particular, when we have multiple explanatory variables in the model we should remove them successively and refit the (reduced) model after each parameter has been removed.

- A statistical analysis begins with a starting model to describe the response variable as a function of the explanatory variable(s) and then a series of steps to try to simplify the model as much as possible. Parameter estimates and conclusions are based on the final model.

Example 8.8. Model parameterizations. This example presents a slightly different way to think about parameters and models, and it should illustrate how different statistical models correspond to different parameterizations for two-way analysis of variance models. The four models described below are illustrated graphically in Figure 8.6, albeit with four categories for the variable B to illustrate the difference between models more clearly.

The additive two-way analysis of variance model is given by

$$y = A + B$$

where A and B are the two variables. Let us assume that A has two categories and that B has three categories. We can specify the mean of each of the observations in the 3×2 table:

		A	
		1	2
	1	μ	$\mu + \alpha$
B	2	$\mu + \beta_2$	$\mu + \alpha + \beta_2$
	3	$\mu + \beta_3$	$\mu + \alpha + \beta_3$

Here, μ is the expected value for the reference group, α is the contrast between the two categories for variable A, and the parameters β_2 and β_3 represent the contrasts (relative to category 1) for variable B. We can see that we have four effective parameters for this model — namely μ, α, β_2, and β_3 —

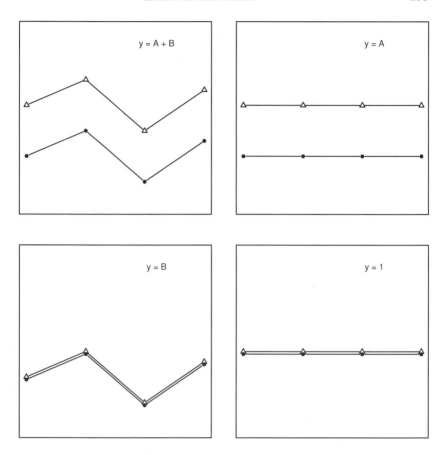

Figure 8.6: Graphical illustration of four different statistical models. The points show the expected values for different combinations of two categorical variables A (represented by the two different symbols, • and △) and B (the change along the x-axis). Upper left has an additive effect of both A and B (y = A + B). In the upper right panel there is only an effect of A, y = A, while the lower left figure corresponds to the model y = B. The lower right panel is the model with no effect of A or B, y = 1.

and that the number of parameters matches the formula $1 + (k_1 - 1) + (k_2 - 1) = 1 + 2 + 1 = 4$ we saw on p. 217.

The hypothesis that there is no difference between the two levels of A corresponds to the model

$$y = B.$$

This hypothesis sets $\alpha = 0$ and gives the design

		A	
		1	2
	1	μ	μ
B	2	$\mu + \beta_2$	$\mu + \beta_2$
	3	$\mu + \beta_3$	$\mu + \beta_3$

If we compare the two previous designs, we see that the model is reduced to a simple one-way analysis of variance, where we have exactly one parameter for each level of B regardless of which category of A the observations belong to. Likewise, the model y = A results in the following design of expected means:

		A	
		1	2
	1	μ	$\mu + \alpha$
B	2	μ	$\mu + \alpha$
	3	μ	$\mu + \alpha$

where we have two parameters — one for the expected value of the first category and one for the contrast.

Finally, we can look at the model

$$y = 1$$

where we have no explanatory variables (sometimes it is also written as "y = " because there are no explanatory variables). This model is the same as the one-sample situation since we have the same expected mean level for every possible combination of categories.

		A	
		1	2
	1	μ	μ
B	2	μ	μ
	3	μ	μ

Notice how the hypotheses correspond to a change from one design (or table of expected values) to another design and how the hypotheses place restrictions on the parameters. If we, for example, start with the model y = A + B, then the hypothesis H_0 corresponds to setting each of the contrasts for A equal to zero. Thus, restrictions placed by specific hypotheses influence the expected values for different combinations of the categorical explanatory variables. □

8.4 Interactions between variables

Until now we have solely discussed additive analysis of variance models, where the contrast between two levels from one variable is the same regardless of the value of the other explanatory variables. In certain situations, however, it might be reasonable to assume that the effect of one variable might depend on the value of another explanatory variable.

We know from one-way analysis of variance that we need more than one observation for each level in order to estimate the within-group variation. The same is true for the levels defined by the interaction term, so in order to fit a model which includes an interaction term we need to have at least two observations for each level of the interaction variable. If we only have one observation for each level of the interaction, then we can only fit the additive model.

8.4.1 Interactions between categorical variables

Assume that we wish to analyze the blood pressure of men and women at ages 40, 50, 60, and 70 years. We can model this as an additive two-way analysis of variance with two explanatory variables: gender (with two categories) and age (with four categories),

```
blood pressure = gender + age
```

This additive model assumes that the expected difference in blood pressure between men and women is the same for every age group. However, we might wish to employ a model where the difference in blood pressure between men and women depends on the age group, such that the difference can increase (or decrease) with age. Then the additive model is not adequate and we need to use a model that allows the effect of age to depend on the gender.

In general, a set of explanatory variables is said to interact if the effect of one variable depends on the level(s) of the other variable(s) in the set. In the example above we believe that the difference between men and women (*i.e.,* the change in effect between men and women) might depend on the value of age.

We can extend the additive model by adding an *interaction* term, which for the two-way analysis of variance model is written as

$$y_i = \alpha_{g(i)} + \beta_{h(i)} + \gamma_{g(i),h(i)} + e_i, \quad i = 1, \ldots, n. \tag{8.19}$$

Here the term $\gamma_{g(i),h(i)}$ is the effect of the interaction between level $g(i)$ of the first explanatory variable and level $h(i)$ of the second explanatory variable. In total there are $k_1 \cdot k_2$ different γ-parameters, since every level of the first explanatory variable is combined with every possible level of the second explanatory variable. The variables gender and age have 2 and 4 categories, respectively, so there are $2 \cdot 4 = 8$ parameters in the model that allow for an interaction between them.

The interaction model specified by (8.19) is highly overparameterized, but the parameters introduced by the interaction term "replace" the parameters from the additive model and we can disregard any contributions from the main effects as long as an interaction term exists in the model. The model with interaction allows for $k_1 \cdot k_2$ different expected values — with no restrictions between the expected values. This is exactly described by the γ parameters, so the effective number of parameters in the model is $k_1 \cdot k_2$. In particular, the α and β parameters corresponding to the additive model are redundant since the main effects of the variables are embedded in the interaction term. The interaction model (8.19) is thus highly overparameterized and we may simply write the model as

$$y_i = \gamma_{g(i),h(i)} + e_i, \quad i = 1, \dots, n.$$

This is identical to (8.19) except for the interpretation of the γ's. Thus in the example with blood pressure for different ages and gender, we have a total of 8 parameters in the model — one for each possible combination of gender and age.

We specify interactions in model formulas by "multiplying" variables together with an asterisk:

```
blood pressure = gender + age + gender*age
```

and we say that we model blood pressure as a function of the combined group of age and gender such that the effect of gender depends on the given age (or conversely, that the effect of age depends on the gender).

Note that there needs to be replications for each combination of the two variables that comprise the interaction in order to be able to fit the model. Otherwise, if there is only one observation per cell, then we get a fit where the observed and the fitted values for each cell are identical, so all the residuals end up being equal to zero and we are unable to estimate the standard deviation σ.

Example 8.9. Model parameterizations (continued from p. 232). If we consider the interaction model

```
y = A + B + A*B
```

we will then have the parameterization

		A	
		1	2
	1	γ_{11}	γ_{12}
B	2	γ_{21}	γ_{22}
	3	γ_{31}	γ_{32}

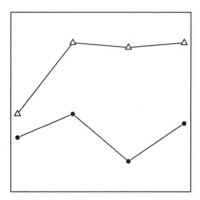

Figure 8.7: Graphical example of the expected values from an interaction between two categorical variables A and B, y = A + B + A*B. The interaction model shown here can be compared to the additive models shown in Figure 8.6.

with a total of $2 \cdot 3 = 6$ parameters. γ_{11} describes the expected mean value for observations that belong to category 1 from A and category 1 from B. γ_{21} is a parameter that describes the expected mean value for observations that belong to category 2 from B and category 1 from A and so forth. We have one parameter for each of the six combinations of A and B and there are no restrictions on these parameters, so essentially the interaction model can be viewed as a one-way analysis with six groups.

Notice that although the model says A + B + A*B, we have only 6 parameters — whenever an interaction term appears in a model, the main effects (A + B) will not introduce any additional parameters since the model is over-parameterized. We can write the same design in the following way, where we extend the two-way additive design we saw in Example 8.8.

		A	
		1	2
	1	μ	$\mu + \alpha$
B	2	$\mu + \beta_2$	$\mu + \alpha + \beta_2 + \gamma_{22}$
	3	$\mu + \beta_3$	$\mu + \alpha + \beta_3 + \gamma_{32}$

Figure 8.7 illustrates how an interaction corresponds to a model where the change in level of one variable depends on the category of another variable. The effects of B are clearly different for the two symbols, • and △ (corresponding to the two groups of A), in Figure 8.7. Alternatively, we can say the

difference in effects represented by the two symbols, • and △, depends on the value of B.   □

We can think of the interaction term as a single categorical explanatory variable where the categories are defined by the combination of terms that make up the combination of the two variables. If we view the interaction terms in this light then we can use the same techniques as in Chapters 5 and 7 to provide inferences about the interaction.

Example 8.10. Pork color over time (continued from p. 221). If we look at the left panel of Figure 8.4 we see that the lines (which each represent a pig) do not appear to be particularly parallel. That suggests that there could be an interaction between pig and day. The interpretation would be that the brightness of pork chops from some pigs changes faster than it does for other pigs.

Unfortunately, because we have only one observation for each of the 30 categories defined by the interaction between the 3 days and 10 pigs, we cannot fit the model

```
colour score = day*pig
```

so we are not able to test if a model with an interaction between pig and day fits better than the additive model we used earlier.   □

Least squares estimation is used to estimate the parameters of (8.19) and to identify the effective number of parameters. The algorithms in statistical software packages are very effective in determining the effective number of parameters, degrees of freedom, *etc.*, and we will not go into more detail here. However, it is worth emphasizing that we now have an extremely flexible toolbox that enables us to specify and make statistical inferences for very complex models.

8.4.2 Hypothesis tests

The same techniques as previously are used for statistical tests of hypotheses involving interactions. If we consider model (8.19), then a null hypothesis related to the interaction term can be written as

$$H_0 : \gamma_{11} = \gamma_{12} = \cdots = \gamma_{k_1 k_2} = 0,$$

where we specify the hypothesis that all interaction terms are the same (and zero); *i.e.*, that the interaction does not change the model. Hence, the null model corresponds to the additive model.

To test this hypothesis of no interaction, we calculate

$$F_{\text{obs}} = \frac{\text{MS}_{\text{interaction}}}{\text{MS}_e}$$

$$= \frac{\left(\text{SS}_{\text{no interaction}} - \text{SS}_{\text{with interaction}}\right) / \left(\text{df}_{\text{no interaction}} - \text{df}_{\text{with interaction}}\right)}{\text{SS}_{\text{with interaction}} / \text{df}_{\text{with interaction}}},$$

which follows an F distribution and is based on the same approach as we saw in (6.9), with subscripts denoting the null model (without interaction) and the full model (with interaction). The numerator and denominator degrees of freedom are easily calculated by a statistical software package. We reject the hypothesis of "no interaction" if the p-value is small and we fail to reject it if the p-value is large.

The only new issue we should be aware of when testing hypotheses in models that include interactions is the *hierarchical principle*:

Infobox 8.2: Hierarchical principle

If a model contains an interaction between categorical variables, then we must always keep lower order interactions and main effects of variables associated with the interaction in the model as well.

Basically, the hierarchical principle tells us that we cannot test a hypothesis about a main effect as long as that main effect is still part of an interaction in the model. Consider the blood pressure model we discussed previously:

```
blood pressure = age + gender + age*gender
```

If $\alpha_{40}, \alpha_{50}, \alpha_{60}$, and α_{70} are the parameters defined by the levels of the age categorical variable, then we are not allowed to test the hypothesis

$$H_0 : \alpha_{40} = \cdots = \alpha_{70}$$

as long as we still have the interaction age*gender in the model. The hypothesis simply does not make sense as long as the interaction is still present, and we can test only H_0 in the additive model

```
blood pressure = age + gender
```

It is not too hard to realize that the hypothesis H_0 has no meaning in the interaction model. Essentially, H_0 states that there is no effect of age. However, at the same time we allow for an interaction term in the model that says that not only is there an effect of age but that this effect also depends on gender. So on the one hand, we are trying to remove age from the model while we still keep it in the model on the other hand. Hence the importance of the hierarchical principle.

Example 8.11. Cucumber disease (continued from p. 217). We wish to analyze the cucumber data with interaction between climate and dose to allow for the possibility that the effect of dose might depend on the climate. Figure 8.8 shows the interaction plot, and it seems that the average difference between the levels of climates A and B changes as the dose increases. If we fit the model with an interaction between dose and climate,

```
disease spread = dose + climate + dose*climate
```

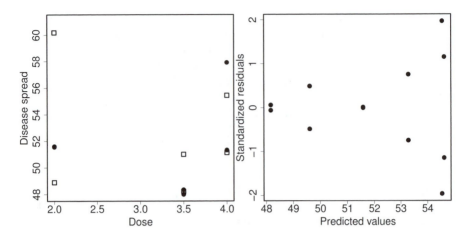

Figure 8.8: Interaction plot, with climate "A" represented by filled circles and climate "B" by open squares (left panel), and standardized residual plot for the interaction model (right panel) for the cucumber disease data.

we get a residual plot (right panel of Figure 8.8) that possibly shows a tendency of increasing variance for larger predicted values; but that tendency is based on a single pair of observations (the difference we see for climate "B" at dose 2.0) and the log-transformed data do not improve the result substantially (figure not shown). We choose to continue with the interaction model and want to test the hypothesis

$$H_0 : \text{no interaction between climate and dose.}$$

We fit the model using a statistical software package and get an analysis of variance table (Table 8.6). The analysis of variance table shows that the in-

Table 8.6: Analysis of variance table for the cucumber data where we include an interaction between dose and climate

	SS	df	MS	F_{obs}	p-value
Dose	58.426	2	29.213	1.7772	0.2477
Climate	3.127	1	3.127	0.1902	0.6780
Dose*climate	9.477	2	4.738	0.2883	0.7594
Residuals	98.627	6	16.438		

teraction between dose and climate is not significant ($p = 0.7594$), and we fail to reject the null hypothesis of no interaction between climate and dose. The variation within each combination of dose and climate is larger than the changes between climates that we observed as the dose increased. We conclude that *if* there is an effect of either dose or climate then that effect is additive. Our next step would then be to analyze the model without interaction between dose and climate,

```
disease spread = dose + climate
```

which has analysis of variance table

	SS	df	MS	F_{obs}	p-value
Dose	58.426	2	29.213	2.1619	0.1776
Climate	3.127	1	3.127	0.2314	0.6434
Residuals	108.104	8	13.513		

From the analysis of variance table we conclude that there appears to be no effect of climate and no effect of dose on disease spread in cucumbers. We remove "Climate", refit the model, and obtain a p-value of 0.1496 for the test of the hypothesis of no effect of dose. Hence we can also remove dose from the model and conclude that there is no difference whatsoever among climates or doses on disease spread in cucumber.

One thing worth noting here is that several statistical software packages list hypotheses and corresponding p-values that do *not* follow the hierarchical principle. If we look at the output listed in Table 8.6, we see that there are test statistics and p-values listed for both dose and climate, although the interaction between dose and climate is still in the model. We cannot stress enough that *these p-values cannot be used for drawing any conclusions about the main effect as long as the interaction is still present!* □

8.4.3 Interactions between categorical and quantitative variables

In the previous section we discussed interactions between two categorical variables. Here we will discuss interactions between a categorical and a quantitative variable.

Interactions between two categorical variables are used to specify a model where the effects of one variable depend on the level of the other variable. When we talk about interactions between categorical and quantitative variables we seek to specify the same kind of dependency. The effect of the quantitative variable is described through the partial slope parameter, so an interaction means that the partial slope corresponding to the quantitative variable depends on the level of the categorical variable.

Estimates and hypothesis tests are calculated and interpreted as for interactions between two categorical variables.

Consider our example from before with blood pressure, but let us now assume that we view age as a quantitative variable instead of a categorical variable. We still write the model as

```
blood pressure = age + gender + age*gender
```

but this model essentially specifies two regression lines with two different intercepts and two different slopes — one for men and one for women. If we test the hypothesis "no interaction between gender and age", then we test the hypothesis that the slope for men is identical to the slope for women. The interaction between age and gender is expressed by a change in slopes, and hence no interaction means that there is the same effect of age for both sexes. A test of

$$\alpha_{men} = \alpha_{women}$$

corresponds to a model where the two regression lines are forced to have the same intercept but can have different slopes. All of these models are illustrated in Figure 8.9.

Example 8.12. Birth weight of boys and girls. Dobson (2001) presented data from a study that was undertaken to investigate how the sex of the baby and the age of the fetus influence birth weight during the last weeks of the pregnancy. Specifically, do baby boys develop differently from baby girls as the fetus gets older?

Figure 8.10 shows the data and the residual plot for the model

```
birth weight = sex + age + sex*age
```

where we consider sex as a categorical variable with two levels and where we view age as a quantitative variable. That means we are initially fitting a regression line to the baby boys and another regression line to the baby girls. There is nothing in the residual plot that raises an alarm, so we continue with the data analysis.

Let us first test the hypothesis that there is no interaction between age and sex. That means that the two regression lines are parallel (the slopes are identical), so the age dependence is the same for both boys and girls. The analysis of variance table is

	SS	df	MS	F_{obs}	p-value
Sex	9575.33	1	9575.33	0.2935	0.5940
Age	1099182.35	1	1099182.35	33.6953	<0.0001
Sex*age	6346.22	1	6346.22	0.1945	0.6639
Residuals	652424.52	20	32621.23		

The p-value for the hypothesis test of no interaction is 0.6639, so we fail to reject the hypothesis. Thus we do not need the two regression slopes to be different, and we can remove the interaction term from the model. We refit the model

```
birth weight = sex + age
```

now with one intercept for boys, one intercept for girls, and a common slope. This model yields the analysis of variance table

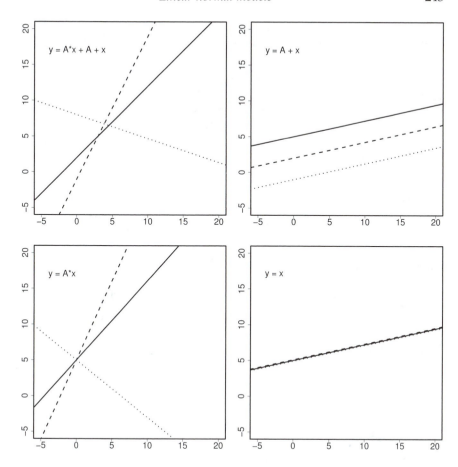

Figure 8.9: Illustration of the possible types of models we can achieve when we have both a categorical, A, and a quantitative variable, x. The upper left figure shows an interaction between the A and x (*i.e.*, the model y = A + x + A*x), where the interaction allows for different slopes and intercepts according to the level of A. The upper right panel shows three parallel lines (*i.e.*, they have the same slope) but with different intercepts, which corresponds to the model y = A + x. The lines in the lower left panel have identical intercepts but different slopes (*i.e.*, y = A*x) while the lines coincide on the lower right figure, so y = x.

	SS	df	MS	F_{obs}	p-value
Sex	157303.68	1	157303.68	5.0145	0.0361
Age	1094939.92	1	1094939.92	34.9040	<0.0001
Residuals	658770.75	21	31370.04		

where we see that both age and sex are significant. The common slope is significantly different from zero (p-value less than 0.0001), and there is a

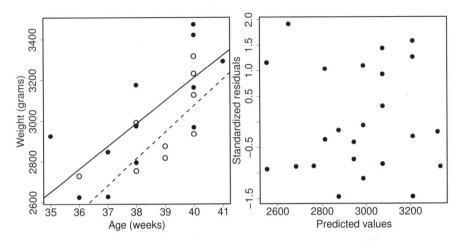

Figure 8.10: Scatter plot of birth weight against age for baby boys (solid dots) and girls (circles). The two lines show the fitted regression lines for boys (solid line) and girls (dashed line). The right panel shows the residual plot for a model with an interaction between sex of the baby and age.

significant difference in levels between boys and girls since the hypothesis $H_1 : \alpha_{\text{boys}} = \alpha_{\text{girls}}$ has a *p*-value of 0.0361. Hence we reject H_1 and we end with the following model to describe our data.

$$\texttt{birth weight = sex + age}$$

This corresponds to

$$y_i = \alpha_{\text{boys}} + \beta_{\text{age}} \cdot \text{age}_i + e_i, \quad y_i = \alpha_{\text{girls}} + \beta_{\text{age}} \cdot \text{age}_i + e_i,$$

for boys and girls, respectively. The estimates for the final model are shown in the table below:

Parameter	Estimate	SE	T_{obs}	*p*-value
α_{girls}	-1773.32	794.59	-2.232	0.0367
$\alpha_{\text{boys}} - \alpha_{\text{girls}}$	163.04	72.81	2.239	0.0361
β_{age}	120.89	20.46	5.908	<0.0001

This regression model states that the birth weight depends on the sex of the baby and the age but that the effect of age is the same for both boys and girls. The estimated regression slope in this model for age is $\hat{\beta}_{\text{age}} = 120.89$ and the estimate for the difference in levels between boys and girls is $\hat{\alpha}_{\text{boys}} - \hat{\alpha}_{\text{girls}} = 163.04$. Thus we can conclude from this study that the birth weight on average increases 120.89 grams per week near the end of the pregnancy and that boys on average weigh 163.04 grams more than girls. □

Finally, consider the following example with a non-standard application of regression analysis.

Table 8.7: Mixture proportions and optical densities for 10 dilutions of serum from mice

Mixture proportion, m	50	150	450	1350	4050	12150
Optical density, y	1.67	1.41	1.05	0.62	0.12	0.03

Example 8.13. ELISA experiment (continued from p. 126). In the ELISA experiment from Example 5.10, the optical density was also measured for 6 dilutions of serum from mice. The results are listed in Table 8.7, and the complete data, consisting of the data from both the standard dissolution and the mice serum, are plotted in the left part of Figure 8.11.

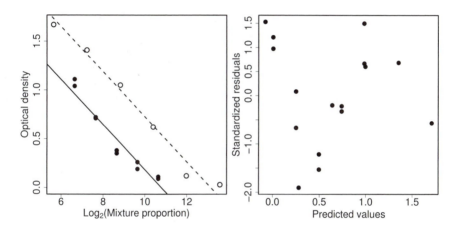

Figure 8.11: Scatter plot of the optical density against the mixture proportions for the standard dissolution (solid dots) and for the mice serum (circles). The regression lines are the fitted lines in the model with equal slopes for the dissolution types. The right panel shows the residual plot for the model where the slopes are allowed to differ.

Similarly to Example 5.10 (p. 126), we use the logarithmic mixture proportions (with base 2) as the continuous explanatory variable. As the initial model for the optical density we use a regression model where both the intercept and the slope are allowed to differ between the two types of dissolutions. This corresponds to model formula

```
optical density = type + log2(mixture) + type*log2(mixture)
```

where `mixture proportion` has been abbreviated to `mixture`. The right part of Figure 8.11 shows the residual plot for this model, which gives us no worries. It turns out that the slopes are not significantly different ($F = 0.47$, $p = 0.50$), so the model can be reduced to the model with a common slope for the two types. This corresponds to the model formula

```
optical density = type + log2(mixture)
```

and we can also write it in terms of intercept and slope parameters:

$$y_i = \alpha_{std} + \beta \cdot \log_2(m_i) + e_i$$
$$y_i = \alpha_{serum} + \beta \cdot \log_2(m_i) + e_i$$

for the two types, respectively. The estimates in this model are summarized as follows:

Parameter	Estimate	SE
α_{std}	2.494	0.111
α_{serum}	3.037	0.125
β	−0.231	0.012

One of the purposes of the analysis was to estimate the ratio between the ubiquitin concentration for the two types of dissolutions. Denote the concentrations C_{std} and C_{serum}, respectively. Assume now that mixture proportions m_{std} and m_{serum} are such that they lead to the same optical density. This means that the concentrations of ubiquitin are the same in the standard dilution with mixture proportion m_{std} and the serum dilution with mixture proportion m_{serum}; *i.e.*,

$$\frac{C_{std}}{m_{std}} = \frac{C_{serum}}{m_{serum}} \quad \text{or} \quad \frac{C_{serum}}{C_{std}} = \frac{m_{serum}}{m_{std}}.$$

The expected optical density is the same for the two mixture proportions, meaning that

$$\alpha_{std} + \beta \cdot \log_2(m_{std}) = \alpha_{serum} + \beta \cdot \log_2(m_{serum}).$$

This is equivalent to

$$\log_2\left(\frac{m_{serum}}{m_{std}}\right) = \frac{\alpha_{std} - \alpha_{serum}}{\beta},$$

which is estimated to $(2.494 - 3.037)/(-0.231) = 2.351$. Hence we estimate the ratio between the ubiquitin concentrations to

$$\frac{C_{serum}}{C_{std}} = \frac{m_{serum}}{m_{std}} = 2^{2.351} = 5.10$$

and conclude that the ubiquitin concentration is five-fold larger in the mice serum compared to the standard dissolution. $\square$

8.5   R

In this section we will see just how powerful and flexible the lm() function in R is. Example 8.1 on p. 210 was about the data from 31 cherry trees. This dataset is part of the R distribution and can be loaded directly with the data() function. A multiple regression model is fitted with lm() simply by writing all the explanatory variables on the right-hand side of the model formula in the call to lm():

```
> data(trees)          # Load the tree dataset from R
> attach(trees)        # and attach it
> model <- lm(Volume ~ Height + Girth)   # Multiple regression
> plot(predict(model), rstandard(model)) # Make residual plot
```

The residual plot generated by the last command is the one seen in the left panel of Figure 8.2. If we wish to analyze the model suggested by the cone geometric shape, (8.5), we can make the transformation directly in the call to lm(). Alternatively, we can define new vectors that contain the transformed variables.

```
> conemodel <- lm(log(Volume) ~ log(Height) + log(Girth))
> summary(conemodel)

Call:
lm(formula = log(Volume) ~ log(Height) + log(Girth))

Residuals:
      Min        1Q    Median        3Q       Max
-0.168561 -0.048488  0.002431  0.063637  0.129223

Coefficients:
            Estimate Std. Error t value Pr(>|t|)
(Intercept) -6.63162    0.79979  -8.292 5.06e-09 ***
log(Height)  1.11712    0.20444   5.464 7.81e-06 ***
log(Girth)   1.98265    0.07501  26.432  < 2e-16 ***
---
Signif. codes:  0 '***' 0.001 '**' 0.01 '*' 0.05 '.' 0.1 ' ' 1

Residual standard error: 0.08139 on 28 degrees of freedom
Multiple R-squared: 0.9777,     Adjusted R-squared: 0.9761
F-statistic: 613.2 on 2 and 28 DF,  p-value: < 2.2e-16
```

In Example 8.3 on p. 215 with strength of paper for paper bags, we used quadratic regression to model the relationship between paper strength and hardwood contents. If we assume that the dataset is read into the data frame paperstr in R, then we can model the quadratic regression as follows:

```
> attach(paperstr)
> paper <- lm(strength ~ hardwood + I(hardwood^2))
> plot(predict(paper), rstandard(paper))    # Make residual plot
> summary(paper)

Call:
lm(formula = strength ~ hardwood + I(hardwood^2))

Residuals:
    Min      1Q  Median      3Q     Max
-5.8503 -3.2482 -0.7267  4.1350  6.5506

Coefficients:
              Estimate Std. Error t value Pr(>|t|)
(Intercept)   -6.67419    3.39971  -1.963   0.0673 .
hardwood      11.76401    1.00278  11.731 2.85e-09 ***
I(hardwood^2) -0.63455    0.06179 -10.270 1.89e-08 ***
---
Signif. codes:  0 '***' 0.001 '**' 0.01 '*' 0.05 '.' 0.1 ' ' 1

Residual standard error: 4.42 on 16 degrees of freedom
Multiple R-squared: 0.9085,     Adjusted R-squared: 0.8971
F-statistic: 79.43 on 2 and 16 DF,  p-value: 4.912e-09
```

We use the inhibit interpretation function, I(), in the call to lm() since the operator ^ already has a meaning in model formulas, and we want it to be interpreted as the power function so the hardwood variable is squared. The I() function can also be used to add two variables inside the formula in lm() if we wish to do that: for the formula ~ z + I(x+y) the first "+" is interpreted as addition of a term in the linear model, while the second "+" is the usual arithmetic operator. Alternatively, we could write

```
> hardwoodsqr <- hardwood^2
> paper <- lm(strength ~ hardwood + hardwoodsqr)
```

The residual plot produces the graph seen in the right-hand panel of Figure 8.3.

We previously used lm() to fit and analyze one-way analysis of variance models in R. Just like we could extend the simple linear regression to multiple regression by adding more explanatory variables in the formula for lm(), we can extend one-way analysis of variance to multi-way analysis of variance in the exact same way: by supplying additional (categorical) explanatory variables to the formula in lm().

In Example 8.6 on p. 221 about meat quality of pork, we have two explanatory variables that both can be analyzed as being categorical: the individual pig and the day.

```
> pig <- rep(1:10, 3)                          # Input pig no.
> day <- rep(c(1, 4, 6), times=c(10, 10, 10))  # Input day no.
> brightness <- c(51.6, 53.4, 59.7, 52.6, 59.6, 62.5, 52.6,
+ 54.1, 49.3, 54.5, 56.3, 54.0, 61.5, 52.2, 59.3, 60.1, 52.7,
+ 56.6, 52.3, 56.3, 54.8, 53.9, 59.9, 51.0, 64.1, 61.4, 55.6,
+ 63.9, 51.6, 58.0)
> pig <- factor(pig)    # R should interpret the pig and day
> day <- factor(day)    # number as categorical variables
> quality <- lm(brightness ~ pig + day)
> summary(quality)

Call:
lm(formula = brightness ~ pig + day)

Residuals:
     Min       1Q    Median       3Q      Max
-2.91000 -1.15667 -0.06333  0.79917  4.46000

Coefficients:
            Estimate Std. Error t value Pr(>|t|)
(Intercept)  53.0433     1.2611  42.060  < 2e-16 ***
pig2         -0.4667     1.6281  -0.287 0.777670
pig3          6.1333     1.6281   3.767 0.001411 **
pig4         -2.3000     1.6281  -1.413 0.174811
pig5          6.7667     1.6281   4.156 0.000593 ***
pig6          7.1000     1.6281   4.361 0.000377 ***
pig7         -0.6000     1.6281  -0.369 0.716783
pig8          3.9667     1.6281   2.436 0.025451 *
pig9         -3.1667     1.6281  -1.945 0.067570 .
pig10         2.0333     1.6281   1.249 0.227707
day4          1.1400     0.8918   1.278 0.217358
day6          2.4300     0.8918   2.725 0.013895 *
---
Signif. codes:  0 '***' 0.001 '**' 0.01 '*' 0.05 '.' 0.1 ' ' 1

Residual standard error: 1.994 on 18 degrees of freedom
Multiple R-squared: 0.8558,    Adjusted R-squared: 0.7676
F-statistic: 9.708 on 11 and 18 DF,  p-value: 1.761e-05
```

Note that summary() provides information only on estimates and standard errors for each parameter in the model but does not provide information on the hypothesis that tests the overall effect of any of the factors. As usual, the estimates for the categorical explanatory variable from summary() are contrasts relative to a reference level, as described in Section 8.3.2. In the output above, the intercept is the estimated average for pig 1 (since pig no. 1 is the reference pig) measured at day 1 (since day 1 is the reference level for the day variable). Also, the estimate found for, say, day4, is the estimated *contrast* between factor level 4 (for the variable day) and the reference level (*i.e.,* day 1). We can therefore conclude that the brightness score on day 4 is on average 1.14 points higher than on day 1, as seen in the table at the end of Example 8.6. Likewise, we can see from the estimates that the brightness of the meat from pig 10 is on average 2.033 points higher than the brightness measured on pig 1.

To test the hypothesis that there is no difference among the levels for a categorical explanatory variable, we use the drop1(), function which tests the overall effect of each single term in the model. If we add the option test="F" to drop1(), then R will automatically calculate the proper *p*-value from the F distribution.

```
> drop1(quality, test="F")
Single term deletions

Model:
brightness ~ pig + day
        Df Sum of Sq    RSS    AIC F value     Pr(F)
<none>                 71.57  50.08
pig      9    395.05 466.63  88.33 11.0395 1.133e-05 ***
day      2     29.56 101.13  56.46  3.7174   0.04452 *
---
Signif. codes:  0 '***' 0.001 '**' 0.01 '*' 0.05 '.' 0.1 ' ' 1
```

From the call to drop1() we see that there is a significant difference among the 10 pigs (with a *p*-value less than 0.0001), while the differences in average effect among the three days is only borderline significant (*p*-value of 0.04452). We can also compare models using anova() instead of drop1(). For example, to compare a two-way additive model with a model with no effect of day, we type

```
> noday <- lm(brightness ~ pig)
> anova(noday,quality)
Analysis of Variance Table

Model 1: brightness ~ pig
Model 2: brightness ~ pig + day
  Res.Df     RSS Df Sum of Sq      F  Pr(>F)
1     20 101.133
2     18  71.571  2    29.562 3.7174 0.04452 *
---
Signif. codes:  0 '***' 0.001 '**' 0.01 '*' 0.05 '.' 0.1 ' ' 1
```

which yields the same result as before.

8.5.1 Interactions

In Example 8.12 on p. 242, where birth weight was modeled as a function of gender and length of pregnancy, we suggested an interaction between gender and length of pregnancy since that would allow for different effects of length of pregnancy for boys and girls. We would analyze this model in the following way:

```
> bw <- read.table("birthweight.txt", header=TRUE) # Read data
> summary(bw)                                       # Summary
      sex           age             weight
 female:12   Min.   :35.00   Min.   :2412
 male  :12   1st Qu.:37.00   1st Qu.:2785
             Median :38.50   Median :2952
             Mean   :38.54   Mean   :2968
             3rd Qu.:40.00   3rd Qu.:3184
             Max.   :42.00   Max.   :3473


> model <- lm(weight ~ sex + age + sex*age, data=bw)
> summary(model)


Call:
lm(formula = weight ~ sex + age + sex * age, data = bw)

Residuals:
    Min      1Q  Median      3Q     Max
-246.69 -138.11  -39.13  176.57  274.28

Coefficients:
             Estimate Std. Error t value Pr(>|t|)
(Intercept) -2141.67    1163.60   -1.841 0.080574 .
sexmale       872.99    1611.33    0.542 0.593952
age           130.40      30.00    4.347 0.000313 ***
sexmale:age   -18.42      41.76   -0.441 0.663893
---
Signif. codes:  0 '***' 0.001 '**' 0.01 '*' 0.05 '.' 0.1 ' ' 1

Residual standard error: 180.6 on 20 degrees of freedom
Multiple R-squared: 0.6435,     Adjusted R-squared:  0.59
F-statistic: 12.03 on 3 and 20 DF,  p-value: 0.0001010
```

We find the estimates for the parameters in the model in the output from summary(). Note that female is the reference level for the categorical variable sex.

The effect for age lists the estimated increase in weight per week for the reference level, so on average baby girls gain 130.40 grams per week of pregnancy (at the end of the last trimester, as that is the only period covered by the data). The interaction term shows the contrast in slopes between the boys and girls, so on average boys gain $130.40 - 18.42 = 111.98$ grams per week while girls gain 130.40 grams.

To test for any overall effects of gender or age, we again use the drop1() with option test="F":

```
> drop1(model, test="F")
Single term deletions

Model:
weight ~ sex + age + sex * age
        Df Sum of Sq    RSS    AIC F value  Pr(F)
<none>                656425    253
sex:age  1     6346 658771    251  0.1945 0.6639
```

In this case we get only a single test, since drop1() very strictly follows the hierarchical principle and will not test any main effects that are part of an interaction (even though it is still possible in this case to test for the main effect of sex, although the hypothesis tested may be of little biological relevance). Based on the analysis we conclude that there is no interaction between gender and length of pregnancy, so the boys and girls have the same average weight increase per week. Thus we remove the interaction from the model and fit the reduced model to the data:

```
> model2 <- lm(weight ~ sex + age, data=bw)
> summary(model2)

Call:
lm(formula = weight ~ sex + age, data = bw)

Residuals:
    Min      1Q  Median      3Q     Max
-257.49 -125.28  -58.44  169.00  303.98

Coefficients:
            Estimate Std. Error t value Pr(>|t|)
(Intercept) -1773.32     794.59  -2.232   0.0367 *
sexmale       163.04      72.81   2.239   0.0361 *
age           120.89      20.46   5.908 7.28e-06 ***
---
Signif. codes:  0 '***' 0.001 '**' 0.01 '*' 0.05 '.' 0.1 ' ' 1

Residual standard error: 177.1 on 21 degrees of freedom
```

```
Multiple R-squared:  0.64,        Adjusted R-squared:  0.6057
F-statistic: 18.67 on 2 and 21 DF,   p-value: 2.194e-05
```

We then check if we can remove any of the explanatory variables from this simpler model:

```
> drop1(model2, test="F")
Single term deletions

Model:
weight ~ sex + age
        Df Sum of Sq      RSS      AIC F value       Pr(F)
<none>                  658771      251
sex      1   157304   816074      254  5.0145    0.03609 *
age      1  1094940  1753711      273 34.9040 7.284e-06 ***
---
Signif. codes:  0 '***' 0.001 '**' 0.01 '*' 0.05 '.' 0.1 ' ' 1
```

Age is highly significant and gender is barely significant, so we keep both variables as main effects in the model and the final model is

```
birth weight = sex + age
```

We can find the estimates for the parameters from the call to summary(), so the average weight increase per week is 120.89 grams for both boys and girls and boys are on average 163.04 grams heavier than girls.

It is worth noting that the *p*-values found for both the call to summary() and drop1() are identical, but that is generally not the case. We get identical results only because both explanatory variables for these data have one degree of freedom associated with them and because there are no interactions in the final model. We need to use the drop1() function to test the overall hypothesis of no effect of an explanatory variable if the explanatory variable is categorical with more than two categories.

8.6 Exercises

8.1 ℝ **Brightness of pork chops.** Consider the dataset on brightness of pork chops from Example 8.6 in the text. We analyzed those data using an additive two-way analysis of variance. Try analyzing the same data as a one-way analysis of variance where we do not take pig into account. What is the conclusion about the effect of days in this model? Discuss any differences in conclusions between the two models.

The pork brightness data are found in the external pork.txt file.

8.2 ® **Yield of corn after fertilizer treatment.** Imagine a field experiment with 8 plots (geographical areas). Two varieties of corn (A and B) were randomly assigned to the 8 plots in a completely randomized design so that each variety was planted on 4 plots. Suppose 4 amounts of fertilizer (5, 10, 15, and 20 units) were randomly assigned to the 4 plots in which variety A was planted. Likewise, the same four amounts of fertilizer were randomly assigned to the 4 plots in which variety B was planted. Yield in bushels per acre was recorded for each plot at the end of the experiment. The data are provided below and can also be found in the external file cornyield.txt:

Yield	Variety	Amount of fertilizer
134	A	5
140	A	10
146	A	15
153	A	20
138	B	5
142	B	10
145	B	15
147	B	20

1. Specify a model for the experiment and list the possible hypotheses.

2. Analyze the data and determine if there is an effect of variety and/or fertilizer dosage. If you find any significant effects then be sure to quantify the results.

8.3 ® **Clutch size of turtles.** Ashton et al. (2007) examined the effect of turtle carapace length on the clutch size of turtles. The data can be found in the file turtles.txt and contain information on 18 turtles.

1. Specify the following three statistical models: clutch size as a linear regression, a quadratic regression, and a cubic regression of carapace length.

2. Import the data and fit the three models in R.

3. Which model should you use to describe clutch size as a function of carapace length?

4. Consider the results from the cubic regression model. How would you test if the following model could describe the relationship between clutch size and carapace length?

$$y_i = \alpha + \beta_1 x_i + \beta_3 x_i^3 + e_i$$

Test if this model fits the data just as well as the standard cubic model.

8.4 Ⓡ **Drugs in rat's livers.** An experiment was undertaken to investigate the amount of drug present in the liver of a rat (Weisberg, 1985). Nineteen rats were randomly selected, weighed, placed under a light anesthetic, and given an oral dose of the drug. It was believed that large livers would absorb more of a given dose than a small liver, so the actual dose given was approximately determined as 40 mg of the drug per kilogram of body weight. After a fixed length of time, each rat was sacrificed, the liver weighed, and the percent dose in the liver was determined.

The file `ratliver.txt` contains information on the nineteen rats on the response and on three covariates:

`BodyWt` is the body weight of each rat in grams.

`LiverWt` is the weight of each liver in grams.

`Dose` is the relative dose of the drug given to each rat as a fraction of the largest dose.

`DrugInLiver` is the proportion of the dose in the liver.

1. Specify a statistical model for the experiment and estimate the parameters of the model.
2. Specify the hypotheses you are able to test with this model.
3. Test the hypotheses from question 2 on the original dataset and state the conclusions.
4. State the final model and the corresponding parameter estimates.
5. Look at the residual plot for the initial model. Do you see any potential problems? If so, how might they affect the results? What can be done to correct this?
6. Test the hypotheses from question 2 *after* you have corrected any problems with the model fit.
7. Do the conclusions change after you have corrected any problems with the model fit?

8.5 **[M] Least squares estimates for quadratic regression.** Look at the following variant of a quadratic regression model:

$$y_i = \beta_1 x_i + \beta_2 x_i^2 + e_i,$$

where we assume that the intercept $\alpha = 0$. Derive the formulas for the least squares estimate of β_1 and β_2.

8.6 Ⓡ **Weight gain of rats.** The file `ratweight.txt` contains the weight gain for rats fed on four different diets: combinations of protein source (beef or cereal) and protein amount (low and high). Data are from Hand et al. (1993).

1. Read the data into R and make reasonable plots to get a feel for the data.

2. Specify a statistical model where weight gain is a function of protein source and protein amount such that an interaction is included in the model to allow the effect of protein amount to depend on protein source.

3. State the hypotheses you can test in this model.

4. Test the hypotheses and state the conclusion. In particular, the investigators are interested in quantifying the differences found between the two protein sources as well as the differences found between the two protein amounts.

8.7 ℝ **Butterfat and dairy cattle.** The file `butterfat.txt` contains information on the average butterfat content (percentages) for random samples of 20 cows (10 two year olds and 10 mature (greater than four years old)) from each of five breeds. The data are from Canadian records of pure-bred dairy cattle and from Hand et al. (1993).

There are 100 observations on two age groups (two years and mature) and five breeds, and the dataset has three variables: `Butterfat` (the average butterfat percentages), `Breed` (the cattle breed: one of Ayrshire, Canadian, Guernsey, Holstein-Fresian or Jersey), and `Age` (the age of the cow: either "Mature" or "2years").

1. Analyze the data (basically, go through the same steps as for Exercise 8.6). The researchers are particularly interested in determining if the change in butterfat contents over age is the same for the five breeds.

8.8 ℝ **Body fat in women.** It is expensive and cumbersome to determine the body fat in humans as it involves immersion of the person in water. The data in the file `bodyfat.txt` provide information on body fat, triceps skinfold thickness, thigh circumference, and mid-arm circumference for twenty healthy females aged 20 to 34 (Neter et al., 1996). It would therefore be very helpful if a regression model with some or all of these predictor variables could provide reliable predictions of the amount of body fat, since the measurements needed for the predictor variables are easy to obtain. The dataset contains the following four variables: `Fat` (body fat), `Triceps` (triceps skinfold measurement), `Thigh` (thigh circumference), and `Midarm` (mid-arm circumference).

Find the model that best predicts body fat from (some of) the three explanatory variables that are easy to measure.

Chapter 9

Probabilities

When an experiment is performed multiple times, we are not guaranteed to get the exact same result every time. If we roll a die we do not get the same number of pips every time, and when milking a cow we end up with slightly different amounts of milk daily even if we treat and feed the cow the same way every day. The likeliness of different outcomes is described by probabilities, and this chapter introduces probability concepts and rules for calculations with probabilities.

The *probability* of a random event is the limit of its relative frequency in a large number of experiments. In other words, if we are interested in the probability of an event we can use the relative frequency of the event as an estimate of the probability. In the simplest situation, we register whether or not an event occurs (for example, if a regular die rolls an even number of pips). Let us call the event A and consider the number of times where the event A occurs, n_A, out of n repetitions of the experiment (*i.e.*, we roll the die n times). The fraction n_A/n is the *relative frequency* of the event A, just as in Section 1.5. Recall also that the relative frequency stabilizes as the number of experiments, n, tends towards infinity (Section 1.5), so we interpret the probability of the event A as the relative frequency of A for an infinite number of experiments. We will write that more formally in the next section. Skovgaard et al. (1999, Chapter 1) provide a more thorough coverage of probabilities, and their work was used as inspiration for this chapter.

9.1   Outcomes, events, and probabilities

The set of all possible outcomes from an experiment is called the *sample space* and we denote it U. A possible *outcome*, u, is an element in the sample space, so $u \in U$. The possible outcomes when we throw a die are $1, 2, \ldots, 6$, so the sample space is the set $U = \{1, 2, \ldots, 6\}$.

An *event*, A, is a subset of the sample space, and we write $A \subseteq U$ and say that the event A has occurred if the outcome of the experiment belongs to the set A. For example, if we are interested in the event of rolling an odd number of pips on a single die, then the event A is the set $\{1, 3, 5\}$. The empty set,

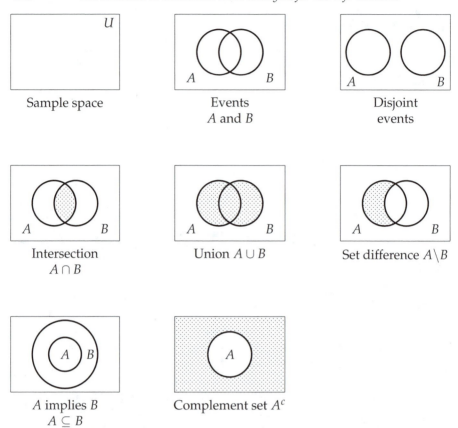

Figure 9.1: Relationships between two events A and B.

$\emptyset$, and the entire sample space, U, are events which occur never and always, respectively.

We can use set theory to combine two or more sets. If A and B are events, then we define (see Figure 9.1):

- Intersection: $A \cap B$ is the event that both A and B occur.

- Union: $A \cup B$ is the event that either A or B or both occur.

- Complement: A^c is the event that A does not occur.

A is said to *imply* B if A is a subset of B ($A \subseteq B$) since B must occur if A occurs. A and B are said to be disjoint if $A \cap B = \emptyset$. That means that they cannot both occur at once. The terminology is illustrated by the Venn diagrams in Figure 9.1.

Example 9.1. Die throwing. If we throw a die, the sample space is $U = \{1, 2, \ldots, 6\}$. The two events "odd number of pips" and "at least 5 pips" are

$$A = \{1, 3, 5\}, \qquad B = \{5, 6\}.$$

For these two events we have that

$$A \cap B = \{5\} \qquad A \cup B = \{1,3,5,6\}$$
$$A \backslash B = \{1,3\} \qquad A^c = \{2,4,6\}.$$

□

If n_A and n_B are the frequencies of events A and B each occurring out of n experiments, then

- $0 \leq n_A/n \leq 1$, since $0 \leq n_A \leq n$.

- $n_U/n = 1$, since $n_U = n$.

- $n_{A \cup B}/n = n_A/n + n_B/n$ if A and B are disjoint (*i.e.* $A \cap B = \varnothing$), since $A \cup B$ occurs $n_A + n_B$ times when A and B cannot both occur at the same time.

A probability is interpreted as a relative frequency in an infinitely large number of experiments, so if we want to assign probabilities to events then these probabilities should fulfill the same requirements as the relative frequencies. Motivated by the properties from relative frequencies, we can now present the definition of a probability distribution and the corresponding probabilities.

Infobox 9.1: Definition of probability

Let U denote the sample space for an experiment. A *probability distribution* on U is a function P which assigns a number, $P(A)$, between zero and one to any event $A \subseteq U$ such that

$$P(A \cup B) = P(A) + P(B) \text{ if } A \cap B = \varnothing, \qquad (9.1)$$

and so that $P(U) = 1$.

The probability distribution P describes how the total probability mass of 1 should be distributed on the sample space. The definition from Infobox 9.1 ensures that the probability of an event cannot be less than 0 or greater than 1 (corresponding to 0% and 100%, respectively) and requires that the total probability mass is 1 (*i.e.*, the probabilities add up to 100%). Formula (9.1) says that the probability of either event A or event B occurring is equal to the sum of the probabilities of the individual events if we assume that it is impossible to observe both A and B at the same time. Formula (9.1) is commonly called the *addition rule* since it allows us to add probabilities together if they belong to non-overlapping events.

Note how the definition determines which probability distributions are allowed on the sample space U, but the definition says nothing about which distributions are adequate for a given experiment. For example, if we consider the die throwing example, the definition has no requirement that each of the six sides of the die should be equally probable. Keep in mind that the probability of an event A is the relative frequency of the event A. Hence, a reasonable estimate of a probability of an event A based on a sample is the relative frequency of the event A in the sample.

Example 9.2. Die throwing (continued from p. 258). When we consider a regular die it is reasonable to think of all six possible outcomes as equally likely,

$$P(\{1\}) = P(\{2\}) = \cdots = P(\{6\}).$$

If we let #A denote the number of elements (possible outcomes) from an event A, then this corresponds to the probability distribution given by

$$P(A) = \frac{\#A}{6}.$$

It is easy to show that this definition of a probability distribution fulfills the criteria in Infobox 9.1. □

Basic probability rules can be derived from the criteria in Infobox 9.1 by using the relationships shown in Figure 9.1. Infobox 9.2 lists some of the probability rules.

Infobox 9.2: Probability rules

Let A and B be events from the sample space U. Then

1. $P(A) \leq P(B)$ if $A \subseteq B$.

2. $P(A^c) = 1 - P(A)$.

3. $P(A \cup B) = P(A) + P(B) - P(A \cap B)$.

4. $P(A_1 \cup \cdots \cup A_k) = P(A_1) + \cdots + P(A_k)$ if $A_1, \ldots, A_k$ are pairwise disjoint events, $A_i \cap A_j = \emptyset$ for all $i \neq j$.

The properties found in Infobox 9.1 and Infobox 9.2 form the fundamental probability rules. Some of the rules may seem so evident from an intuitive point of view that we would not hesitate to use them. As a matter of fact, we already used rule 2 for computation of p-values in the previous chapters. However, rules 1 to 4 all follow from the definition in Infobox 9.1. We will not go through all rules here but just illustrate rule 3 in the die example.

Example 9.3. Die throwing (continued from p. 260). Let us consider the probability of observing an odd number of pips or rolling at least 4 with a single die roll (or both). We use the natural probability distribution from Example 9.2. We are interested in the events

$$A = \{1, 3, 5\}, \qquad B = \{4, 5, 6\}$$

and wish to calculate $P(A \cup B)$. We can calculate this probability by "counting elements": The union event $A \cup B = \{1, 3, 4, 5, 6\}$ has 5 elements and the probability distribution yields the probability $5/6$. Alternatively, we can use rule 3, which gives

$$P(A \cup B) = \frac{3}{6} + \frac{3}{6} - \frac{1}{6} = \frac{5}{6},$$

since the intersection of A and B, $A \cap B = \{5\}$, has exactly one element. Had we only computed $P(A) + P(B) = \frac{3}{6} + \frac{3}{6}$ we would have counted $\{5\}$ twice, but rule 3 remedies this by subtracting $P(\{5\})$ once. □

9.2 Conditional probabilities

In this section we define the concept of conditional probability: the probability of an event given that another event has already occurred.

Consider the events A, B and $A \cap B$ in an experiment with sample space U. Imagine that we make n independent replications of the experiment and let n_A, n_B and $n_{A \cap B}$ denote the number of times that the three events occur, respectively. Let us look more closely at the relative frequency of

- the event A among all the experiments (*i.e.*, n_A/n), and

- the event A among the experiments where B occurred (*i.e.*, $n_{A \cap B}/n_B$).

The first relative frequency approaches the probability of event A as n tends towards infinity. The second relative frequency approaches the *conditional probability of A given B*, which should be viewed as the probability of A among those outcomes where B occurs.

Infobox 9.3: Conditional probability

Let A and B be events and assume that $P(B) > 0$. The *conditional probability* of A given B (*i.e.*, that the event A occurs when we already know that event B has occurred) is written $P(A|B)$ and is defined as

$$P(A|B) = \frac{P(A \cap B)}{P(B)}. \tag{9.2}$$

We will often need to rewrite the conditional probability definition so it has the form

$$P(A \cap B) = P(B)P(A|B),$$

which is obtained directly from (9.2) by multiplying with $P(B)$.

Example 9.4. Specificity and sensitivity. Meat samples are analyzed by a chemical *diagnostic test* for the presence of a specific bacteria. Ideally, the test is positive $(+)$ if the bacteria is present in the meat sample and negative $(-)$ if it is absent. Table 9.1 shows the results from such a test where the presence of *E. coli* bacteria O157 is investigated.

Table 9.1: Presence of *E. coli* O157: number of positive and negative test results from samples with and without the bacteria

	Positive	Negative
With bacteria	57	5
Without bacteria	4	127

We can see from Table 9.1 that the test is not perfect. We sometimes incorrectly identify samples as positive although they do not have the bacteria, and in some cases we also fail to identify the bacteria in samples where it is present. We shall use the following two conditional probabilities to describe the test:

$$P(+|\text{bacteria}) = 1 - P(-|\text{bacteria})$$
$$P(+|\text{no bacteria}) = 1 - P(-|\text{no bacteria}).$$

The event $\{+ \cap \text{no bacteria}\}$ when a sample without bacteria is identified as positive is called a *false positive* result, and the event $\{- \cap \text{bacteria}\}$ where a sample with bacteria tests negative is called a *false negative* result. In Table 9.1 we have 4 false positives and 5 false negatives.

The *sensitivity* of the test method is defined by the probability $P(+|\text{bacteria})$, and $P(-|\text{no bacteria})$ is called the *specificity*. The sensitivity is the probability that the diagnostic test will show the correct result (*i.e.*, that the test is positive) if bacteria are indeed present, and the specificity is the probability that the diagnostic test will show the correct result (*i.e.*, that the test is negative) if bacteria are not present in the sample. In general, we prefer diagnostic tests which have both very high sensitivity and very high specificity.

Often we use these two conditional probabilities to describe the efficiency of the diagnostic test. As an analogy, we think of the shepherd who cried wolf as having high sensitivity but little specificity, whereas the guard dog who sleeps through every burglary has very little sensitivity. In the example from Table 9.1, we estimate the sensitivity to $57/62 = 0.92$ and the specificity to $127/131 = 0.97$. □

We can use Bayes' theorem if we wish to "invert" the conditional probability. If we know the conditional probability $P(B|A)$ and the two marginal probabilities $P(A)$ and $P(B)$, then we can calculate the inverse conditional probability $P(A|B)$.

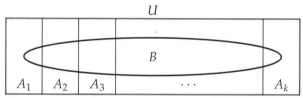

Figure 9.2: Partition of the sample space U into disjoint events $A_1, \ldots, A_k$. The event B consists of the disjoint events $A_1 \cap B, \ldots, A_k \cap B$.

Infobox 9.4: Bayes' theorem

Bayes' theorem applies to events A and B with $P(A) > 0$ and $P(B) > 0$:

$$P(A|B) = \frac{P(B|A)P(A)}{P(B)}. \tag{9.3}$$

We can generalize Bayes' theorem if we partition the total sample space into k disjoint sets, $A_1, \ldots, A_k$, such that $A_1 \cup \cdots \cup A_k = U$, as shown in Figure 9.2. We assume here that we already know or have a model for the conditional probabilities of B given each of the A_i's and seek to reverse the conditional probabilities so we can calculate the conditional probability of each A_i given the event B.

This might be relevant in a situation where we already have observed the event B and then wish to say something about how probable each of the different A_i's are. A common example that matches this setup is disease diagnostics. Let the A_i's denote different possible disease types and let B be an event corresponding to certain symptoms; for example, fever or rash. The conditional probabilities $P(B|A_i)$ are the probabilities of the symptoms for individuals with disease A_i. We wish to calculate $P(A_i|B)$ for the different diseases so we can provide the best possible diagnosis; *i.e.*, tell which disease is most likely given the symptoms we observe. We can do that using a generalized version of Bayes' theorem and the law of total probability.

Infobox 9.5: Law of total probability

Let $A_1, \ldots, A_k$ be a partition of the full sample space U into pairwise disjoint events (*i.e.*, $A_1 \cup \cdots \cup A_k = U$), and assume also that $P(A_i) > 0$ for all i. For any event B, we have the *law of total probability*

$$P(B) = P(B|A_1)P(A_1) + \cdots + P(B|A_k)P(A_k), \qquad (9.4)$$

and if $P(B) > 0$, we have the generalized Bayes' theorem:

$$P(A_i|B) = \frac{P(B|A_i)P(A_i)}{P(B|A_1)P(A_1) + \cdots + P(B|A_k)P(A_k)} = \frac{P(B|A_i)P(A_i)}{P(B)}.$$
$$(9.5)$$

9.3   Independence

Independence is an important concept in probability theory and statistics. In some of the previous chapters we have already discussed the intuitive understanding about how experiments or observations can be independent but the mathematical concept of independence assigns a precise meaning to the intuitive understanding.

Infobox 9.6: Independence

Two events, A and B, are said to be *independent* (under the probability distribution P) if
$$P(A \cap B) = P(A) \cdot P(B). \qquad (9.6)$$

We can generalize (9.6) and say that n events, $A_1, \ldots, A_n$, are said to be mutually independent if

$$P(A_{i_1} \cap \ldots \cap A_{i_k}) = P(A_{i_1}) \cdots P(A_{i_k})$$

for all possible subsets $A_{i_1}, \ldots, A_{i_k}$ selected among $A_1, \ldots, A_n$. As a special case, we have that

$$P(A_1 \cap \ldots \cap A_n) = P(A_1) \cdots P(A_n). \qquad (9.7)$$

The property (9.6) is also commonly referred to as the *multiplication rule*.

Note that the probability distribution should be specified *before* it makes sense to ask if events are independent. We should also note that Infobox 9.6

ensures that two events A and B are independent if and only if $P(A|B) = P(A)$. In other words, knowing that event B has occurred does not change the probability that A occurs.

Example 9.5. Two dice. As an example of independent events, think of the situation where we throw two regular dice. In this case it would be natural to assume that the result of one die has no impact on the result of the other die; *i.e.*, that an observed roll of 6 on the first die would not influence the probability of rolling 6 on the second die.

The sample space for this experiment has 36 elements, one for each combination of the two dice. If we assign probability $1/36$ to each of the elements, then the outcomes of the two dice are independent. For example, rolling 6 on the first die and at the same time rolling 6 on the second die is $\frac{1}{36} = \frac{1}{6} \cdot \frac{1}{6}$. Likewise, rolling at least 5 with the first die and even with the second die is $\frac{2}{6} \cdot \frac{3}{6} = \frac{6}{36}$, corresponding to the 6 possible outcomes (5,2), (5,4), (5,6), (6,2), (6,4), and (6,6).

When every possible element in the sample space has equal probability then we call the probability distribution the *uniform distribution*. A uniform distribution assigns equal probability to all possible outcomes; *i.e.*, all $6 \cdot 6 = 36$ possible outcomes have the same probability $1/36$ in the situation where we roll two dice. $\square$

Example 9.6. Card games and independent events. Draw a card at random from a standard deck of 52 playing cards. Are the events $A = \{$hearts$\}$ and $B = \{$court card$\}$ independent?

The sample space consists of 52 possible outcomes, and we assume a uniform distribution of those so each of the 52 outcomes are equally probable. The probability of getting a hearts card is $P(A) = 13/52 = 1/4$ and the probability of a court card is $P(B) = 12/52 = 3/13$ since there are 3 court cards of each suit. We also know that there are 3 court cards in the hearts suit, so

$$P(A \cap B) = \frac{3}{52} = P(A) \cdot P(B).$$

The two events A and B are therefore independent. This example shows that two events can be independent in the mathematical sense defined by (9.6) even though they are not physically independent — it is the same card we are talking about when we look at A and B. $\square$

Until now in this chapter we have discussed independence of events, but in the previous chapters we required independent observations. If Y is a random variable, for example $Y \sim N(\mu, \sigma^2)$, and a is a number, then we can consider the event that Y is less than a; *i.e.*, $\{Y \leq a\}$. We can assign probabilities to these types of outcomes and that allows us to discuss independence. Two random variables X and Y are independent if and only if for any numbers a and b the events $\{X \leq a\}$ and $\{Y \leq b\}$ are independent events. Thus two variables are independent if

$$P(X \leq a, Y \leq b) = P(X \leq a) \cdot P(Y \leq b)$$

for all numbers a and b (compare to formula (4.8)), or equivalently, if

$$P(X \le a | Y \le b) = P(X \le a)$$

for all numbers a and b where $P(Y \le b) > 0$. In particular, the outcome of one variable does not change the probability of the outcomes of the other variable, reflecting our intuitive understanding of independence.

Throughout the text we have discussed replications of experiments. Replication is one of the fundamental concepts in statistics and in experimental scientific research. Physical and biological results which cannot be reproduced have little scientific value. When we mention replications of experiments we assume that the sample space and the probability distribution are the same from experiment to experiment and that the experiments are independent. Independence of experiments is defined exactly as in (9.6) since we require that events from the different experiments are independent.

The purpose of the following example is to illustrate how the definitions and rules from this chapter can be used.

Example 9.7. Cocaine users in the USA. Consider the following table from Daniel (1995), where a sample of 111 cocaine users supposedly representative of the population of adult cocaine users in the USA who are not incarcerated and not undergoing treatment are classified according to their lifetime frequency of cocaine use:

Lifetime frequency of cocaine use	Male (M)	Female (F)	Total
1–19 times (A)	32	7	39
20–99 times (B)	18	20	38
100 + times (C)	25	9	34
Total	75	36	111

We are interested in properties concerning the population of cocaine users rather than this particular sample, but we use the relative frequencies in the sample to estimate the population probabilities. Assume that we randomly pick an adult cocaine user. Estimate the probability (*i.e.*, compute the corresponding relative frequency in the sample) that this person

1. is a male (event M)?

2. is a female (F)?

3. a) is a male (M) *and* a person who has used cocaine more that 100 times (C)?

   b) is a male (M) *or* a person who has used cocaine more than 100 times (C) (or both)?

4. has used cocaine more than 100 times (C) if it is known that the person was a male (M)?

5. is a male (M) if we know that it is a person who has used cocaine more than 100 times (C)?

6. is a male (M) (try using the law of total probability (9.4))?

Now assume that we pick two adult cocaine users at random with replacement; *i.e.*, such that an individual can be sampled twice. Estimate the probability (compute the corresponding relative frequency) that

7. the second individual was a male (M2) given that the first one was a male (M1)?

8. both are males (M1 and M2)?

The answers to the questions above can be found in various ways. The solutions shown below are not always the most straightforward but are used to show how the probability definitions and results from this chapter can be used.

1. We can read this result directly from the table by looking at the relative frequency:

$$P(M) = \frac{75}{111} = 0.6757.$$

2. M^c is the complement to M and that corresponds to the set F, so from question 1 we get

$$P(F) = P(M^c) = 1 - P(M) = 1 - 0.6757 = 0.3243.$$

3. a) We can get the result directly from the table as well as by counting the individuals in sets M and C:

$$P(M \cap C) = \frac{25}{111} = 0.2252.$$

b) The probability of C can be calculated as in question 1, and if we combine that with the results from questions 1 and 3a and probability rule 3 we get:

$$
\begin{aligned}
P(M \cup C) &= P(M) + P(C) - P(M \cap C) \\
&= \frac{75}{111} + \frac{34}{111} - \frac{25}{111} = 0.675 + 0.306 - 0.225 = 0.756.
\end{aligned}
$$

4. The conditional probability of C given M defined from the intersection and the previous results gives that:

$$P(C|M) = \frac{P(C \cap M)}{P(M)} = \frac{0.2252}{0.6757} = 0.3333.$$

which we also could have found as $25/75$ directly by looking at the numbers from the table.

5. The probability of the reverse conditional probability is

$$P(M|C) = \frac{P(M)}{P(C)}P(C|M) = \frac{0.6757}{0.3063} \cdot 0.3333 = 0.7353.$$

6. We can split up the event M according to three possible events A, B, or C:

$$P(M) = P(M|A)P(A) + P(M|B)P(B) + P(M|C)P(C)$$

and then calculate the conditional probabilities as above. It is also possible to get the conditional probabilities directly from the table. For example, there are 25 males out of 34 individuals in the C event, which yields $25/34 = .7353$ — same result as in question 5. We get

$$
\begin{aligned}
P(M) &= \left(\frac{32}{39}\right)\left(\frac{39}{111}\right) + \left(\frac{18}{38}\right)\left(\frac{38}{111}\right) + \left(\frac{25}{34}\right)\left(\frac{34}{111}\right) \\
&= 0.8205 \cdot 0.3514 + 0.4737 \cdot 0.3423 + 0.7353 \cdot 0.3063 \\
&= 0.2883 + 0.1622 + 0.2252 = 0.6757.
\end{aligned}
$$

7. The first and second sample are independent since the two selections have nothing to do with each other. Therefore, we get

$$P(M2|M1) = P(M2) = 0.6757.$$

8. The probability is

$$P(M2 \cap M1) = P(M2)P(M1) = 0.6757 \cdot 0.6757 = 0.4566.$$

$\square$

9.4   Exercises

9.1 Abortion. The following table from Daniel (1995) shows the results from a questionnaire about free abortion. The survey was conducted in an American city and comprised 500 individuals partitioned according to the area of the city where they lived (A, B, C, or D):

	"How do you feel about free abortion?"			
Area	For (F)	Against (I)	Undecided (U)	Total
A	100	20	5	125
B	115	5	5	125
C	50	60	15	125
D	35	50	40	125
Total	300	135	65	500

1. Estimate the probability that a random individual from the sample of 500

   - was for free abortion.
   - was against free abortion.
   - was undecided.
   - lived in area B.
   - was for free abortion given that he/she lived in area B.
   - was undecided or lived in area D.

2. Estimate the following probabilities:

   - $P(A \cap U)$
   - $P(I \cup D)$
   - $P(D^c)$
   - $P(I|D)$
   - $P(B|U)$
   - $P(F)$

9.2 Probability distribution of fair die. Prove the assumption from Example 9.2 that

$$P(A) = \frac{\#A}{6}$$

is a probability distribution; *i.e.*, that it fulfills the requirements from Infobox 9.1.

9.3 Cheese shop. A cheese shop receives 65% of its cheeses from dairy factory A, 25% of its cheeses from dairy factory B, and the remaining 10% from dairy factory C. Assume that the prevalence of *listeria* bacteria is 1% (from factory A), 4% (from factory B), and 2% (from factory C).

1. What is the probability that *listeria* is found in a randomly chosen cheese from the shop?

2. How big a percentage of the cheeses where *listeria* is present are from dairy factory A, B, and C, respectively?

9.4 Germination of seeds. Assume that the probability of a random seed from some plant to germinate and turn into a sprout is 0.6. We plant three seeds and assume that the events that they germinate are independent of each other. What is the probability that

1. all three seeds germinate and turn into sprouts?

2. none of the seeds germinate and turn into sprouts?

9.5 Pollen and flower color of sweet peas. In an experiment with sweet pea plants (*Lathyrus odoratus*), Bateson et al. (1906) looked at the offspring plants and classified them according to the color of their flowers (red or purple) and whether or not they had long or round pollen

	Long pollen	Round pollen
Purple flower	4831	390
Red flower	393	1338

1. Estimate the probability of observing a purple flower. Estimate the probability of observing a long pollen. Is it reasonable to assume that they both are 3/4? (A proper way to decide this can be found in the next chapter, so for now you should just see if the estimate is close to 3/4 or not.)

2. Determine the conditional probability of purple flowers given a) long pollen and b) round pollen. Is it reasonable to assume that flower color and pollen form are segregated independently in sweet peas? (This can also be answered more precisely using methods from the next chapter.)

9.6 Equine protozoal myeloencephalitis. EPM is a neurological disease in horses. Protozoal parasites infect and invade the central nervous system, causing damage to the brain and spinal cord and resulting in symptoms such as muscle atrophy and respiratory problems. The prevalence of EPM in the horse population is 1%, and there exists a clinical method to diagnose horses even if they show no symptoms. The test method provides one of two results: positive (the horse is infected) or negative (the horse is not infected).

We prefer diagnostic tests where the sensitivity and the specificity are both as high as possible since that means that the diagnostic test is likely to provide the correct result for both healthy as well as diseased animals.

Assume that a diagnostic test for EPM has a sensitivity of 89% and a specificity of 92%.

1. What is the probability that a randomly selected horse will test positive on the diagnostic test (*i.e.*, the test shows that the horse is infected) if the horse is healthy?

2. What is the probability that a randomly selected horse (from the total population of both healthy and diseased horses) will test positive on the diagnostic test?

Often researchers are interested in the *positive predictive value* of a diagnostic test. This is defined as the probability that the animal is diseased given that the diagnostic test is positive. Typically, it is the positive predictive value that is of interest for a physician or veterinarian because the positive predictive value helps diagnose the individual given the results of the diagnostic test.

   3. Assume that we examine a horse which has a positive diagnostic test. Calculate the positive predictive value (*i.e.*, the probability that the horse examined is indeed diseased). Is this the result you would expect? Why/why not?

9.7 Game show. Imagine you are a contestant in a television game show where a car is the top prize. There are three closed doors. The car is located behind one of the doors and there are goats behind the other two doors.

When the show starts you must select a door at random but you are not allowed to open it yet. The host of the game show then opens one of the remaining two doors and reveals a goat. You now have the opportunity to stay with your initial choice and open the first door you selected or open the third door instead. Which strategy should be used to have the biggest chance of winning the car?

Chapter 10

The binomial distribution

Binary variables are variables with only two possible outcomes ("success" and "failure"). The statistical models from the previous chapters are not meaningful for such response variables because the models used the normal or Gaussian distribution to describe the random variation; in particular, the response was assumed to be continuous. In the remaining chapters we will discuss statistical models for binary data and count data where the response is discrete.

10.1   The independent trials model

First we will present the design behind the *independent trials model* or *Bernoulli trials*, which describes a sequence of trials or experiments. Data for the independent trials model for n trials should fulfill the following criteria:

- There are two possible outcomes for each trial: "success" and "failure".

- Each trial has the same probability of success, p.

- The outcome of one trial does not influence the outcome of any of the other trials (the independence property).

The independent trials model is clearly different from the linear models we have looked at thus far. There are only two possible outcomes for the response, "success" and "failure", and there is a single probability, p, of each trial being a success. This probability is an unknown parameter in the model.

The outcome consists of all possible combinations of successes and failures, and if we use the independence property we can easily calculate the corresponding probabilities using the multiplication rule (9.7).

Example 10.1. Independent trials. Assume, for example, that we have $n = 3$ trials and that the probability of success is p. There are $2^3 = 8$ possible ordered outcomes that we can observe:

$$SSS \quad SSF \quad SFS \quad FSS$$
$$SFF \quad FSF \quad FFS \quad FFF$$

where S represents success and F is failure. The result FSF means that the first trial was a failure, the second a success, and the third a failure. We also know that each trial has the same probability of success, p, and that they are independent; so, for example, the probability of outcome FSF is

$$P(FSF) = (1 - p)p(1 - p) = p(1 - p)^2.$$

We can summarize the possible outcomes and their respective probabilities in a *probability tree*, as shown in Figure 10.1. Each node in the tree represents a trial and for every trial we can either observe a success (follow the line away from the node towards the top) or a failure (follow the line towards the bottom of the figure). Note that the probability of each of the eight possible outcomes can be written in the form $p^{n_s}(1 - p)^{n_f}$, where n_s and n_f are the number of successes and failures for the outcome, respectively. It is worth emphasizing that because of the independence assumption, the probabilities depend only on the number of successes and failures but *not* on the order in which they occur. An outcome like SSF has the same probability as the outcome FSS simply because the number of successes and failures is the same.

We can summarize the results from the probability tree from Figure 10.1 in the following table, since we can observe from 0 to 3 successes from an independent trial experiment with $n = 3$:

Number of successes	Frequency	Probability of each outcome	Total probability
0	1	$(1 - p)^3$	$(1 - p)^3$
1	3	$p(1 - p)^2$	$3p(1 - p)^2$
2	3	$p^2(1 - p)$	$3p^2(1 - p)$
3	1	p^3	p^3

Thus the probability of observing, say, 2 successes and 1 failure is $3p^2(1 - p)$ from an experiment with $n = 3$. The reason we can add up the probabilities in the "Total probability" column is that each of the outcomes is disjoint; so Infobox 9.1 allows us to add up the individual probabilities.   □

10.2   The binomial distribution

Assume we have data from an experiment that match the independent trials setup and that we are interested in the number of successes out of n trials. We can use the multiplication rule (9.6) to calculate the probability of each possible outcome and then use (9.1) from Infobox 9.1 to add up the probabilities corresponding to the same number of successes, just like in Example 10.1. Note that we are interested in the *number* of successes out of the n trials and *not* in whether or not a particular trial was a success or a failure.

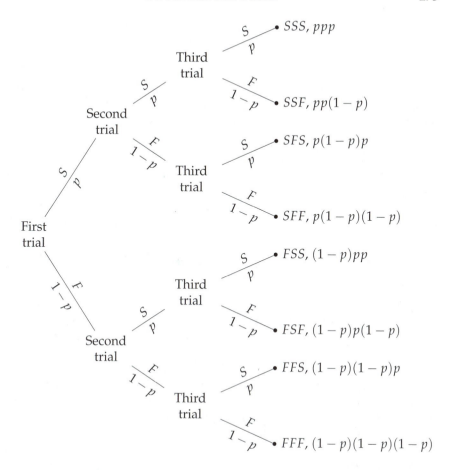

Figure 10.1: Probability tree for an independent trials experiment with $n = 3$ and probability of success p. S and F correspond to success and failure, respectively.

The *binomial model* or *binomial distribution* with size n and probability p models the number of successes, Y, out of n trials, and the probability of observing j successes is defined as

$$P(j \text{ successes}) = P(Y = j) = \binom{n}{j} \cdot p^j \cdot (1-p)^{n-j}. \tag{10.1}$$

The term p^j corresponds to the success probability for each of the j successes, and the term $(1-p)^{n-j}$ corresponds to each of the $n-j$ failures.

The *binomial coefficient* is defined as

$$\binom{n}{j} = \frac{n!}{j!(n-j)!},$$

where $n!$ ("n-factorial") is defined for any positive integer as

$$n! = n \cdot (n-1) \cdot (n-2) \cdots 3 \cdot 2 \cdot 1$$

and where $0! = 1$. The binomial coefficients count all the possible outcomes with a certain number of successes. In other words, $\binom{n}{j}$ is the number of possible outcomes (out of n experiments) where we observed exactly j successes (and therefore $n - j$ failures). For example, $\binom{3}{2} = \frac{3!}{2! \cdot 1!} = 3$, so there are three possible outcomes with 2 successes and 1 failure out of 3 trials (see Figure 10.1). Likewise, $\binom{20}{8} = \frac{20!}{8! \cdot 12!} = 125970$ is the number of possible outcomes with 8 successes out of 20 trials.

If Y represents the number of successes out of n independent trials with the same probability of success, we say that Y follows a binomial distribution with size n and probability of success p, and we write that as $Y \sim \text{bin}(n, p)$. The only parameter in the binomial distribution is the success probability, p, as the number of observations, n, is fixed.

Figure 10.2 shows the probabilities of the possible outcomes calculated from (10.1) for four binomial distributions of size 20 with different success probabilities: 0.1, 0.25, 0.5, and 0.8. We can see that the distribution is close to symmetric except when the success parameter is close to the edge; *i.e.*, when p is close to zero or one.

Example 10.2. Germination. Assume that seeds from a particular plant have a probability of 0.60 of germination. We plant three seeds and let Y denote the number of seeds that germinate. If we assume that the seeds germinate independently of each other, then we can use the binomial distribution; so $Y \sim \text{bin}(3, 0.6)$, and we get that, *e.g.*, the probability that two out of the three seeds germinate is

$$P(Y = 2) = \binom{3}{2} \cdot (0.6)^2 \cdot (1 - 0.6)^{3-2} = 0.432.$$

We can also calculate the probability that, say, at least one of the seeds germinate,

$$
\begin{aligned}
P(Y \geq 1) &= P(Y \in \{1,2,3\}) = P(Y = 1) + P(Y = 2) + P(Y = 3) \\
&= \binom{3}{1} 0.6^1 \cdot 0.4^{3-1} + \binom{3}{2} 0.6^2 \cdot 0.4^{3-2} + \binom{3}{3} 0.6^3 \cdot 0.4^{3-3} \\
&= 0.936,
\end{aligned}
$$

where the second equality follows from the fact that $\{1\}, \{2\}, \{3\}$ are disjoint sets. We could also have calculated this as $P(Y \geq 1) = 1 - P(Y = 0) = 1 - 0.4^3$ since the complement of $\{1,2,3\}$ is $\{0\}$. □

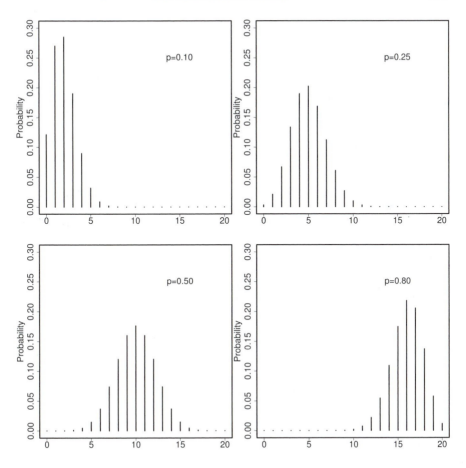

Figure 10.2: Probability distributions for four binomial distributions all with $n = 20$ and with probability of success $p = 0.1, 0.25, 0.50,$ and 0.80.

10.2.1   Mean, variance, and standard deviation

A variable Z is said to follow a *Bernoulli distribution* with parameter p if it follows a binomial distribution with size 1; *i.e.*, $Z \sim \text{bin}(1, p)$. There are two possible outcomes for a Bernoulli variable — zero or one — and the mean (or expected) value is

$$P(Z = 1) \cdot 1 + P(Z = 0) \cdot 0 = p \cdot 1 + (1 - p) \cdot 0 = p.$$

The mean is a weighted average of the possible values (zero and one) with weights equal to the corresponding probabilities. We look at all possible outcomes and the corresponding probabilities of observing them.

We can write a binomial variable, $Y \sim \text{bin}(n, p)$, as a sum of n independent Bernoulli variables, since this corresponds to the outcome of n independent trials that all have the same probability of success, p. Thus

$$Y = \sum_{i=1}^{n} Z_i \quad \text{where } Z_i = \begin{cases} 1 & \text{if trial } i \text{ was a "success"} \\ 0 & \text{if trial } i \text{ was a "failure"} \end{cases}.$$

The Z_i's are independent and all have the same mean p, so the expected value of Y will be the sum of the means of each Z^*. The expected value of Y is

$$\sum_{i=1}^{n} p = np.$$

If we follow the same approach, we can calculate the variance of a binomial distribution by starting with the variance of a Bernoulli variable, Z. The variance of a variable is defined as the average squared deviation from its mean, so we can calculate that as a weighted average of the squared deviances with weights given as the probability of each observation:

$$\begin{aligned} \text{Var}(Z) &= (1-p)^2 \cdot P(Z=1) + (0-p)^2 \cdot P(Z=0) \\ &= (1-p)^2 p + p^2(1-p) = p(1-p). \end{aligned}$$

The variance of a sum of independent variables equals the sum of the individual variances, so

$$\text{Var}(Y) = \text{Var}(\sum_{i=1}^{n} Z_i) = \sum_{i=1}^{n} \text{Var}(Z_i) = np(1-p), \tag{10.2}$$

and since the standard deviation is the square root of the variance, we get that

$$\text{sd}(Y) = \sqrt{np(1-p)}.$$

It is worth noting that the one parameter for the binomial distribution, p, defines both the mean and the variance, and we can see that once we have the mean we can directly calculate the variance and standard deviation. Notice that the calculations for Z are completely identical to the calculations we saw in Example 4.8 on p. 86.

Figure 10.3 shows the variance of a Bernoulli variable as a function of the parameter p. We can see that there is very little deviation for a binomial variable when p is close to zero or one. That makes sense since if most of the independent trials are identical then there is very little uncertainly left. On the other hand, the largest variance is found when $p = 0.5$; *i.e.*, when there is the same chance of observing zeros and ones.

---

*Strictly speaking, we have not seen a result that states that we in general can simply add expectations. However, we saw in Infobox 4.2(a) that the expected value of the sum of two independent, normally distributed variables with means μ_1 and μ_2, respectively, equals the sum, $\mu_1 + \mu_2$. Likewise, the variance of the sum equals the sum of the variances. Actually, it can be proven that this result is true for any variable — not just for those that follow a normal distribution.

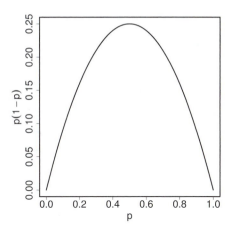

Figure 10.3: Variance of a Bernoulli variable for different values of the parameter p.

Example 10.3. Germination (continued from p. 276). In our example with germination of seeds we have $n = 3$ and $p = 0.60$. Thus on average we expect $np = 3 \cdot 0.6 = 1.8$ seeds to germinate. The standard deviation is

$$\text{sd}(Y) = \sqrt{np(1-p)} = \sqrt{3 \cdot 0.6 \cdot 0.4} = 0.8485.$$

□

The binomial distribution has some nice properties. We have already seen from Figure 10.2 that the binomial distribution is close to symmetric and the general shape resembles a normal distribution. Furthermore, we get that the binomial distribution is retained for sums of binomial variables that have the same parameter p. If Y_1 and Y_2 are independent, and if $Y_1 \sim \text{bin}(n_1, p)$ and $Y_2 \sim \text{bin}(n_2, p)$, then

$$Y_1 + Y_2 \sim \text{bin}(n_1 + n_2, p). \tag{10.3}$$

In particular, the mean and standard deviation of $Y_1 + Y_2$ are given by $(n_1 + n_2)p$ and $\sqrt{(n_1 + n_2)p(1-p)}$, respectively.

The formula (10.3) follows from the fact that Y_1 counts the number of successes in n_1 trials and Y_2 counts the number of successes in n_2 trials; so it is intuitively clear that $Y_1 + Y_2$ counts the number of successes out of the $n_1 + n_2$ trials.

Example 10.4. Germination (continued from p. 276). Assume now that we plant an additional 17 seeds of the same type. Out of the $3 + 17$ seeds, we expect an average of $20 \cdot 0.6 = 12$ to germinate, and the standard deviation is $\sqrt{(17 + 3) \cdot 0.6 \cdot 0.4} = \sqrt{4.8} = 2.19$. □

10.2.2   Normal approximation

The binomial distribution (10.1) allows us to calculate the probability of different outcomes; but there can be practical numerical problems in actually calculating the probabilities when the number of trials, n, is large, as the next example shows.

Example 10.5. Blood donors. Denmark has the largest consumption of blood-products per capita in Europe but also the highest number of blood donors. This very high usage is largely due to the high quality of blood-products (and associated low risks connected with use), tradition, the low cost, the continued availability of a sufficient number of donors, and the high acceptance of blood donation in the Danish population.

The main aim of the blood donor associations in Denmark is to ensure Danish self-sufficiency in blood-products, and in 2007 the number of bleedings in the greater Copenhagen area was 125,754. The frequency of blood donors with blood type 0 rhesus negative is 6%.

Imagine now that the blood donor association wants to compute the probability that they get at least 7500 blood-units of type 0 rhesus negative blood next year if the number of bleedings is the same. Thus we seek to calculate

$$P(Y \geq 7500),$$

where we assume that $Y \sim \text{bin}(125754, 0.06)$. In principle, we could calculate this value by using the probabilities defined by (10.1) repeatedly for the values $7500, 7501, 7502, \ldots, 125,754$. However, we might encounter numerical difficulties evaluating each of the individual probabilities in (10.1) because of the large n. □

Recall that the binomial distribution has mean np and variance $np(1-p)$ and the form is symmetric and resembles that of a normal distribution provided p is not too close to zero or one. We can therefore try to approximate the binomial distribution with a normal distribution with the same mean and variance, $N(np, np(1-p))$. In Figure 10.5 we plot the density of the approximate normal distribution on top of four different binomial distributions and see that in each case the normal distribution is a very good approximation to the binomial distribution. Thus, it may not be unreasonable to calculate probabilities for the binomial distribution using the normal distribution as follows:

$$P(Y \leq y) \approx \Phi\left(\frac{(y+0.5) - np}{\sqrt{np(1-p)}}\right). \tag{10.4}$$

Notice how we add 0.5 to y in the numerator of (10.4). This is because the normal distribution is a continuous distribution, whereas the binomial distribution is a discrete distribution. If we wish to approximate the probability that a binomial variable results in a single value, say 3, then we say that we get the best approximation if we compare that value to the interval $(2.5, 3.5)$

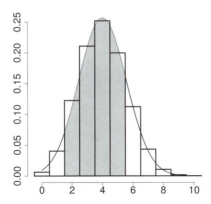

Figure 10.4: Approximation of the binomial probability $P(2 \leq Y \leq 5)$ with a normal distribution. To get the best possible approximation, we use the interval from 1.5 to 5.5 when we calculate the area under the normal density curve.

for the continuous distribution; *i.e.*,

$$
\begin{aligned}
P(Y = 3) \quad &= \quad P(Y \leq 3) - P(Y \leq 2) \\
&\approx \quad \Phi\left(\frac{3.5 - np}{\sqrt{np(1-p)}}\right) - \Phi\left(\frac{2.5 - np}{\sqrt{np(1-p)}}\right).
\end{aligned}
$$

The same idea is illustrated in Figure 10.4.

Example 10.6. Blood donors (continued from p. 280). In the blood type example, we want to calculate

$$P(Y \geq 7500)$$

when $Y \sim \text{bin}(125754, 0.06)$. We approximate this with the normal distribution with mean $125754 \cdot 0.06 = 7545.24$ and variance $125754 \cdot 0.06 \cdot 0.94 = 7092.526$. We get

$$
\begin{aligned}
P(Y \geq 7500) \quad &= \quad 1 - P(Y \leq 7499) \approx 1 - \Phi\left(\frac{7499 + 0.5 - 7545.24}{\sqrt{7092.526}}\right) \\
&= \quad 1 - \Phi(-0.543) = 0.7064.
\end{aligned}
$$

We conclude that there is a probability of 70.64% of receiving at least 7500 type 0 rhesus-negative blood units the following year if we assume that the number of bleedings is the same. □

Throughout this section we have mentioned that n should be sufficiently large for the normal approximation to hold. A rule-of-thumb states that the normal approximation is valid if both $np \geq 5$ and $n(1 - p) \geq 5$. The rule essentially provides the criteria for when the parameter p is far enough away from zero and one.

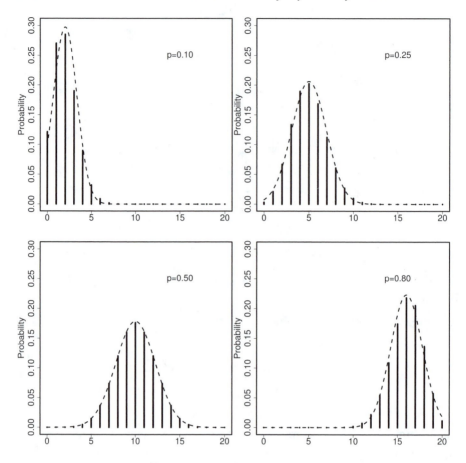

Figure 10.5: Probability distributions for four binomial distributions, all with $n = 20$ and with probability of success $p = 0.1, 0.25, 0.50$, and 0.80 and corresponding normal distribution approximations (dashed curves).

10.3   Estimation, confidence intervals, and hypothesis tests

There is just a single parameter for a binomial distribution, the probability p, and the obvious estimate for that is obtained by counting the number of successes and then dividing by the number of trials,

$$\hat{p} = \frac{\text{number of successes}}{\text{number of trials}}.$$

This result is usually found through maximum likelihood estimation (see Section 5.2.7), but we get the same result through the least squares method.

Our observation is y and the expected value of Y is np. The squared deviation between the observed value and the expected value to be minimized is thus

$$Q(p) = (y - np)^2.$$

The minimum of this function is zero, which is obtained exactly when $y = np$. Thus we set $\hat{p} = y/n$, corresponding to the proportion of successes to the total number of trials.

The standard deviation of a binomial variable Y is $\text{sd}(Y) = \sqrt{n \cdot p \cdot (1 - p)}$, so the standard error of $\hat{p} = \frac{y}{n}$ becomes

$$\text{SE}(\hat{p}) = \sqrt{\frac{\hat{p}(1 - \hat{p})}{n}}. \tag{10.5}$$

We can use the approach from (5.20) to construct a confidence interval for p. In (5.20) we multiplied the standard error by a quantile from the t distribution, and we used this t quantile because the data were assumed to be normally distributed. In our situation, we know that the data are *not* normally distributed, but the central limit theorem described in Section 4.4 allows us to use the normal approximation for y/n provided the number of observations, n, is large. The standard error is determined from $\hat{p}$ alone (there is no extra variance parameter in this model). As a consequence we should use a quantile from the normal distribution rather than the t distribution. The 95% confidence interval for p then becomes

$$\hat{p} \pm 1.96 \cdot \sqrt{\frac{\hat{p}(1 - \hat{p})}{n}}. \tag{10.6}$$

The confidence interval defined by (10.6) has a serious "flaw": if n is small and p is close to the boundaries zero or one, then we might end up with a confidence interval that contains negative values or values above one. Clearly, values below zero or above one do not make sense, as probabilities cannot have values outside the interval $[0, 1]$. When we report confidence intervals, we should restrict them so that the lowest possible value is zero and the highest is one. However, we should also be aware that when we "cut off" one end of the confidence interval, it will not have the desired coverage. We consider an improved confidence interval in Section 10.3.1.

Example 10.7. Apple scab. Twenty apples of the variety "Summer red" were collected at random from an old apple tree. Seven of the twenty apples had signs of the black or grey-brown lesions associated with apple scab. We want to estimate the probability that an apple picked at random from this tree has apple scab.

We assume that the number of apples with apple scab among 20 sampled apples is binomial, with size 20 and parameter p, and observe $y = 7$. Our estimate of the proportion of apples infected with apple scab from this tree is thus

$$\hat{p} = \frac{7}{20} = 0.35.$$

The 95% confidence interval for the proportion of apples that are infected with scab is

$$\frac{7}{20} \pm 1.96 \cdot \sqrt{\frac{\frac{7}{20}(1 - \frac{7}{20})}{20}} = 0.35 \pm 1.96 \cdot 0.1067 = (0.1410, 0.5590).$$

Hence we are 95% confident that the interval 14.10% to 55.90% contains the proportion of apples from the tree that are infected with apple scab.   □

The confidence interval contains the values of p that are in accordance with the observed data, and it can be used informally to test a hypothesis

$$H_0 : p = p_0$$

for some known value p_0. We simply check if p_0 is in the confidence interval for p or not.

If we want to make a formal test of H_0 for the binomial distribution, we use the same approach as in Chapter 6. We choose a test statistic and compute the corresponding p-value. Assume we want to test the hypothesis $H_0 : p = p_0$ and denote the observed value Y_{obs}. Under the null hypothesis, we know that $Y \sim \text{bin}(n, p_0)$ and we can calculate the probability of observing Y_{obs}:

$$P(Y = Y_{obs}) = \binom{n}{Y_{obs}} \cdot p_0^{Y_{obs}} \cdot (1 - p_0)^{n - Y_{obs}}.$$

We will use Y_{obs} itself as our test statistic. Recall that the p-value is defined as the probability of observing something that is *as extreme or more extreme* — *i.e.*, is less in accordance with the null hypothesis — than our observation. Outcomes y with probabilities less than $P(Y = Y_{obs})$ are more extreme than Y_{obs}, so we must add the probabilities of all possible outcomes that have probability smaller than or equal to $P(Y = Y_{obs})$ under the null hypothesis:

$$p\text{-value} = \sum_{y:P(Y=y)\leq P(Y=Y_{obs})} P(Y = y). \tag{10.7}$$

Thus we only add the probabilities that are less likely (under the null) than what we have observed.

Figure 10.6 shows an example on how we do this in practice for a variable $Y \sim \text{bin}(8, p)$ where we have observed the value 1. Assume we wish to test the hypothesis that $H_0 : p = 0.35$. The figure shows the distribution for a binomial variable with size 8 and parameter 0.35; the dotted horizontal line is the probability corresponding to the observed value. The p-value corresponds to the sum of the outcomes that are at least as "extreme" as our observation; *i.e.*, have probability less than or equal to the probability of the observation. The solid vertical lines in Figure 10.6 correspond to those outcomes and probabilities. If we add these probabilities, we get the total probability of observing something at least as extreme as our original observation if the null hypothesis is true. Thus the *p*-value would be

$$p\text{-value} = P(Y = 0) + P(Y = 1)$$
$$+ P(Y = 5) + P(Y = 6) + P(Y = 7) + P(Y = 8) = 0.2752.$$

If *n* is large, we can approximate the calculations for the *p*-value with the normal approximation, as described in Section 10.2.2.

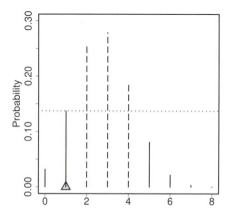

Figure 10.6: Probability distribution under the null hypothesis $Y \sim \text{bin}(8, 0.35)$. The gray triangle shows the observation, the dashed horizontal line is the probability of the observation (under the null hypothesis), and the solid vertical lines represent the probabilities of the outcomes that are used for calculating the *p*-value. The dashed lines are the probabilities of the outcomes that are not contradicting the null hypothesis.

Example 10.8. Apple scab (continued from p. 283). Assume that we wish to test the hypothesis

$$H_0 : p = 0.25$$

that the proportion of apples with scab is 25%. We observed 7 apples out of 20 with scab, and if H_0 is true then the probability of observing 7 apples with scab is $P(Y = 7) = 0.1124$. We can test the hypothesis using the confidence interval calculated in Example 10.7, and since the value 0.25 is contained in the confidence interval we would not reject H_0. The *p*-value becomes $P(Y \leq 2) + P(Y \geq 7) = 0.3055$. Thus, we do not reject the hypothesis that the proportion of apples with scab is 25%. □

10.3.1   Improved confidence interval

The confidence interval (10.6) is derived by approximating the binomial distribution with a normal distribution. This approximation is justified by the central limit theorem but may not be appropriate when the sample size is low and p is close to zero or one. In those situations, the binomial distribution is asymmetric, as we saw in Figure 10.2, and the normal approximation is not adequate.

Agresti and Coull (1998) proposed an improved approximate confidence interval for a binomial proportion p. The center of the confidence interval is shifted slightly towards 0.5 such that the interval becomes asymmetric relative to the estimate $\hat{p}$. To be specific, the center of the improved $1 - \alpha$ confidence interval is

$$\tilde{p} = \frac{y + \frac{1}{2}z_{1-\alpha/2}^2}{n + z_{1-\alpha/2}^2}, \tag{10.8}$$

where y is the observed number of successes, n is the number of trials, and $z_{1-\alpha/2}$ is the $1 - \alpha/2$ quantile for the normal distribution. The improved approximative $1 - \alpha\%$ confidence interval for a proportion p is defined as

$$\tilde{p} \pm z_{1-\alpha/2} \cdot \sqrt{\frac{\tilde{p}(1 - \tilde{p})}{n + z_{1-\alpha/2}^2}}. \tag{10.9}$$

For a 95% confidence interval, $z_{1-\alpha/2}$ becomes the familiar value 1.96. The modified confidence interval defined by (10.9) is less likely than the traditional confidence interval (10.6) to contain values outside the range from zero to one; it resembles the form of the binomial distribution better when the dataset is small, and unlike the traditional confidence interval (10.6), it can calculate confidence intervals when the number of successes is zero or n (although the coverage may not be at the level desired).

Notice that since $z_{1-0.05/2} = 1.96 \approx 2$, we get that

$$\frac{y + 2}{n + 4} \pm 1.96 \cdot \sqrt{\frac{\frac{y+2}{n+4}\left(1 - \frac{y+2}{n+4}\right)}{n + 4}}$$

is the 95% confidence interval for p. Thus the 95% confidence interval for a proportion p corresponds to adding 2 successes and 2 failures to the dataset.

Example 10.9. Apple scab (continued from p. 283). For the apple scab data, we observed 7 apples with scab out of 20. If we use the simple approximation for the improved 95% confidence interval, then we get that it is centered around

$$\tilde{p} = \frac{7 + 2}{20 + 4} = 0.375.$$

The improved 95% confidence interval becomes

$$\tilde{p} \pm 1.96\sqrt{\frac{\tilde{p}(1 - \tilde{p})}{n + 4}} = 0.375 \pm 1.96\sqrt{\frac{0.375 \cdot (1 - 0.375)}{20 + 4}} = (0.1813; 0.5687).$$

If we were interested in the 90% confidence interval for p, we would use $z_{0.95} = 1.6449$ and we get

$$\tilde{p} = \frac{7 + 0.5 \cdot 1.6449^2}{20 + 1.6449^2} = 0.3679,$$

so the improved 90% confidence interval for p is

$$0.3679 \pm 1.6449 \sqrt{\frac{0.3679 \cdot (1 - 0.3679)}{20 + 1.6449^2}} = (0.2014; 0.5343).$$

$\square$

10.4 Differences between proportions

Section 5.2.5 showed how we could estimate the contrast between two group levels with an analysis of variance. Likewise, we might be interested in estimating the difference between the probabilities of two binomial variables Y_1 and Y_2 corresponding to two independent samples. Thus we have $Y_1 \sim \text{bin}(n_1, p_1)$ and $Y_2 \sim \text{bin}(n_2, p_2)$ and we are interested in the hypothesis $H_0 : p_1 = p_2$, which is equivalent to the hypothesis $H_0 : p_1 - p_2 = 0$.

We already know that

$$\hat{p}_1 = \frac{y_1}{n_1} \quad \text{and} \quad \hat{p}_2 = \frac{y_2}{n_2},$$

and naturally the estimate for $p_1 - p_2$ is simply the difference between the two estimates,

$$\widehat{p_1 - p_2} = \hat{p}_1 - \hat{p}_2 = \frac{y_1}{n_1} - \frac{y_2}{n_2}. \tag{10.10}$$

The estimates $\hat{p}_1$ and $\hat{p}_2$ are independent, as they are computed from observations from two independent populations, and we get that

$$\text{Var}(\hat{p}_1 - \hat{p}_2) = \text{Var}(\hat{p}_1) + \text{Var}(\hat{p}_2) = \frac{\hat{p}_1(1 - \hat{p}_1)}{n_1} + \frac{\hat{p}_2(1 - \hat{p}_2)}{n_2},$$

where we use the fact that we can add variances for independent variables for the first equality and we use (10.5) twice for the second equality. We get the standard error of the difference if we take the square root of the variance, so

$$\text{SE}(\hat{p}_1 - \hat{p}_2) = \sqrt{\frac{\hat{p}_1(1 - \hat{p}_1)}{n_1} + \frac{\hat{p}_2(1 - \hat{p}_2)}{n_2}}. \tag{10.11}$$

We can now construct the 95% confidence interval for the difference between two proportions, $p_1 - p_2$,

$$\hat{p}_1 - \hat{p}_2 \pm 1.96 \cdot \sqrt{\frac{\hat{p}_1(1 - \hat{p}_1)}{n_1} + \frac{\hat{p}_2(1 - \hat{p}_2)}{n_2}}. \qquad (10.12)$$

We can use this confidence interval to test the hypothesis

$$H_0 : p_1 = p_2,$$

and if the confidence interval excludes zero, we reject the hypothesis; whereas we fail to reject the hypothesis if the interval contains zero. More about this in Chapter 11.

Example 10.10. Smelly pets. Two studies by Wells and Hepper (2000) and Courtney and Wells (2002) were undertaken to see if dog owners and cat owners are able to identify their own pet from the smell alone. For each animal, a blanket was used to pet the animal 50 times before the blanket was stored in a plastic bag. Each owner was then presented with two plastic bags — one contained the blanket from their own animal while the other contained a blanket from another animal of a similar breed, sex, and age. After smelling the two blankets, the owners were asked to identity which of the blankets came from their own pet. The results can be seen below:

	Identified correctly	Misidentified
Dog owners	23	3
Cat owners	13	12

We have two populations — dog owners and cat owners — and we can model the number of correctly identified pets as a binomial variable for each of the two populations — $Y_d \sim \text{bin}(26, p_d)$ and $Y_c \sim \text{bin}(25, p_c)$ — since 26 dog owners and 25 cat owners participated in the study. From the data, we get that

$$\hat{p}_d = \frac{23}{26} = 0.8846$$

and

$$\hat{p}_c = \frac{13}{25} = 0.52.$$

First, we can test if dog owners are able to identify their own pets. Here we should realize that if they were not able to identify their pets by the smell, then they would essentially be picking one of the two blankets at random, which corresponds to a probability of 50% for correctly identifying their own pet. Thus, if we wish to test if they are able to smell their own pet, we should look at the null hypothesis

$$H_0 : p_d = 0.5.$$

If we calculate the 95% confidence interval for p_d, we get

$$\hat{p}_d \pm 1.96 \cdot \sqrt{\frac{\hat{p}_d(1 - \hat{p}_d)}{n_d}} = 0.8846 \pm 1.96 \cdot 0.0626 = (0.7619; 1.000).$$

Since 0.5 is not in the 95% confidence interval, we reject the null hypothesis and conclude that the dog owners to some extent are able to identify their own pet from the smell.

If we wish to investigate if dog owners and cat owners are equally good at identifying their own pets, we seek to test the hypothesis

$$H_0 : p_d = p_c,$$

which we test by looking at the 95% confidence interval for the difference $p_d - p_c$. The confidence interval is

$$0.8846 - 0.52 \pm 1.96 \cdot \sqrt{\frac{0.8846(1 - 0.8846)}{26} + \frac{0.52(1 - 0.52)}{25}} =$$
$$0.3646 \pm 1.96 \cdot 0.1179 = (0.1334, 0.5958).$$

Since zero is not included in the confidence interval, we conclude that the probability for dog owners is significantly different from the probability for cat owners, and from the estimates we see that dog owners fare better than cat owners. □

10.5 R

The `choose()` function in R can be used to calculate the binomial coefficients, and `factorial()` is used for calculating factorials.

```
> choose(3,2)    # Ways to choose 2 out of 3 possible outcomes
[1] 3
> choose(20,8)   # Ways to choose 8 out of 20 possible outcomes
[1] 125970
> factorial(20) / (factorial(8) * factorial(12))
[1] 125970
```

Just like for the normal distribution, there exists a number of commands in R that are useful for computations with the binomial distribution. We shall use the `dbinom()` and `pbinom()` functions, which give the density and cumulative distribution function, respectively.

In Example 10.2 on p. 276 we looked at the germination of seeds that are assumed to have a probability of germination of 0.60. R can calculate the probability of observing exactly 2 of the 3 seeds germinating, $P(Y = 2)$:

```
> dbinom(2, size=3, p=.60)
[1] 0.432
```

In the call to dbinom(), we set the number of trials with the size option and the probability of success with the p option.

To calculate the probability that at least one of the seeds germinates (*i.e.*, $P(Y \geq 1)$), we can use the cumulative distribution function pbinom():

```
> 1 - pbinom(0, size=3, p=.60)
[1] 0.936
```

where we have used the fact that $P(Y \geq 1) = 1 - P(Y < 1) = 1 - P(Y \leq 0)$.

The rbinom() function generates observations from a binomial distribution. The first argument is the number of random variables to sample, and the size and p options should be specified as above for dbinom().

```
> # Generate 10 Bernoulli variables to simulate 10 coin tosses
> rbinom(10, size=1, p=.5)
 [1] 1 0 0 0 0 1 0 0 1 0
> rbinom(20, size=1, p=0.85) # 20 tosses with biased coin
 [1] 1 1 1 0 1 0 1 1 0 1 1 1 0 1 1 1 1 1 1 1
> # Make 10 variables from a binomial distr. with n=4 and p=0.6
> rbinom(10, size=4, p=0.6)
 [1] 2 2 1 3 1 3 3 4 2 1
```

The prop.test() can be used to test a hypothesis about a single proportion, $H_0 : p = p_0$. In the apple scab example (Example 10.7 on p. 283), we observed 7 apples with scab out of 20 apples. If we wish to test the hypothesis that the true proportion of apples with scab is 0.25, we can use the following command:

```
> prop.test(7, 20, correct=FALSE, p=0.25)

        1-sample proportions test without continuity correction

data:  7 out of 20, null probability 0.25
X-squared = 1.0667, df = 1, p-value = 0.3017
alternative hypothesis: true p is not equal to 0.25
95 percent confidence interval:
 0.1811918 0.5671457
sample estimates:
   p
0.35
```

The first argument to prop.test() is the observed number of successes and the second is the number of trials. By default, R uses a continuity correction in the call to prop.test(), which is slightly more sophisticated than the method we have explained in the text, and we should disregard the correction (by adding the option `correct=FALSE`) if we wish to obtain the same result as in the text. From the output we can read that the null hypothesis is $p = 0.25$ that the estimate is $\frac{7}{20} = 0.35$ and that the p-value for testing the hypothesis is 0.3017. By default, R uses the improved confidence interval presented in Section 10.3.1, albeit it uses a slight modification. Note that the p-value is slightly different from the p-value found on p. 285. This is because `prop.test()` uses a chi-square test statistic when calculating the p-value (chi-square test statistics will be explained in Chapter 11). The p-value presented in the text can be found by using the `binom.test()` function:

```
> binom.test(7, 20, p=0.25)

        Exact binomial test

data:  7 and 20
number of successes = 7, number of trials = 20, p-value
= 0.3055
alternative hypothesis: true probability of success is not
equal to 0.25
95 percent confidence interval:
 0.1539092 0.5921885
sample estimates:
probability of success
                0.35
```

Example 10.10 on p. 288 considered the proportion of dog and cat owners who could identify their own animal by smell alone. We can calculate a confidence interval for the difference in proportions:

```
> pdog <- 23/26        # Proportion of dog owners
> pcat <- 13/25        # Proportion of cat owners
> pdiff <- pdog - pcat # Difference in proportions
> pdiff
[1] 0.3646154
> sepdiff <- sqrt(pdog*(1-pdog)/26 + pcat*(1-pcat)/25)
> sepdiff              # SE of difference in proportions
[1] 0.1179398
> pdiff + 1.96*sepdiff # Upper confidence limit for difference
[1] 0.5957775
> pdiff - 1.96*sepdiff # Lower confidence limit for difference
[1] 0.1334533
```

It is also possible to use `prop.test()` to test if the difference between two proportions is equal to zero. To test the hypothesis $H_0 : p_d = p_c$, we enter vectors of successes and trials in the call to `prop.test()`.

```
> prop.test(c(23, 13), c(26, 25), correct=FALSE)

        2-sample test for equality of proportions without
        continuity correction

data:  c(23, 13) out of c(26, 25)
X-squared = 8.1613, df = 1, p-value = 0.004279
alternative hypothesis: two.sided
95 percent confidence interval:
 0.1334575 0.5957732
sample estimates:
   prop 1    prop 2
0.8846154 0.5200000
```

The estimates for the two proportions $\hat{p}_d = 0.88$ and $\hat{p}_c = 0.52$ are found at the end of the output, and the 95% confidence interval for the difference is also printed. R also provides output for a test that the two proportions are identical (with a p-value of 0.004279), so in this case we would reject the hypothesis that the two proportions are equal. Hypothesis tests for proportions are discussed in Chapter 11.

10.6   Exercises

10.1 Gender of bunnies. It can be difficult to determine the sex of baby rabbits younger than 2 months. A rabbit owner has just had a litter with 5 bunnies and because he seeks to sell them he wishes to estimate the probability that at least 4 of the 5 are males.

   1. Specify a statistical model that can accommodate the data from the rabbit breeder. Assume for this exercise that it is known that the probability of bucks is 0.55.

   2. What is the probability of observing 3 males and 2 females?

   3. Estimate the probability that the litter consists of at least 4 males.

10.2 Salmonella and egg shells. In an experiment it was examined if the bacteria *Salmonella enteritidis* is able to penetrate a chicken egg shell under various circumstances. A total of 44 eggs with minor cracks in the shells were examined (so-called high-risk eggs). The salmonella bacteria was administered to all eggs using the same procedure, but half of the eggs were kept for 3 hours at room temperature first to check the effect of this "heat treatment". After eight days, 14 of the 22 eggs with heat treatment were found to contain *Salmonella enteri-*

tidis, while 17 of the 22 normally treated eggs were found to contain *Salmonella enteritidis*.

1. Specify a statistical model for this experiment.

2. Which values of the difference in infection risk between the two groups would we choose not to reject?

3. Is there any significant effect of the heat treatment on the risk that an egg will be infected with *Salmonella enteritidis*?

4. In the same experiment, 71 "healthy eggs" (without cracks) were examined with the heat treatment too, and only 3 eggs were found to be infected after eight days. Is there a significant difference in the infection risk for healthy and cracked eggs? And if yes, how large is this difference?

5. How many eggs would you have to examine in a future study to obtain a 95% confidence interval for the infection risk for heat-treated healthy eggs that was not wider than 0.04?

10.3 **Effectiveness of new treatment.** A company claims that they have a new treatment for a disease in trees and that the treatment improves the condition of 70% of diseased trees that receive the treatment. In order to verify this claim, a researcher applied the treatment to 84 diseased trees and saw that 50 of the trees had improved condition after 2 weeks.

1. Calculate a 99% confidence interval for the proportion of diseased trees that improves their condition after treatment.

2. Test at significance level 1% if the proportion of trees that improves their condition can be said to be 70%.

Chapter 11

Analysis of count data

This chapter is about comparing different groups of categorical data. The problems are similar to the problems that occur in analysis of variance, but now we are interested in the relative frequency or *number of observations* for each category as opposed to the average value of some quantitative variable for each category.

Categorical response data are also called *count data, tabular data,* or *contingency tables,* and we can present the data in a table with rows and/or columns representing the various categories of our categorical variables and where each cell in the table contains the number of observations for the given combination of the categorical variables.

An example of tabular data is shown in Table 11.1, where the leg injuries of coyotes caught by 20 traps of type Northwood and 20 traps of type Soft-Catch were classified according to the severity of leg damage.

The numbers in the table correspond to the number of observed coyotes. Notice for this dataset that the numbers in each row sum to 20, which was the number of traps of each type, and this number was determined by the design of the experiment. An interesting hypothesis for this type of data is if the *distribution* of coyotes in the three categories is the same for the two trap types.

Let us start by looking at the chi-square goodness-of-fit method for a single categorical variable.

11.1 The chi-square test for goodness-of-fit

Let us assume that we have classified n individuals into k groups. We wish to test a null hypothesis H_0 that completely specifies the probabilities of each of the k categories; *i.e.,*

$$H_0 : p_1 = p_{01}, p_2 = p_{02}, \ldots, p_k = p_{0k}.$$

Here $p_1, \ldots, p_k$ are the unknown parameters that describe the probability of each category; *i.e.,* p_i is the probability that a randomly chosen individual will belong to category i and $p_{01}, \ldots, p_{0k}$ are the corresponding pre-specified

Table 11.1: Leg injuries of coyotes caught by two different types of traps. Injury category I is little or no leg damage, category II is torn skin, and category III is broken or amputated leg

	Injury		
Type of trap	I	II	III
Northwood	9	7	4
Soft-Catch	11	3	6

probabilities under the null hypothesis. Note that we must have $\sum_{i=1}^{k} p_i = 1$ and $\sum_{i=1}^{k} p_{0i} = 1$ such that both the probabilities given by the parameters and the probabilities given by the null hypothesis sum to 1. This simply means that each observation must be in one of the k categories. Furthermore, the probabilities for the null hypothesis are all assumed to be between zero and one; *i.e.*, $0 < p_{01} < 1$, $0 < p_{02} < 1,\ldots,0 < p_{0k} < 1$, so we cannot have categories where we expect none or all of the observations.

Once we have the probabilities specified by the null hypothesis, we can calculate the expected number of observations for each category. Thus, for category 1 we would expect $n \cdot p_{01}$ observations if the null hypothesis is true, and so forth.

Example 11.1. Mendelian inheritance. Gregor Mendel published his results from his experiments on simultaneous inheritance of pea plant phenotypes in 1866 (Mendel, 1866). He examined 556 plants and classified them according to their form (wrinkled or round) and color (green or yellow):

Class	Number
Round, yellow	315
Round, green	108
Wrinkled, yellow	101
Wrinkled, green	32

According to Mendel's theory, these phenotypes should appear in the ratio 9:3:3:1, since the two phenotypes are believed to be independent and because both phenotypes show dominance — yellow is dominant over green and round is dominant over wrinkled. Thus, Mendel's model specifies the expected probabilities of the four groups:

$$H_0 : p_{ry} = \frac{9}{16}, \quad p_{rg} = \frac{3}{16}, \quad p_{wy} = \frac{3}{16}, \quad p_{wg} = \frac{1}{16},$$

where "r", "w", "y", and "g" in the subscripts denote round, wrinkled, yellow, and green, respectively. In the experiment, he examined 556 pea plants, and if we believe Mendel's model to be true (*i.e.*, we look at the distribution under H_0), then we would expect $556 \cdot \frac{9}{16} = 312.75$ plants out of the 556 plants to have round and yellow peas. If we do the same calculation for every group, we can summarize the observed data and the expected data in a table.

Table 11.2: Observed and expected values for Mendel's experiment with pea plants

Class	Observed	Expected
Round, yellow	315	312.75
Round, green	108	104.25
Wrinkled, yellow	101	104.25
Wrinkled, green	32	34.75
Total	556	556

The table suggests that the observed and expected numbers seem to be fairly close to each other. □

Previously, we have compared our observations to the expected values given a statistical model (for example, when we looked at residuals). We wish to do something similar for tabular data, so we compare the number of individuals we have observed for a given category with the number of individuals we would expect to observe if the model is correct.

If we wish to test the null hypothesis, then we should compare $\hat{p}_1$ with p_{01}, $\hat{p}_2$ with p_{02}, etc. However, if we let Y_i denote the number of observations in category i, then we know that $\hat{p}_i = \frac{Y_i}{n}$, so

$$Y_i = n\hat{p}_i.$$

The expected number of individuals in category i is np_{i0}. Hence, comparison of p_i and p_{i0} is equivalent to comparison of the observed values Y_i and the expected number np_{i0}.

To test a hypothesis for tabular data, we can use the *chi-square statistic*, which is defined as

$$X^2 = \sum_i \frac{(\text{observed}_i - \text{expected}_i)^2}{\text{expected}_i}. \tag{11.1}$$

Here the summation is over all possible categories. The numerator in the ith term of (11.1) contains the squared difference between the observed and expected number of observations in category i, so it is close to zero if the two nearly coincide and it becomes large if there is a large discrepancy between the two. Since we square the difference (and the expected number of observations in the denominator is always positive), the contribution from category i to the chi-square test statistic is non-negative and close to zero if the data and the model are in agreement for the ith category. Likewise, large discrepancies for the ith category are critical. In particular, the test statistic is close to zero if the data and the model are in agreement in all the categories, while it is far from zero if the disagreement is too large for at least one of the categories.

We need to decide how large a value of X^2 we can accept before we reject the hypothesis. It turns out that the chi-square test statistic X^2 approximately follows a $\chi^2(r)$ *distribution*, where r is the number of degrees of freedom if H_0 is true. For the situation in this section where the probabilities under the null hypothesis are pre-specified, the degrees of freedom is

$$df = \text{number of categories} - 1.$$

The degrees of freedom measure the number of free parameters when the probabilities vary freely compared to the number of free parameters under the null hypothesis. Under the full model, we have $k - 1$ free parameters (one for each of the k categories); but once we have specified the first $k - 1$ probabilities, the last one is fully determined since the probabilities must sum to one. Under the null hypothesis we have no free parameters we need to estimate, since the probability of each category is given by the null hypothesis. Thus we have $k - 1 - 0 = k - 1$ degrees of freedom.

The left panel of Figure 11.1 shows the density for the $\chi^2(r)$ distribution, with 1, 5, and 10 degrees of freedom. We test the hypothesis H_0 by comparing the chi-square test statistics X^2 to the appropriate $\chi^2(r)$ distribution. Like previously, the p-value is the probability (given that the null hypothesis is true) of observing a value that is more extreme (*i.e.*, further away from zero) than what we have observed. In other words, the critical value for X^2 at significance level α is the value such that the right-most area under the χ^2 distribution is exactly α. The white area in the right panel of Figure 11.1 corresponds to a mass of 5%, so the critical value for a $\chi^2(5)$ distribution is 11.07.

Example 11.2. Mendelian inheritance (continued from p. 296). We wish to test the hypothesis

$$H_0 : p_{ry} = \frac{9}{16}, \quad p_{rg} = \frac{3}{16}, \quad p_{wy} = \frac{3}{16}, \quad p_{wg} = \frac{1}{16}.$$

The X^2 test statistic becomes

$$X^2 = \frac{(315 - 312.75)^2}{312.75} + \frac{(108 - 104.25)^2}{104.25} + \frac{(101 - 104.25)^2}{104.25} + \frac{(32 - 34.75)^2}{34.75}$$
$$= 0.470.$$

We compare this value to a $\chi^2(3)$ distribution because we have 4 categories, which results in 3 degrees of freedom. The critical value for a $\chi^2(3)$ distribution when testing at a significance level of 0.05 is 7.81 (see, for example, the statistical tables on p. 399). Since $X^2 = 0.470$ is less than 7.81, we fail to reject the null hypothesis. Actually, we can look up the exact p-value from

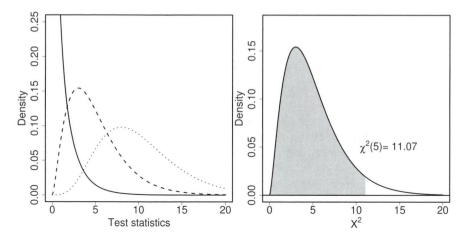

Figure 11.1: The left panel shows the density for the $\chi^2(r)$ distribution with $r = 1$ (solid), $r = 5$ (dashed), as well as $r = 10$ (dotted). The right panel illustrates the density for the $\chi^2(5)$ distribution. The 95% quantile is 11.07, as illustrated by the gray area which has area 0.95.

a $\chi^2(3)$ table or computer package and we get that the p-value is 0.9254. In other words, these data do not contradict Mendel's hypothesis. □

Notice how we divide by the expected values in (11.1) to normalize the differences in the numerators. The reason is that we want the differences to be on the same scale — a difference of $(1001 - 1000)$ is much less severe than the difference $(2 - 1)$. In the first case, we have a single unit change when we normally expect 1000, so the relative difference is very small. In the second case, we expect a value of 1 but observe 2, which is twice as large. We need to normalize all of the categories to prevent one category with a large number of observations to "overshadow" the effects of the categories with fewer observations.

Note also that unlike the distributions we have seen earlier, the distribution of the test statistic is not directly affected by the sample size, n. However, the size of n will influence the test statistic. If we, for example, double the sample size but keep the same relationship between the observed and expected number of individuals, then the differences in the numerators will be twice as large. Since we square each numerator, we will get that each of them will be four times as large (two squared). The denominators, however, are not squared, so they will only be twice as large. Therefore, when we double the number of observations, the chi-square test statistic becomes twice as large too (provided there is the same relationship between observed and expected numbers). This yields a smaller p-value, as we would expect: a discrepancy between the data and the model is easier to detect if we have more observations available.

A final remark about the use of the chi-square test statistic to test hypotheses: the chi-square test statistic only approximately follows a $\chi^2(r)$ distribution. We have the following rule-of-thumb: We believe that we can approximate the distribution of the chi-square test statistic with a $\chi^2(r)$ distribution only if the expected number of each category is at least 5. We should always check that

$$n \cdot p_{01} \geq 5, \quad n \cdot p_{02} \geq 5, \ldots, \quad n \cdot p_{0k} \geq 5.$$

If the requirements from the rule-of-thumb are not fulfilled, then we do not trust the p-values we obtain from the $\chi^2(r)$ distribution. More advanced models or non-parametric methods are needed in those situations, but that is beyond the scope of this book.

11.2 2×2 contingency table

A typical situation is that observations are classified or tabulated according to two categorical variables and not just a single categorical variable, as we saw in the previous section. We shall start by looking at the simplest two-sided contingency table, where we have two categorical variables each with two categories, and then later extend the results to more general $r \times k$ contingency tables, where we have two categorical variables — one with r categories and the other with k categories.

11.2.1   Test for homogeneity

Let us start by assuming that we have independent samples from two populations and that for each individual we have observed a single dichotomous variable. We can summarize the data in a 2×2 table as shown in Table 11.3. In the table, n_1 is the sample size for population 1 and n_2 is the sample size for population 2. y_{11} of the observations in sample 1 are in category 1 of the dichotomous response, while the remaining $y_{12} = n_1 - y_{11}$ are in category 2.

Table 11.3: A generic 2×2 table

	Response 1	Response 2	Total
Sample 1	y_{11}	y_{12}	$n_1 = y_{11} + y_{12}$
Sample 2	y_{21}	y_{22}	$n_2 = y_{21} + y_{22}$
Total	$c_1 = n_{11} + n_{21}$	$c_2 = n_{12} + n_{22}$	$n = n_1 + n_2 = c_1 + c_2$

For the first population, we can look at the probability of observing response 1. If p_{11} is the probability of observing response 1 for an individual from population 1, then we have from Chapter 10 that

$$\hat{p}_{11} = \frac{y_{11}}{n_1}.$$

We can estimate the corresponding probability for population 2,

$$\hat{p}_{21} = \frac{y_{21}}{n_2}.$$

It is natural to be interested in the *test for homogeneity*

$$H_0 : p_{11} = p_{21},$$

which means that the probability of response 1 for population 1 is the same as response 1 for population 2; *i.e.*, the probability of observing a response 1 is the same for the two populations.

We can model the results from each population with a binomial model, since we view the 2×2 table as the outcomes from two independent samples and we have a response with exactly two possible outcomes. Hence, if we wish to test H_0, we could calculate a confidence interval for the difference $p_{11} - p_{21}$ from the two binomial distributions, as we did in Section 10.4. While that is a perfectly valid method for testing H_0, we will instead extend the chi-square goodness-of-fit test to include two-sided contingency tables because it turns out that the method also works when there are more than 2 populations or when the number of response categories is greater than 2.

Example 11.3. Avadex. The effects of the fungicide Avadex on pulmonary cancer in mice was studied by Innes et al. (1969). Sixteen male mice were continuously fed small concentrations of Avadex (the treatment population), while 79 male mice were given the usual diet (the control population). After 85 weeks, all animals were sacrificed and examined for tumors, with the results shown below:

	Tumor present	No tumor	Total
Treatment group	4	12	16
Control group	5	74	79
Total	9	86	95

If we let p_t and p_c denote the probabilities that a mouse from the treatment population and the control population, respectively, develops a tumor, then we can estimate those probabilities by

$$\hat{p}_t = \frac{4}{16} = 0.25 \quad \hat{p}_c = \frac{5}{79} = 0.0633.$$

Likewise, we could estimate the probabilities of *not* observing tumors for the two populations by $12/16 = 0.75$ and $74/79 = 0.9367$.

We wish to test the hypothesis

$$H_0 : p_t = p_c,$$

which states that the mice from the two groups develop tumors with the same probability.

Since there are exactly two categories for tumor status (presence and absence), this is a comparison of two binomial distributions and we can compute the confidence interval for $p_t - p_c$ as in Section 10.4. That only works because there are exactly two categories for tumor status: presence and absence. Thus we can model the number of tumors for each population as a binomial distribution, where $Y_t \sim \text{bin}(16, p_t)$ and $Y_c \sim \text{bin}(79, p_c)$. Formula (10.12) gave the confidence interval for $p_t - p_c$, which can be used to test H_0. The 95% confidence interval for $p_t - p_c$ becomes

$$(-0.0321, 0.4056),$$

which just includes zero, so we fail to reject the null hypothesis that the probability of tumors is the same in the treatment and control groups.   □

Just as in Example 11.3, we will compare the observed values to the expected values under the null hypothesis. We wish to compute the expected value of each cell in the contingency table under the assumption that H_0 is true. If we assume that the probability of observing response 1 is the same for both populations, then we can use the data from both populations to estimate that probability. Under H_0, there is no difference between the populations, so our combined estimate of the probability of observing response 1, p, would be

$$\hat{p} = \frac{c_1}{n} = \frac{y_{11} + y_{21}}{n}.$$

The estimate simply takes all the observations from both samples that had response 1 and divides by the total number of observations. Once we have the combined estimate for response 1, we can use this to compute the expected number of observations for each population. There are n_1 observations in population 1. If H_0 is true we would expect $\hat{p} \cdot n_1 = \frac{c_1 \cdot n_1}{n}$ of them to result in response 1. Likewise, we would expect $\frac{c_2 \cdot n_1}{n}$ of the individuals from population 1 to result in response 2 if H_0 is true. For population 2 we get $\frac{c_1 \cdot n_2}{n}$ and $\frac{c_2 \cdot n_2}{n}$ for responses 1 and 2, respectively. Notice how the expected value for a given cell in the table is calculated as the row total times the column total divided by the total number of observations. We can summarize this result as follows:

The expected number of observations in cell (i, j) for a contingency table is given by

$$E_{ij} = \frac{(\text{Row total for row } i) \cdot (\text{Column total for column } j)}{n} \qquad (11.2)$$

Now that we have the expected number of observations for each cell under H_0, we can carry out a test of H_0 using the same type of chi-square test statistic as in Section 11.1, where large values still are critical. When calculating the test statistic

$$X^2 = \sum_i \frac{(\text{observed}_i - \text{expected}_i)^2}{\text{expected}_i},$$

we sum over all possible cells in the table, so for a 2×2 table we would get four elements in the sum.

The value for the X^2 test statistic should be compared to a χ^2 distribution, where the degrees of freedom is

$$\text{df} = 1.$$

This may seem counter-intuitive since we had df = number of categories -1 in the previous section, but remember that we are testing two different hypotheses. For the full underlying model of the 2×2 table we have two free parameters: p_{11} for sample 1 and p_{21} for sample 2. p_{21} and p_{22} are not free since the sum of the parameters for each sample must be exactly 1, and once p_{11} is given then p_{21} must be fixed at $1 - p_{11}$. Thus, the full model contains two free parameters. Under the null hypothesis, we only have a single free parameter since h_0 restricts our two parameters to be identical. The difference in parameters between the full model and the model under the null hypothesis is $2 - 1 = 1$, which is why we always use 1 degree of freedom for a 2×2 table.

In Section 11.1 we had a rule-of-thumb that stated we should have an expected value of at least 5 for each category. We have the same rule-of-thumb for the 2×2 table: each of the expected values should be at least 5 in order for us to use the $\chi^2(1)$ distribution to compute the p-value.

Example 11.4. Avadex (continued from p. 301). Under the hypothesis H_0 : $p_t = p_c$, the probability of developing a tumor is not affected by population (treatment group). If we wish to estimate the probability of observing a tumor, p, based on both samples, then we get

$$\hat{p} = \frac{4+5}{16+79} = 0.0947.$$

If H_0 is true, we would expect $0.0947 \cdot 16 = 1.52$ of the mice from the treatment group and $0.0947 \cdot 79 = 7.48$ from the control group to develop tumors. We can calculate all four expected values and enter them in another 2×2 table:

	Tumor present	No tumor	Total
Treatment group	1.52	14.48	16
Control group	7.48	71.52	79
Total	9	86	95

When testing H_0, we calculate the test statistic

$$X^2 = \frac{(4 - 1.52)^2}{1.52} + \frac{(12 - 14.48)^2}{14.48} + \frac{(5 - 7.48)^2}{7.48} + \frac{(74 - 71.52)^2}{71.52}$$
$$= 5.4083,$$

which we should compare to a $\chi^2(1)$ distribution. This yields a p-value of 0.020, so we reject the null hypothesis and conclude that the relative frequencies of tumors are different in the two populations. By looking at the estimates $\hat{p}_t$ and $\hat{p}_c$, we conclude that Avadex appears to increase the cancer rate in mice.

Note that this conclusion is different from the conclusion we got in Example 11.3 on p. 301. When we computed the 95% confidence interval for the difference in tumor risk, we found that zero was included in the confidence interval; *i.e.*, we failed to reject the hypothesis of equal risks of tumor. We get different results from the two methods because the normal and χ^2 approximations are not identical for small sample sizes. However, we found that zero was barely in the 95% confidence interval and got a p-value for the chi-square test statistic of 0.020. These two results are really not that different, and when we get a p-value close to 0.05 or get a confidence interval where zero is barely inside or outside, then we should be careful with the conclusions.   □

11.2.2   Test for independence

So far we have viewed 2×2 tables as a way to classify two independent samples with one dichotomous variable. However, we can also obtain a 2×2 table if we have a *single sample* where we have observed two dichotomous variables, as indicated by Table 11.4. In Section 11.2 we tested the hypothesis of homogeneity since we had two independent populations. When we have a single sample scored for two categorical variables, we can make a *test of independence* between the two variables. If we let $p_{11}, p_{12}, p_{21},$ and p_{22} denote the probabilities of the four possible cells in the 2×2 table, then

$$\sum_{i=1}^{2} \sum_{j=1}^{2} p_{ij} = 1,$$

as the probabilities should add up to one. If the null hypothesis is true, then there is independence between the two variables, and the multiplication rule from Infobox 9.6 gives us

$$H_0 : P(\{\text{row } i\} \cap \{\text{column } j\}) = P(\{\text{row } i\}) \cdot P(\{\text{column } j\}).$$

In other words, the test of independence

$$H_0 : \text{The two variables are independent}$$

Table 11.4: A generic 2×2 table when data are from a single sample measured for two categorical variables (category 1 and category 2 are the two possible categories for variable 1, while category A and category B are the two possible categories for variable 2)

	Category A	Category B	Total
Category 1	y_{11}	y_{12}	$n_1 = y_{11} + y_{12}$
Category 2	y_{21}	y_{22}	$n_2 = y_{21} + y_{22}$
Total	$c_1 = y_{11} + y_{21}$	$c_2 = y_{12} + y_{22}$	$n = n_1 + n_2 = c_1 + c_2$

is equivalent to

$$H_0 : p_{ij} = p_i \cdot q_j \quad \text{for all } i \text{ and } j,$$

where the p_i's are the marginal probabilities of the first variable (*i.e.*, n_1/n and n_2/n) and the q_j's are the marginal probabilities of the second variable (*i.e.*, c_1/n and c_2/n). Note how the null hypothesis H_0 defines the same expected number of observations for each cell in the table, as we saw in formula (11.2).

It turns out that the test statistic for the 2×2 contingency table in this context is exactly the same as the chi-square test statistic (11.1) for the homogeneity test, and that we can approximate the test statistic with a $\chi^2(1)$ distribution provided the expected number for each cell is at least 5. Thus, the only difference between the two situations (two samples, one variable or one sample, two variables) is how we formulate the null hypothesis and how to interpret the result and state the conclusion. Let us look at an example.

Example 11.5. Mendelian inheritance (continued from p. 296). In example 11.1, we looked at the four possible phenotypic groups for pea plants. However, we can view those data as the result from a single sample with 556 observations and two dichotomous variables. We summarize the results in a 2×2 table:

	Yellow	Green	Total
Round	315	108	423
Wrinkled	101	32	133
Total	416	140	556

We are interested in testing the hypothesis of independence between pea color and pea shape, which we can write in different ways:

$$H_0 : \text{pea shape and pea color are independent}$$

or column-wise as

$$H_0 : P(\text{round} \mid \text{yellow}) = P(\text{round} \mid \text{green})$$

or row-wise as

$$H_0 : P(\text{yellow} \mid \text{round}) = P(\text{yellow} \mid \text{wrinkled}).$$

The last two hypotheses also suggest independence between shape and color, and all three hypotheses are equivalent, so any population that satisfies one of them also satisfies the others. If we calculate the chi-square test statistic, we get

$$X^2 = 0.1163,$$

which yields a p-value of $p = 0.733$ when compared to a $\chi^2(1)$ distribution*. Thus we fail to reject the null hypothesis and we conclude that pea color and pea shape occur independently of each other.   □

It may not always be obvious whether a 2×2 experiment should be viewed as two independent populations examined for one variable or as a single sample with two variables. However, this distinction is not critical for the test statistic and p-value, as both setups use the exact same procedure for hypothesis testing. If we have two independent groups we would normally test for homogeneity, but if we have two variables with interchangeable roles then we will usually interpret it as a test for independence, but the distinction is not always clear cut.

11.2.3   Directional hypotheses for 2×2 tables

The chi-square test statistic does not provide any information about the directionality of the test. If we find that a test is significant, then we need to compute and compare the estimated probabilities to conclude which of the groups has the highest probability.

In Section 6.1 we discussed directional alternative hypotheses for hypothesis tests. For a 2×2 table, the alternative can be non-directional (*i.e.*, $H_A : p_{11} \neq p_{21}$, like discussed above) or directional, such as

$$H_A : p_{11} > p_{21}$$

or

$$H_A : p_{11} < p_{21}.$$

p-values for directional alternatives are calculated using the following two-step procedure:

1. Estimate the probabilities for both populations.

---

*Notice how the chi-square test in Section 11.1 had 3 degrees of freedom, while it has only 1 degree of freedom here even though we are looking at the exact same data. This discrepancy occurs because we test very different hypotheses. In Section 11.1 we were given the probabilities from somewhere (*i.e.*, from Mendel's law in the example), while in this section we do not have any information about the probabilities from "outside" the experiment but have to estimate them from the data. This estimation costs in degrees of freedom, which is why there is only 1 in the 2×2 table.

2. If the estimates match the direction of the alternative hypothesis, then calculate the p-value using the procedure outlined above but *divide the resulting p-value by 2*. That is, compute the X^2 test statistic, compare the result to a $\chi^2(1)$ distribution, and divide the result by two.

If the estimated probabilities *do not* match the alternative hypothesis, then set the p-value to 0.5.

The reason behind this two-step procedure is as follows: if we use a directional hypothesis and the estimated proportions do not match the alternative, then the null hypothesis is most likely to be true. As an example, look at a hypothesis like

$$H_0 : p_{11} = p_{21} \quad \text{vs.} \quad H_A : p_{11} > p_{21}.$$

If we get estimates such that $\hat{p}_{11} < \hat{p}_{21}$, then the estimates (and hence the data) match the null hypothesis more than they resemble the alternative hypothesis. Hence, we fulfill the last condition in step 2 above. If the computed estimates match the alternative hypothesis, then we divide the p-value by two because we are interested in deviations from the null hypothesis in only one direction, but the test statistic automatically takes deviations in both directions into account: the X^2 test statistic squares every difference between observed and expected values and hence it weighs differences in both directions equally.

Example 11.6. Neutering and diabetes. The effect of early neutering of mice on the development of diabetes was tested on non-obese diabetic mice (NOD mice). The researchers suspect that neutering increases the prevalence of diabetes. One hundred mice were randomly split into two groups of the same size, and the 50 mice in one group were neutered the day after birth. Twenty-six mice from the neutered group were found to have diabetes after 112, days while only 12 of the non-neutered mice had diabetes. We can summarize the results in the 2 × 2 table:

	Diabetes	No diabetes	Total
Neutered group	26	24	50
Non-neutered group	12	38	50
Total	38	62	100

If we let p_1 and p_2 denote the probability of diabetes for the neutered and non-neutered groups, respectively, then we wish to test the hypothesis

$$H_0 : p_1 = p_2 \quad \text{vs.} \quad H_A : p_1 > p_2.$$

Note that the alternative hypothesis is directional since the researchers were only interested in examining whether or not castration increased the risk of diabetes.

To test the hypothesis, we first check whether the estimates correspond to the alternative hypothesis:

$$\hat{p}_1 = \frac{26}{50} = 0.52 \quad \hat{p}_2 = \frac{12}{50} = 0.24.$$

$\hat{p}_1 > \hat{p}_2$, which is in accordance with H_A, so we proceed to calculate the chi-square test statistic using formula (11.1). We get that $X^2 = 8.3192$, which we compare to a $\chi^2(1)$ distribution and get a p-value of 0.003923. Since we have a one-sided hypothesis, we divide this value by 2 to get a final p-value of 0.0020. Thus, we reject the null hypothesis and conclude that neutering does indeed increase the risk of diabetes.   □

11.2.4   Fisher's exact test

For a 2×2 contingency table, we need the expected number of observations in each cell to be at least 5 for the $\chi^2(1)$ approximation to hold. In this section we will briefly discuss an alternative method for hypothesis testing in 2×2 tables when this requirement is not fulfilled. The method is called *Fisher's exact test* and, as the name suggests, it is exact, so it will work for any 2×2 table. The test is exact because it uses the exact *hypergeometric distribution* rather than the approximate chi-square distribution to compute p-values. The method is, however, very computationally intensive and does not scale well to contingency tables much larger than 2×2, so we will give only a brief overview of the idea behind the method and then use a statistical software package to obtain the results.

Fisher's exact test is a statistical test used to determine if there are non-random associations between two categorical variables. The test considers all possible tables that would still result in the same marginal frequencies as the observed table. The p-value for the test of no association is defined as the sum of the probabilities of the tables with probabilities less than or equal to the probability of the observed table. Let us illustrate Fisher's exact test through an example.

Example 11.7. Avadex (continued from p. 301). Of the 95 mice in the experiment, 16 were in the Avadex treatment group and 79 were in the control group.

	Tumor present	No tumor	Total
Treatment group	4	12	16
Control group	5	74	79
Total	9	86	95

The null hypothesis is that the relative frequency of tumors is the same for both the treatment and the control group.

Using combinatorics (and the hypergeometric distribution), it can be shown that the probability of a specific table with fixed margins c_1, c_2, n_1, and n_2, like in Table 11.4, is

$$P(\text{"Table"}) = \frac{\binom{c_1}{y_{11}}\binom{c_2}{y_{12}}}{\binom{n}{n_1}}$$

$$= \frac{c_1!c_2!n_1!n_2!}{y_{11}!y_{12}!y_{21}!y_{22}!n!}$$

The probability of the observed table (under the null hypothesis) is:

$$P\left(\begin{bmatrix} 4 & 12 \\ 5 & 74 \end{bmatrix}\right) = \frac{\binom{9}{4}\binom{86}{12}}{\binom{95}{16}} = 0.0349.$$

Notice that once we change the value of one of the cells in the 2×2 table, the remaining three values will change automatically to keep the margins constant. For example, if we wish to calculate the probability of a table where there are 3 mice from the treatment group who developed tumors, then the probability of the table would have to be

$$P\left(\begin{bmatrix} 3 & 13 \\ 6 & 73 \end{bmatrix}\right) = 0.1325.$$

If we set the number of mice with treatment and tumor to 3, we must have 13 mice with treatment and no tumor if we wish to keep the number of mice in the treatment group constant. So once we change one value in the table, the others must adjust to keep the margins constant. Following the same procedure as above, we can calculate the probability of every possible table that has the same *margins* as our original table.

The p-value for the hypothesis that $H_0 : p_t = p_c$ is the sum of the probabilities of the tables that are at least as unlikely to occur as the table we have observed[†]. In this situation, it turns out that all the tables where there are four or more mice with tumors in the treatment group have a probability as small as our observed table. Consequently, the p-value becomes

$$
\begin{aligned}
p\text{-value} &= P\left(\begin{bmatrix} 4 & 12 \\ 5 & 74 \end{bmatrix}\right) + P\left(\begin{bmatrix} 5 & 11 \\ 4 & 75 \end{bmatrix}\right) + P\left(\begin{bmatrix} 6 & 10 \\ 3 & 76 \end{bmatrix}\right) + \\
&\quad P\left(\begin{bmatrix} 7 & 9 \\ 2 & 77 \end{bmatrix}\right) + P\left(\begin{bmatrix} 8 & 8 \\ 1 & 78 \end{bmatrix}\right) + P\left(\begin{bmatrix} 9 & 7 \\ 0 & 79 \end{bmatrix}\right) \\
&= 0.041.
\end{aligned}
$$

This p-value is slightly larger than the result from the $\chi^2(1)$ distribution. If we look at the table of expected values on p. 303, then we see that one of

---

[†]There are different definitions of the p-value for two-sided tests from Fisher's exact test, and there is no general consensus about which one is the right one to use. We use the same approach here as is used in R.

the expected values is less than 5, which was the rule-of-thumb-level. Since Fisher's exact test is exact and does not require an approximation to the $\chi^2(1)$ distribution, we trust this result better and conclude that we barely reject the hypothesis at the 5% level. $\qquad\qquad\square$

11.3   Two-sided contingency tables

The ideas from Section 11.2 are generalized to cover two-sided contingency tables that are larger than 2×2. In the following we will consider an $r \times k$ contingency table like the one shown in Table 11.5. The $r \times k$ contingency table can stem from either r independent populations, where each observation is from a single categorical variable with k possible categories, or from a single population, where observations are classified according to two categorical variables with r and k categories, respectively.

Table 11.5: A generic $r \times k$ table

	Column 1	Column 2	$\cdots$	Column k	Total
Row 1	y_{11}	y_{12}	$\cdots$	y_{1k}	n_1
Row 2	y_{21}	y_{22}	$\cdots$	y_{2k}	n_2
$\vdots$	$\vdots$	$\vdots$	$\vdots$	$\vdots$	$\vdots$
Row r	y_{r1}	y_{r2}	$\cdots$	y_{rk}	n_r
Total	c_1	c_2		c_k	n

If we view the data as r populations, then we can test for homogeneity, and if we view the data as a single sample with two categorical variables, then we can test for independence; *i.e.*, we wish to test either

$$H_0 : \text{homogeneity for the } r \text{ populations}$$

or

$$H_0 : \text{independence between the two categorical variables.}$$

In both cases we end up with the same chi-square test statistic as we saw in the 2×2 table:

$$X^2 = \sum_i \frac{(\text{observed}_i - \text{expected}_i)^2}{\text{expected}_i}, \qquad (11.3)$$

where the sum is over all possible cells in the $r \times k$ table and where the expected values are calculated like in formula (11.2); *i.e.*,

$$E_{ij} = \frac{(\text{Row total for row } i) \cdot (\text{Column total for column } j)}{n}.$$

We get the same formula for the expected values because we can use the exact same arguments for the $r \times k$ table as we did for the 2×2 table. The degrees of freedom for testing H_0 for an $r \times k$ table is

$$\mathrm{df} = (r-1) \cdot (k-1), \tag{11.4}$$

so we compare the X^2 test statistic to a $\chi^2((r-1) \cdot (k-1))$ distribution to obtain the p-value.

The degrees of freedom are calculated as follows: for the full model we have $r \cdot (k-1)$ parameters. $k-1$ parameters can vary freely since the proportions for the k groups must sum to one and we have r populations. Under the null hypothesis we assume that the probability for each column is the same for all populations, so there we have $k-1$ parameters. The difference is $r \cdot (k-1) - (k-1) = (r-1) \cdot (k-1)$. We can see that the degrees of freedom for the $r \times k$ tables give a value of 1 if we have $r = 2$ and $k = 2$, so we get the same result as we did for the 2×2 table.

Example 11.8. Cat behavior. In a study about cats taken to an animal shelter, it was registered whether the cats were taken to the shelter because of bad behavior and whether other animals were present in the household. A total of 1111 cats were included in this study.

Other animals	Behavioral problems	
	yes	no
Yes	53	502
No	115	410
No information	17	14

We can view this data in two possible ways:

- We have 3 populations (those with other animals in the household, those with no animals in the household, and those we do not know about), and for each population we have registered one categorical variable (behavior status) with two possible categories.

- We have a single sample of 1111 cats and we have subsequently registered two variables — behavior status (problems and no problems) and other animals in household (yes, no, or no information).

Regardless of how we view the statistical design, the expected values become

Other animals	Behavioral problems	
	yes	no
Yes	92.42	462.58
No	87.42	437.57
No information	5.16	25.84

The chi-square test statistic is

$$X^2 = \frac{(53 - 92.42)^2}{92.42} + \cdots + \frac{(14 - 25.84)^2}{25.84} = 63.1804,$$

which we compare to a χ^2 distribution with $(2 - 1) \cdot (3 - 1) = 2$ degrees of freedom. That gives a p-value of 0, so we clearly reject the null hypothesis of independence between other animals in the household and problems with behavior.

When we compare the observed and expected values, we see that cats who lived with no other animals in the household and have behavioral problems are overrepresented in the sample (assuming H_0 is true), and similarly that cats with problems and who lived with other animals are underrepresented. We conclude that other animals in the household appear to prevent problems with bad behavior. $\square$

11.4   R

Example 11.1 on p. 296 concerns the results from Mendel's experiment.

```
> obs <- c(315, 108, 101, 32)        # Input the observed data
> prop <- c(9/16, 3/16, 3/16, 1/16) # and expected proportions
> expected <- sum(obs)*prop          # Calculate expected number
> expected                           # based on the hypothesis.
[1] 312.75 104.25 104.25   34.75
> # Compute chi-square goodness-of-fit statistic
> sum((obs-expected)**2/expected)
[1] 0.470024
```

The chi-square goodness-of-fit test statistic should be compared to a χ^2 distribution with the correct degrees of freedom. The cumulative distribution function for the χ^2 distribution in R is called pchisq(), so we can evaluate the size of the test statistic by the call

```
> 1 - pchisq(0.470024, df=3)
[1] 0.9254259
```

where we have used the rightmost tail of the χ^2 distribution, since large values for the test statistic are critical. The degrees of freedom is specified by the option df.

The matrix() function can be used to enter data directly for a contingency table. Another possibility is to use table() to cross-classify factors and count the number of combinations for each factor level from a data frame. Here we will focus on using matrix() for direct input.

In Example 11.8 on p. 311 we saw data from a 3 × 2 table on cat behavior for 1111 cats.

```
> shelter <- matrix(c(53, 115, 17, 502, 410, 14), ncol=2)
> shelter
     [,1] [,2]
[1,]   53  502
[2,]  115  410
[3,]   17   14
> # Enter the data row-wise instead of column-wise
> matrix(c(53, 502, 115, 410, 17, 14), ncol=2, byrow=TRUE)
     [,1] [,2]
[1,]   53  502
[2,]  115  410
[3,]   17   14
```

The observations are included as a vector for the first argument in the call to matrix(), and the option ncol defines the number of columns in the matrix. R automatically determines the necessary number of rows based on the length of the vector of observations. Another possible option is nrow, which fixes the number of rows (and then R will calculate the number of columns). Note that R by default fills up the matrix column-wise and that the byrow option to matrix() should be used if data are entered row-wise.

To calculate the expected number of observations for each cell in the contingency table, we use formula (11.2), which requires us to calculate the sum of the table rows and columns. We can use the apply() function to apply a single function to each row or column of a table. In our case, we wish to use the sum() function on each row and column to get the row and column totals:

```
> rowsum <- apply(shelter, 1, sum) # Calculate sum of rows
> rowsum
[1] 555 525  31
> colsum <- apply(shelter, 2, sum) # Calculate sum of columns
> colsum
[1] 185 926
```

The first argument to apply() is the name of the matrix or array. The second argument determines if the function specified by the third argument should be applied to the rows (when the second argument is 1) or columns (when the second argument is 2). Once the row and column sums are calculated, we can use (11.2) to find the expected values. We can either do that by hand for each single cell in the table or get R to do all the calculations by using the outer matrix product, %o%, to calculate the product for all possible combinations of the two vectors that contain the row and column sums:

```
> # Calculate the table of expected values
> expected <- rowsum %o% colsum / sum(colsum)
```

```
> expected
          [,1]        [,2]
[1,]  92.416742  462.58326
[2,]  87.421242  437.57876
[3,]   5.162016   25.83798
```

Once we have the expected numbers, it is easy to calculate the chi-square test statistic:

```
> shelter-expected        # Difference between obs. and expected
          [,1]        [,2]
[1,]  -39.41674   39.41674
[2,]   27.57876  -27.57876
[3,]   11.83798  -11.83798
> sum((shelter-expected)**2/expected)   # Chi-square statistic
[1] 63.18041
> 1-pchisq(63.18041, df=2)              # Corresponding p-value
[1] 1.909584e-14
```

If we only wish to test the hypothesis of independence or homogeneity, then we do not need to do the calculations by hand but can use the chisq.test() function:

```
> chisq.test(shelter)

        Pearson's Chi-squared test

data:  shelter
X-squared = 63.1804, df = 2, p-value = 1.908e-14
```

The chisq.test() function produces a warning if any of the expected numbers are below 5 as shown below

```
> new.shelter <- shelter
> new.shelter[3,1] <- 12    # Change element 3,1
> chisq.test(new.shelter)

        Pearson's Chi-squared test

data:  new.shelter
X-squared = 47.6689, df = 2, p-value = 4.455e-11

Warning message:
In chisq.test(new.shelter) : Chi-squared approximation may be
  incorrect
```

R treats 2 × 2 tables slightly differently than larger tables since it by default uses a continuity correction (called Yates' continuity correction) for 2 × 2 tables. To get the exact same results as we have discussed in the text, we need to turn off Yates' continuity correction, which is done with the `correct=FALSE` option. If we look at the Avadex example on p. 301, we find that R produces the following results:

```
> avadex <- matrix(c(4, 5, 12, 74), ncol=2)
> avadex
     [,1] [,2]
[1,]    4   12
[2,]    5   74
> chisq.test(avadex)

        Pearson's Chi-squared test with Yates' continuity
        correction

data:  avadex
X-squared = 3.4503, df = 1, p-value = 0.06324

Warning message:
In chisq.test(avadex) : Chi-squared approximation may be
   incorrect

> chisq.test(avadex, correct=FALSE)

        Pearson's Chi-squared test

data:  avadex
X-squared = 5.4083, df = 1, p-value = 0.02004

Warning message:
In chisq.test(avadex, correct = FALSE) : Chi-squared
   approximation may be incorrect
```

Apart from the fact that both calls to `chisq.test()` produce the warning that at least one of the expected numbers is too low, we can see that the second call with `correct=FALSE` is the one that returns the same chi-square test statistic and *p*-value, as we found in Example 11.4 on p. 303. In this example, we have little confidence in comparing the chi-square test statistic to a χ^2 distribution since some of the expected numbers are too low, as shown by the warning. Instead, we can use Fisher's exact test from the `fisher.test()` function:

```
> fisher.test(avadex)

        Fisher's Exact Test for Count Data

data:  avadex
p-value = 0.04106
```

```
alternative hypothesis: true odds ratio is not equal to 1
95 percent confidence interval:
  0.834087 26.162982
sample estimates:
odds ratio
  4.814787
```

Based on Fisher's exact test, we get a *p*-value of 0.041, reject the null hypothesis (although just barely), and conclude that the relative tumor frequency is not the same for the Avadex and the control group. The output from fisher.test() also lists the odds ratio. The odds ratio will be explained in Section 12.1.

11.5   Exercises

11.1 Cheating gambler. A casino has reintroduced an old betting game where a gambler bets on rolling a '6' at least once out of 3 rolls of a regular die. Suppose a gambler plays the game 100 times, with the following observed counts:

Number of Sixes	Number of Rolls
0	47
1	40
2	11
3	2

The casino becomes suspicious of the gambler and suspects that the gambler has swapped the die for an unfair die that is more likely to roll six. What do they conclude? To answer that question we should

1. Determine the expected number of rolls with 0, 1, 2, and 3 sixes that we would expect if the dice were fair.
2. Compare the observed to the expected values using a chi-square test. State the conclusion in words.

11.2 Veterinary clinic. The veterinarians at a veterinary clinic want to investigate if the number of acute visits to the clinic is equally frequent over all days of the week, or if visits are more frequent on the weekend, when pet owners are not at work. If the visits are equally frequent for all days, then there is probability $\frac{1}{7}$ that a visit is on any given day. The number of visits was registered over a period of one year, and the veterinarians found that 87 visits were on a Saturday, 67 visits were on a Sunday, while 246 visits were on a workday (Monday-Friday).

1. Test if the hypothesis that visits for workdays, Saturdays, and Sundays are proportional to the number of days in each group fits the data.

A proportion of the acute visits to the clinic is about a contagious disease in dogs. The disease is found in 15% of the dogs that turn up at the clinic and the treatment requires that the dogs are isolated for 24 hours. The facilities at the clinic can isolate only one dog at a time.

2. What is the probability that there are not enough isolation rooms if there are 4 dogs that turn up at the clinic on a single day?

3. How many isolation rooms will the clinic need if there should be at least 99% probability that there are a sufficient number of isolation units if 4 dogs visit the clinic in a single day?

11.3 Pet preferences. A researcher wants to examine if there is a relationship between the gender of an individual and whether or not the person prefers dogs or cats. The researcher asked 88 persons which of the two animals they preferred. Of the 46 men in the experiment, 30 preferred a dog while only 15 of the 42 women preferred dogs.

1. Enter the results in a 2 × 2 table and test if there is significant association between a person's sex and whether the person prefers dogs or cats. Write the conclusion in words. In particular, if you find a significant association, then you should make sure to state which preferences the two sexes have.

2. Calculate a confidence limit for the true difference between the proportion of men and women who prefer dogs.

3. A new study is planned where the researchers want to estimate the proportion of men who prefer dogs over cats. How many men should they ask if they want a precision such that a 95% confidence interval has a total width of 0.05?

11.4 ⓡ Comparison of test statistics. A study by Willeberg (1976) tried to examine if the amount of exercise and the amount of food influences the risk of feline lower urinary tract diseases in cats. The table below shows the results for the cats that exercise a lot:

		Urinary tract disease		
		Yes	No	Total
Food	Normal	5	12	17
	Excessive	14	5	19
		19	17	36

1. Estimate the difference in probability of urinary tract disease between cats with normal and excessive amounts of food. Test the hypothesis that this difference is zero.

2. Make a chi-square test for independence in the 2 × 2 table. What is your conclusion?

3. Use `fisher.test()` in R to test independence. What is your conclusion?

4. Compare and discuss the results of the previous 3 questions that essentially test the same thing. Do the four tests always provide the same result?

11.5 Malaria. The incidence of three types of malaria in three tropical regions is listed in the table below:

	South America	Asia	Africa
Malaria strain A	451	313	145
Malaria strain B	532	28	56
Malaria strain C	27	539	456

Test if there is independence between the region and the type of malaria. Remember to quantify any results and write the conclusion in words.

11.6 Fisher's exact test. A sample of 27 university students is asked if they are currently on a diet. The data look like this:

	Men	Women
Dieting	1	8
Not dieting	11	7

Test the hypothesis that the proportion of people on a diet is the same for male and female students. State the conclusion in words.

11.7 ℝ Distribution of boys and girls in large families. Consider the data from Part IV of Case 8 (p. 368). In the case we used an *ad hoc* method to compare the observed and expected values for different combinations of boys and girls from families with 12 children. In this exercise, we shall make a proper test for the binomial model.

1. Calculate a goodness-of-fit test for the binomial model.

   [Hint: Note that the goodness-of-fit model requires that the expected value for each category is at least 5, which is clearly not the case here. One way to remedy this is to group some of the categories together. For example: The outcome "0 boys and 12 girls" has an expected value of 0.9, which is less than 5. We could then group category "0 boys" with "1 boy" and obtain a new category called "0 or 1 boys", which has expected probability $P(Y = 0) + P(Y = 1)$. If the expected value of this new group is at least 5, then we can continue. If not, we need to group more categories with their neighboring categories to increase the expected number of observations.]

2. What is your conclusion about fitting a binomial model to the data?

Chapter 12

Logistic regression

If the response variable is binary, we can use *logistic regression* to investigate how multiple explanatory variables influence the binary response. The situation is similar to that of linear models (Chapter 8), except that the normal distribution is not appropriate when the response is binary rather than quantitative. In logistic regression we use the binomial distribution (Chapter 10) and let the explanatory variables influence the probability of success.

12.1 Odds and odds ratios

It is quite common to test the null hypothesis that two population proportions, p_1 and p_2, are equal. We have already seen two ways to test this. In Section 10.4 we constructed a confidence interval for the difference $p_1 - p_2$ and then checked if zero was included in the interval. In Section 11.2 we saw how we could make a formal test in a 2×2 table by a chi-square test or Fisher's exact test. In this section we shall consider another measure of dependence between two proportions.

The *odds* of an event A are calculated as the probability of observing event A divided by the probability of not observing event A:

$$\text{odds of } A = \frac{P(A)}{1 - P(A)}. \tag{12.1}$$

For example, in humans an average of 51 boys are born in every 100 births, so the odds of a randomly chosen delivery being a boy are:

$$\text{odds of boy} = \frac{0.51}{0.49} = 1.04.$$

If the probability of an event is 0.5, then the odds of that event is 1. If the odds of an event are greater than one, then the event is more likely to happen than not (the odds of an event that is certain to happen is infinite); if the odds are less than one, the event is less likely to happen.

The *odds ratio* is the ratio of two odds from different populations or conditions. If we let p_1 denote the probability of some event A in one population/under one condition and p_2 denote the probability of A under another condition, then

$$\theta = \frac{\frac{p_1}{1-p_1}}{\frac{p_2}{1-p_2}} = \frac{p_1 \cdot (1 - p_2)}{p_2 \cdot (1 - p_1)} \tag{12.2}$$

is the odds ratio of A. An odds ratio of 1 means that event A is equally likely in both populations. An odds ratio greater than one implies that the event is more likely in the first population, while an odds ratio less than one implies that the event A is less likely in the first population relative to the second population. The odds ratio measures the effect size between two binary data values.

The estimate of the odds ratio is

$$\hat{\theta} = \frac{\hat{p}_1 \cdot (1 - \hat{p}_2)}{\hat{p}_2 \cdot (1 - \hat{p}_1)}, \tag{12.3}$$

and it can be proven by using the central limit theorem that *the logarithm* of $\hat{\theta}$ has a standard error which is approximately given by

$$SE(\log \hat{\theta}) = \sqrt{\frac{1}{y_{11}} + \frac{1}{y_{12}} + \frac{1}{y_{21}} + \frac{1}{y_{22}}}, \tag{12.4}$$

where the y's are the observations from the 2×2 table (see Table 11.3). A 95% confidence interval for $\log(\theta)$ is therefore given by

$$\log(\hat{\theta}) \pm 1.96 \cdot SE(\log \hat{\theta}),$$

which we can then back-transform to get a confidence interval for θ on the original scale. Note that we use 1.96 from the normal distribution because the result is based on the central limit theorem, so we believe we have such a large number of observations that the limits from the normal distribution are correct.

Example 12.1. Avadex (continued from p. 301). In the example with Avadex and mice, we have that

$$\hat{p}_t = \frac{4}{16} = 0.25 \quad \hat{p}_c = \frac{5}{79} = 0.0633.$$

The estimated odds of tumors for the Avadex treatment group are

$$\widehat{odds} = \frac{0.25}{0.75} = \frac{1}{3} = 0.3333$$

while the estimated odds for the control group are

$$\widehat{odds} = \frac{0.0633}{0.9367} = 0.06757.$$

The estimated odds ratio is

$$\hat{\theta} = \frac{0.3333}{0.06757} = 4.9333.$$

Thus we estimate that the odds of having tumors are 4.9 times as great for Avadex mice as for control mice. The confidence interval for $\log(\theta)$ is

$$\log(4.9333) \pm 1.96 \cdot \sqrt{\frac{1}{4} + \frac{1}{12} + \frac{1}{5} + \frac{1}{74}} = (0.147, 3.045),$$

and if we transform that back with the exponential function to the original scale, we get a 95% confidence interval for θ:

$$(1.15, 21.02).$$

We say that values between 1.15 and 21.02 are in agreement with the data at the 95% confidence level. Alternatively, we can conclude that we are 95% confident that the odds of tumors are between 1.15 and 21.02 times greater in the Avadex treatment group compared to the control group. $\square$

Some people find it easier to interpret the difference $p_1 - p_2$ than the odds ratio. However, the odds ratio has some nice properties; in particular, there are some statistical designs (*e.g.*, case-control designs) where p_1 and p_2 cannot be estimated, but where it is still possible to calculate the odds ratio. Also, odds and odds ratios are natural for specification of logistic regression models.

12.2 Logistic regression models

Let n be the number of observations in a dataset where we want to model a binary response (*e.g.*, success/failure, alive/dead, or healthy/ill). In addition, we assume that we have information on d explanatory variables. The *logistic regression model* for this experiment is written as

$$Y_i \sim \mathrm{bin}(1, p_i), \quad i = 1, \ldots, n, \tag{12.5}$$

in combination with an expression for p_i. Here Y_i denotes whether observation i was a success ($Y_i = 1$) or a failure ($Y_i = 0$) and where the n observations are assumed to be independent. We assume that each observation follows a binomial distribution, but unlike the setup presented in Chapter 10, we now let the probability of success, p_i, depend on the explanatory variables for observational unit i. As a consequence, each individual observation need not have the same probability of success. The log odds for the event $Y = 1$ are

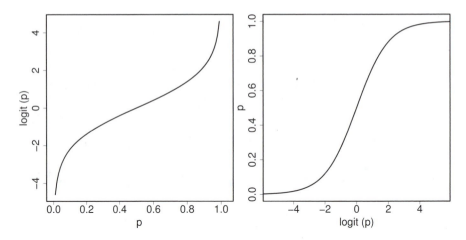

Figure 12.1: The logit transformation for different values of p (left panel) and the inverse function: p as a function of the logit value (right panel).

also called the *logit* of p_i and are defined as

$$\text{logit}\,(p_i) = \log\left(\frac{p_i}{1-p_i}\right). \tag{12.6}$$

The logit function is depicted in Figure 12.1 together with its inverse function. From the right panel in Figure 12.1 we see that no matter which values we have on the logit scale (the x-axis), we keep the probability of success, p, (seen on the y-axis) in the range from 0 to 1. If we had used a linear function, then some x-values would result in values on the y-axis outside the interval from zero to one. These values could not be interpreted as probabilities.

We cannot let all the p_i's vary freely, since we would then have a parameter for every observation in the dataset. Instead, we place some restrictions — in the same fashion as in the linear model — on *how* the p_i's can vary, such that they depend on the explanatory variables:

$$\text{logit}\,(p_i) = \alpha + \beta_1 x_{i1} + \cdots + \beta_d x_{id}, \quad i = 1, \ldots, n. \tag{12.7}$$

Note that we use the explanatory variables to model *the probability* of success for observation i, and that it is linear on the logit scale. This ensures that the probabilities p_i are always between 0 and 1 regardless of the right-hand side of formula (12.7). It is possible to back-transform the results if we desire to present the results as actual probabilities:

$$p_i = \frac{\exp(\alpha + \beta_1 x_{i1} + \cdots + \beta_d x_{id})}{1 + \exp(\alpha + \beta_1 x_{i1} + \cdots + \beta_d x_{id})}. \tag{12.8}$$

The logistic regression model defined by (12.5) and (12.7) can handle both quantitative and categorical explanatory variables even though we have formulated the logistic regression model (12.7) in the same way as we defined the multiple regression model. To include categorical explanatory variables, all we need to do is define the categories through dummy variables, as described in Section 8.2.2. Unlike in linear models, we see from definition (12.7) that there is no error term. This is because the uncertainty about Y_i comes from the binomial distribution and hence is determined directly by the probability of success, as we saw in (10.2).

It is worth emphasizing that all estimates are on the log odds scale. In particular, $\hat{\beta}_j$ is the estimated additive effect on the log odds if variable x_j is increased by one unit, whereas $\exp(\hat{\beta}_j)$ is the estimated change in odds. Likewise, if variable x_j is a dummy variable (*e.g.*, if 1 represents males and 0 females), then $\hat{\beta}_j$ will be the estimated log odds ratio for observing a success for males relative to observing a success for females.

Logistic regression is a massive subject, and readers are referred to Kleinbaum and Klein (2002). The logistic regression model is part of a larger class of models called *generalized linear models*, not to be mistaken for the class of linear models described in Chapter 8, which only allows for normally distributed errors. (The class of linear models, however, is also part of the generalized linear models.) See McCullagh and Nelder (1989) for more details on generalized linear models.

Example 12.2. Moths. Collett (1991) presented a study where groups of 20 male and female moths were exposed to various doses of *trans-cypermethrin* in order to examine the lethality of the insecticide. After three days it was registered how many moths were dead or immobilized. Data are shown in the table below:

	\multicolumn{6}{c|}{Dose (μg)}					
Sex	1	2	4	8	16	32
Males	1	4	9	13	18	20
Females	0	2	6	10	12	16

In this example, we will look only at male moths, and we would like to model the effect of dose on the proportion of moths that die. We use a logistic regression model as defined by (12.5) and (12.7) and state that logit of the probability that moth i dies is given as

$$\text{logit}(p_i) = \alpha + \beta \cdot \text{dose}_i.$$

We have two parameters in this model: α and β. The intercept α should be interpreted as log odds of a male moth dying if it is not exposed to the insecticide (*i.e.*, the dose is zero), and β is the increase in log odds of dying for every unit increase in dose. If we fit a logistic regression model to the data for the male moths, we get

$$\hat{\alpha} = -1.9277 \quad \text{and} \quad \hat{\beta} = 0.2972.$$

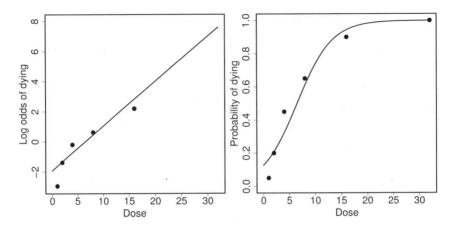

Figure 12.2: The log odds (left) and the probability (right) of male moths dying based on the estimated logistic regression model. Points represent the observed relative frequencies for the six different doses.

We conclude that the odds that a male moth dies without being exposed to the insecticide are $\exp(\hat{\alpha}) = 0.1455$, which corresponds to an estimated probability of $\exp(-1.9277)/(1 + \exp(-1.9277)) = 0.1270$. Moreover, $\exp(\hat{\beta}) = 1.3461$, which is the estimated odds ratio of a male moth dying at dose $x + 1$ relative to a male moth dying at dose x. Thus, we can say that the odds ratio of dying increases by 34.61% every time the dose increases by one unit.

The estimated model can be seen in Figure 12.2. The left-hand side shows the linear relationship between dose and log odds for death, whereas the right-hand side shows the relationship between dose and probability of dying. Notice that there are no observed log odds for dose 32 because all moths died in that group. We have used formula (12.8) to calculate the probability of dying for each dose based on the estimated parameters. For dose 16, formula (12.8) gives that the probability of death is

$$p = \frac{\exp(-1.9277 + 16 \cdot 0.2972)}{1 + \exp(-1.9277 + 16 \cdot 0.2972)} = \frac{16.9032}{1 + 16.9032} = 0.9441,$$

and similarly for the other doses.   □

Example 12.3. Feline urological syndrome. In Exercise 11.4 we used data from a study by Willeberg (1976) to investigate feline urinary tract disease. The investigators looked at both food intake and exercise as possible explanatory variables for the disease. The complete data are shown in Table 12.1.

Here we have two categorical explanatory variables, exercise and food intake. If we use an additive logistic regression model (*i.e.*, a model where there is no interaction between food and exercise), then we have

$$\mathrm{logit}(p_i) = \alpha_{g(i)} + \beta_{h(i)}, \quad i = 1, \ldots, n,$$

Table 12.1: Data on urinary tract disease in cats

Food	Exercise	Tract disease	Total
Normal	Little	3	8
Normal	Much	5	17
Excessive	Little	28	30
Excessive	Much	14	19

and the result for cat i follows a binomial distribution $\text{bin}(1, p_i)$. We get the following estimates:

Parameter	Estimate	Standard error
Intercept	2.2737	0.5400
$\alpha_{normal} - \alpha_{excessive}$	-2.3738	0.6058
$\beta_{much} - \beta_{little}$	-1.0142	0.6140

As usual for categorical explanatory variables, the intercept parameter corresponds to a reference level. In this case the reference is a cat with excessive food intake and little exercise. Cats with excessive food intake and little exercise have odds for urinary tract disease of $\exp(2.2737) = 9.7153$, which corresponds to a probability of urinary tract disease of 0.9067. Note that we can compare this probability based on the model to the observed relative frequency of $28/30 = 0.9333$. The odds ratio for urinary tract disease between normal and excessive food intake is

$$\exp(-2.3738) = 0.0931.$$

Thus, the odds of urinary tract disease for normal food intake is 0.0931 times the odds of urinary tract disease for excessive food intake. Cats with normal food intake and little exercise have odds of urinary tract disease of

$$\exp(2.2737 + (-2.3738)) = \exp(-0.1001) = 0.9047,$$

which corresponds to a probability of

$$\frac{\exp(-0.1001)}{1 + \exp(-0.1001)} = 0.4750.$$

From the parameters, we see that there is a positive effect of normal food intake relative to excessive food intake and that much exercise reduces the odds of urinary tract disease relative to little exercise. $\square$

12.3 Estimation and confidence intervals

For linear models, we estimated the parameters using the method of least squares by minimizing the sum of squared deviations of predicted values from observed values. For logistic regression, least squares estimation is not desirable, since least squares estimation

- assumes that the errors are normally distributed, but binary variables are not quantitative and hence not normally distributed.

- assumes variance homogeneity for all observations, but that is not the case for binary data as the variance depends on the probability of success.

For most of the generalized linear models, the least squares estimation is not a satisfactory technique to estimate the parameters because the results are biased — the estimates are consistently too large or too small. Instead, we use maximum likelihood estimation (see Section 5.2.7) to find the parameter values that best fit the data. Different sets of parameter values are compared through the likelihood function, which for a given set of parameter values describes the probability of the observed data (given that the parameter values in question are the true ones). The maximum likelihood estimates are the parameter values that make this probability as large as possible.

The likelihood function for the logistic regression model is

$$L(\alpha, \beta_1, \ldots, \beta_d; y) = \prod_{i=1}^{n} p_i^{y_i} (1 - p_i)^{1-y_i}, \quad i = 1, \ldots, n, \tag{12.9}$$

where we have used the multiplication rule (9.6), since the observations are assumed to be independent, and where the parameters enter the likelihood function through the p_i's as described by (12.7). The maximum likelihood estimates are the set of parameters that maximize (12.9). A computer is required to calculate the estimates, as we cannot find any closed-form estimates that maximize the likelihood, and an iterative procedure is used, where the current set of parameter estimates are refined until the maximum of the likelihood is obtained. We will not go into more detail about maximum likelihood estimation for logistic regression models but refer interested readers to Mc-Cullagh and Nelder (1989).

12.3.1 Complete and quasi-complete separation

As described above, we use maximum likelihood estimation to fit a logistic regression model and estimate the parameters. However, no estimation technique can handle the problem of *complete separation*, where an explanatory variable or a linear combination of explanatory variables from a model perfectly predicts the response value. This is a common problem in logistic regression since there is very little variation in binary responses — especially if a dataset is small relative to the number of parameters in the model.

For example, consider a dataset where every response is 0 (a failure) if the dose is less than a certain threshold, and where the response is always 1 (a success) if the dose is greater than the threshold. The value of the response can be perfectly predicted for the observational units in the sample simply by checking if the dose is less than or greater than the threshold. In this situation the maximum likelihood estimate for the parameter becomes infinity. As a consequence, statistical software programs often report parameter estimates that are very large (or very small) or extremely large standard errors of parameter estimates. These values do not make sense from a point of view of interpretation.

Quasi-complete separation denotes the situation where separation occurs for only a subset of the observations in the data. Under quasi-complete separation, the parameter estimate for the separating variable (and its standard error) will also be infinite in size, but the estimates for the other explanatory parameters may remain largely unaffected.

If complete or quasi-complete separation is detected, the explanatory variables exhibiting separation should be removed from the analysis.

Example 12.4. Moths (continued from p. 325). Previously we considered dose as a quantitative explanatory variable for the moth dataset, but we also could have included dose as a categorical explanatory variable. Thus we have a model where the response, Y_i, for moth i is modeled as $Y_i \sim \text{bin}(1, p_i)$, and where

$$\text{logit}(p_i) = \alpha_{g(i)}, \ i = 1, \dots, 120, \tag{12.10}$$

and $g(i)$ defines which level of dose was given to moth i. This means that we assign a parameter to each category and that this parameter is the odds of death for the given dose. For example, α_1 is the odds of death for dose 1, and we can estimate it as $\hat{\alpha}_1 = \frac{1/20}{19/20} = 1/19 = 0.0526$. However, the odds for dying for dose 32 become $\hat{\alpha}_{32} = \frac{20/20}{0/20} = 20/0$, which is not well-defined. Thus for model (12.10), we have quasi-complete separation.

Note that if we consider dose as a quantitative explanatory variable, then we do not have quasi-complete separation for this dataset, since there is not a value of dose that separates the deaths from the survivals. □

12.3.2 Confidence intervals

In Sections 5.3 and 10.3 we calculated confidence intervals for parameters by using formula (5.20),

$$\text{estimate} \pm \text{quantile} \cdot \text{SE(estimate)}.$$

We can use the same approach for logistic regression models if we use the quantile from the normal distribution. This is because maximum likelihood estimates are asymptotically normally distributed. The parameter estimates and corresponding standard errors can be calculated by a computer program.

Example 12.5. Moths (continued from p. 325). If we use maximum likelihood to fit the logistic regression model to the moth data, we get the estimates and standard errors shown in the table below:

Parameter	Estimate	SE	z-value	p-value
α	-1.92771	0.40195	-4.796	<0.0001
β	0.29723	0.06254	4.752	<0.0001

We compute the 95% confidence interval for α by

$$-1.9278 \pm 1.96 \cdot 0.40195 = (-2.7156, -1.1400)$$

and consequently the 95% confidence interval for the odds of death when the dose is zero becomes
$$(0.06616, 0.3198)$$

by transformation with the exponential function. Likewise, the 95% confidence interval for β is $0.29723 \pm 1.96 \cdot 0.06254 = (0.1747, 0.4198)$, and if we transform this back we get a 95% confidence interval of

$$(1.1908, 1.5217)$$

for the odds ratio of a unit change in dose.   □

12.4   Hypothesis tests

We cannot use the t and F distributions when testing hypotheses about the parameters from a logistic regression model since these two distributions rely on the assumption that the errors are normally distributed. Hypotheses are instead tested using a Wald test or the extremely general and flexible *likelihood ratio test*.

12.4.1   Wald tests

Maximum likelihood estimates are asymptotically normally distributed; *i.e.*, if we have a large number of observations, then the estimate is normally distributed. Thus, we can use the normal distribution to test a hypothesis about a single parameter, and the *Wald test statistic*,

$$Z_{obs} = \frac{\text{estimate} - \text{true value}}{\text{SE(estimate)}},$$

approximately follows a $N(0, 1)$ distribution. Consequently, if we wish to test a simple hypothesis about a single parameter (*e.g.*, $H_0 : \beta = 0$), then we can compare the Wald test statistic, Z_{obs}, with the true value equal to zero to the standard normal distribution. In particular, we reject the null hypothesis at significance level 0.05 if $|Z_{obs}| \geq 1.96$.

Example 12.6. Moths (continued from p. 330). For the moth data, we may for example test the hypotheses

$$H_0 : \alpha = 0 \text{ vs. } H_A : \alpha \neq 0$$

or

$$H_1 : \beta = 0 \text{ vs. } H_B : \beta \neq 0.$$

The table with the estimates and corresponding standard errors is reproduced below, now including the Wald test statistic and the corresponding p-value:

Parameter	Estimate	SE	z-value	p-value
α	-1.92771	0.40195	-4.796	<0.0001
β	0.29723	0.06254	4.752	<0.0001

The Wald test statistic for the hypothesis H_0 is

$$Z_{obs} = \frac{-1.92771}{0.40195} = -4.796.$$

If we compare this value to the $N(0, 1)$ distribution, we get the p-value

$$p\text{-value} = 2 \cdot P(Z \geq |-4.796|) < 0.0001.$$

Thus, we reject the null hypothesis H_0 and conclude that the log odds for moths that receive dose 0 are significantly different from zero. This means that the odds for moths that receive dose 0 are significantly different from 1 or that the probability of death for dose 0 is significantly different from 0.5.

The test statistic for hypothesis H_1 is

$$Z_{obs} = \frac{0.29723 - 0}{0.06254} = 4.752,$$

which results in a p-value of < 0.0001. Thus there is a significant effect of the insecticide, and since $\hat{\beta} > 0$, we conclude that the proportion of dead moths increases with increasing dose. $\square$

12.4.2 Likelihood ratio tests

The Wald test is useless if we wish to test more complicated hypotheses with restrictions on several parameters — for example, if we have a categorical explanatory variable with more than two categories and we wish to test a hypothesis like $H_0 : \beta_1 = \cdots = \beta_k$. We need a measure of model fit which measures the discrepancy between the model and the data — just like we used the residual sum of squares for normal models — in order to make a general test that can accommodate restrictions on more than one parameter.

In Section 6.4 we discussed hypothesis tests for comparison of nested linear normal models. We used the change in residual sum of squares between two models to evaluate whether or not they are significantly different. We will follow the same idea here but need another measure to compare two nested model. For generalized linear models this measure is called the *deviance*, and we test a hypothesis by comparing the deviance between the two models.

The *log likelihood* is the natural logarithm of the likelihood (12.9), and inserting the maximum likelihood estimates gives us a measure of the model fit. The larger the log likelihood, the better the model fit. Twice the difference between two log likelihoods is called the *likelihood ratio test statistic*, and it can be used to measure the deviance between the two models,

$$\text{LR} = 2 \cdot (\log(L_{\text{full}}) - \log(L_0)), \tag{12.11}$$

where L_0 is the likelihood under the null hypothesis and L_{full} is the likelihood under the full model. Thus we can use a likelihood ratio test to compare the fit of two models. Large values of LR are critical for the hypothesis, whereas small values are in support of the hypothesis. It turns out that the distribution of the likelihood ratio test statistic asymptotically follows a χ^2 distribution with a number of degrees of freedom that is the difference in the number of parameters between the two models. Hence, the p-value is

$$p\text{-value} = P(X^2 \geq \text{LR}),$$

where X^2 is χ^2 distributed with a number of degrees of freedom that equals the difference in number of parameters between the two models.

The χ^2 approximation is more reliable than the normal approximation for Wald test statistic (in the sense that it gives more reliable results) and can also be used to test a hypothesis about a single parameter. However, the Wald test statistic is computationally easy and is given automatically in the output of most software programs.

Example 12.7. Moths (continued from p. 329). In Example 12.4 we considered dose as a categorical explanatory variable for the moth data. Recall from Example 12.4 that we have quasi-complete separation for the moth data when we include dose 32 and consider dose as a categorical variable. Quasi-complete separation presents a problem with parameter estimation but has little impact on the maximum of the likelihood function when we fit a model. This is because observations where p_i is either zero or one do not contribute to the likelihood function (12.9).

The estimates for the log odds for dose 1 as well as the contrasts relative to dose 1 (*i.e.*, log odds ratios) are calculated on a computer and are summarized in the following table:

Parameter	Estimate	SE	z-value	p-value
α_1	-2.944	1.026	-2.870	0.00411
$\alpha_2 - \alpha_1$	1.558	1.168	1.334	0.18234
$\alpha_4 - \alpha_1$	2.744	1.120	2.450	0.01430
$\alpha_8 - \alpha_1$	3.563	1.128	3.159	0.00158
$\alpha_{16} - \alpha_1$	5.142	1.268	4.054	< 0.0001
$\alpha_{32} - \alpha_1$	28.696	52998.405	0.001	0.99957

First, we can notice that the estimate for $\alpha_{32} - \alpha_1$ is very large and has an extremely large standard error. That is a consequence of the quasi-complete separation. We should emphasize that the corresponding Wald test ($p = 0.99957$) is useless, and it cannot be used to test the hypothesis that $\alpha_{32} - \alpha_1 = 0$ because the Wald test is based on the incorrect estimate and standard error. The contrast between doses 8 and 1 is 3.563, so the odds ratio of dying between doses 8 and 1 is $\exp 3.563 = 35.2689$. Since we have a parameter for each dose, we find the observed relative frequencies when we use (12.8). For example, the relative frequency for dose 1 is $\exp(-2.944)/(1 + \exp(-2.944)) = 0.05$ and for dose 8 it is $\exp(-2.944 + 3.563)/(1 + \exp(-2.944 + 3.563)) = 0.65$.

We wish to test the hypothesis that there is no effect of dose, which corresponds to the hypothesis

$$H_0 : \alpha_1 = \cdots = \alpha_{32}.$$

To test H_0, we use a likelihood ratio test. The log likelihood for the full model is $\log(L_{full}) = -7.1735$, and under the null hypothesis the log likelihood becomes $\log(L_0) = -42.7423$. The likelihood ratio test statistic is

$$LR = 2 \cdot (-7.1735 - (-42.7423)) = 71.138,$$

which we should compare to a χ^2 distribution with $6 - 1 = 5$ degrees of freedom. The resulting p-value is < 0.0001, and we reject the null hypothesis that the odds of dying are the same for all doses. □

12.5 Model validation and prediction

There is no obvious way to perform adequate graphical model validation for logistic regression models. The observations attain only one of two possible values, and the variance of the response (and hence the residuals) is a function of the mean value, so some modifications to the techniques used for normal data are needed.

The predicted value for a logistic regression is $\hat{y}_i = \hat{p}_i$, where $\hat{p}_i$ is found by inserting the estimated parameters in (12.8). Note that we are predicting the probability of success. Once we have the predicted value, we can use the same approach for computing residuals as we used for the normal models; *i.e.*, compare the observed data to their expected values.

The *Pearson residuals* are defined as

$$r_i = \frac{y_i - \hat{y}_i}{\sqrt{\hat{y}_i \cdot (1 - \hat{y}_i)}}, \qquad (12.12)$$

so they are just the regular residuals divided by the estimated standard deviation of y_i^*. Just as in Section 7.1.1 on normal models, we should standardize these residuals to make sure that all Pearson residuals have the same unit variance and hence base the residual analysis on the standardized residuals. The standardized Pearson residuals are easily calculated by statistical software programs, so we will not go into detail here.

The residual plot is a plot of the standardized residuals (standardized r_i) against the predicted values, $\hat{y}_i$. Since y_i is either zero or one, the points will fall into two groups: if y_i is zero then the residual is negative, but if y_i is one then the residual is positive. A small value of $\hat{y}_i$ corresponds to a small probability of observing the value one, and there should thus be only a few positive residuals for small predicted values. Similarly, there should be only a few negative residuals for large predicted values. As a rule-of-thumb, observations with a standardized Pearson residual exceeding two in absolute value are worth a closer look.

Example 12.8. Nematodes in mackerel. Roepstorff et al. (1993) investigated if cooling right after catching prevents nematodes (roundworms) from moving from the belly of mackerel to the fillet. A total of 150 mackerels were investigated and their length, number of nematodes in the belly, and time before counting the nematodes in the fillet were registered. The response variable is binary: presence or absence of nematodes in the fillet.

We model this dataset as a logistic regression model, where p_i is the probability of mackerel i having nematodes in the fillet and where

$$\text{logit}(p_i) = \alpha + \beta_1 \cdot \text{length}_i + \beta_2 \cdot \text{nematodes in belly}_i + \beta_3 \cdot \text{time}_i.$$

Figure 12.3 shows the residual plot for the mackerel data. We can see two "curves", where the upper curve corresponds to the successes (the mackerels with nematodes in the fillet) and the lower curve is the failures (the mackerels without nematodes in the fillet). Because the response variable can attain only two possible values, we get these two "collections" of points, and it can be difficult to determine if there is any pattern in the standardized residuals.

---

*The reason we point out that these residuals are Pearson residuals is because it is also possible to define a slightly different type of residual, the *deviance residuals*, which can be used with different types of generalized linear models. The deviance residuals for the ith observation are defined by their contribution to the deviance score. However, we will look only at Pearson residuals here.

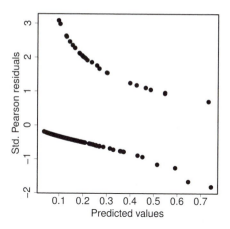

Figure 12.3: Residual plot for the mackerel data.

From the residual plot in Figure 12.3, it appears as if there are too many successes for small predicted values, since quite a few observations have standardized Pearson residuals numerically greater than 2. This suggests that there are some observations which the model is not able to capture.
□

An alternative to graphical model validation is to compare observed and fitted proportions. This is almost pointless with continuous explanatory variables, since each observation may have a unique value, but if we group the continuous explanatory variables we end up with grouped data where we are able to compare the observed frequencies in each group with the corresponding expected frequencies. Then we can use the technique from Chapter 11 and calculate the chi-square goodness-of-fit test statistic (11.1) as a measure of model fit. Hence, we compute

$$X^2 = \sum_{j=1}^{J} \frac{(\text{observed}_j - \text{expected}_j)^2}{\text{expected}_j} = \sum_{j=1}^{J} \frac{(\text{observed}_j - n_j \hat{p}_j)^2}{n_j \hat{p}_j}, \quad (12.13)$$

where the sum is over all possible groups, J, of the response and explanatory variables, n_j is the number of observations in group j, and $\hat{p}_j$ is the estimated relative frequency for group j. The chi-square test statistic follows a chi-square distribution with $J/2 - r$ degrees of freedom, where r is the number of parameters in the model and where $J/2$ is the number of possible groups defined by the explanatory variables. Hence, we use (12.13) as a model validation tool by grouping continuous explanatory variables (if any), refitting the logistic regression model with the grouped explanatory variables, and then calculating the chi-square test statistic. Because the accuracy of the chi-square test statistic is poor when the expected number of observations is low, we need to group the explanatory variables in groups that are not too sparse.

Example 12.9. Moths (continued from p. 325). In the moth example, we used dose as a continuous explanatory variable in the logistic regression model and found estimates $\hat{\alpha} = -1.9277$ and $\hat{\beta} = 0.2972$. The observed data for male moths are shown in the table below:

	\multicolumn{6}{c}{Dose (μg)}					
	1	2	4	8	16	32
Survive	1	4	9	13	18	20
Die	19	16	11	7	8	0

If we use formula (12.8) and the maximum likelihood estimates, we can calculate the expected value for each group. Based on the logistic regression model, the probability of survival at dose 1 is $\exp(\hat{\alpha} + \hat{\beta})/(1 + \exp(\hat{\alpha} + \hat{\beta})) = \exp(-1.9277 + 0.2972)/(1 + \exp(-1.9277 + 0.2972)) = 0.1638$. Hence, we expect $20 \cdot 0.1638 = 3.275$ moths to survive at dose 1 and $20 - 3.275$ to die. Similar computations for the other doses will show that the expected values are

	\multicolumn{6}{c}{Dose (μg)}					
	1	2	4	8	16	32
Survive	3.275	4.172	6.465	12.212	18.883	19.990
Die	16.725	15.828	13.535	7.788	1.117	0.010

The chi-square test statistic becomes

$$X^2 = \frac{(1 - 3.275)^2}{3.275} + \cdots + \frac{(0 - 0.010)^2}{0.010} = 4.2479$$

with a contribution for each of the 12 entries in the table. The value should be compared to the χ^2 distribution with $6 - 2 = 4$ degrees of freedom. This yields a p-value of 0.3735, so we do not reject the logistic regression model.

Note that for these data it would be pointless to use graphical validation of the model. There are only six different doses, so we have only six possible predicted values and we will not be able to see any patterns on a residual plot. □

Example 12.10. Nematodes in mackerel (continued from p. 334). For the mackerel data, we have three continuous explanatory variables: number of nematodes in the belly, length, and time. For each of the three quantitative variables, we define new categorical variables with, say, three categories of equal size for length and number of nematodes in the belly and two categories for time. Thus, to make model validation, we fit the logistic regression model

$$\text{logit}(p_i) = \alpha_{\text{length}_i} + \beta_{\text{nematodes in belly}_i} + \gamma_{\text{time}_i},$$

where the explanatory variables are now categorical factors with three or two levels each. For example, the nematodes in the belly variable ranges from 0 to 41, and we make cuts at 1 and 3 to split it into categories [0, 1], (1, 3], and (3, 41] that contain roughly the same number of observations.

The chi-square test statistic becomes $X^2 = 104.9104$, with $3 \cdot 3 \cdot 2 - 6 = 18 - 6 = 12$ degrees of freedom; 18 because there are 18 combinations of the categorical variables and 6 because the additive model has 6 parameters. This results in a p-value of 0, so we clearly reject the logistic regression model. This corresponds to the conclusion we would draw from the residual plot shown in Figure 12.3.

It is worth pointing out that the residual plot shown in Figure 12.3 and the chi-square test actually validate different models. In Example 12.8, we used a model where we assumed that all three explanatory variables were continuous and made a residual plot based on that model. In this example, we have assumed that the explanatory variables were categorical for the chi-square test. It may seem counter-intuitive that we take a continuous variable and make it less informative by grouping it into a few categories if we really want to validate a model where it is continuous. However, instead of making no model validation at all, it may be better to validate a slightly less informative model; but that also means that we should be wary of putting too much emphasis on the exact p-value. □

12.6   R

There are several ways to specify the input for logistic regression analysis in R, but they all make use of the glm() function (glm for generalized linear model) to fit the model and estimate the parameters. One way is to code the response variable as a matrix with two columns — the first column denotes the number of successes and the second column is the number of failures. This representation is especially useful if there are several observations that share the same set of values or conditions for the explanatory variables.

We will illustrate the glm() function with the moth data from Example 12.2 (p. 325):

```
> dose <- c(1, 2, 4, 8, 16, 32)
> malemoths <- matrix(c(1, 4, 9, 13, 18, 20,
+ 19, 16, 11, 7, 2, 0), ncol=2)
> malemoths
     [,1] [,2]
[1,]    1   19
[2,]    4   16
[3,]    9   11
[4,]   13    7
[5,]   18    2
```

```
[6,]    20    0
```

```
> logreg <- glm(malemoths ~ dose, family=binomial)
> logreg
```

```
Call:  glm(formula = malemoths ~ dose, family = binomial)
```

```
Coefficients:
(Intercept)          dose
   -1.9277        0.2972
```

```
Degrees of Freedom: 5 Total (i.e. Null);   4 Residual
Null Deviance:       71.14
Residual Deviance: 4.634        AIC: 22.98
```

Note that each row of the `malemoths` matrix contains information on 20 moths that were all exposed to the same dose. Hence, the number of rows corresponds to the length of the `dose` vector, and the sum of the elements in each row is 20. We find the estimates $\hat{\alpha} = -1.9277$ and $\hat{\beta} = 0.2972$. The `summary()` function is used to extract more information and to get standard errors and Wald test statistics in the same way that we used `summary()` for linear models estimated with `lm()`.

```
> summary(logreg)
```

```
Call:
glm(formula = malemoths ~ dose, family = binomial)
```

```
Deviance Residuals:
      1         2         3         4         5         6
-1.5729   -0.0954    1.1798    0.3631   -0.7789    0.1426
```

```
Coefficients:
            Estimate Std. Error z value Pr(>|z|)
(Intercept) -1.92771    0.40195  -4.796 1.62e-06 ***
dose         0.29723    0.06254   4.752 2.01e-06 ***
---
Signif. codes:  0 '***' 0.001 '**' 0.01 '*' 0.05 '.' 0.1 ' ' 1
```

```
(Dispersion parameter for binomial family taken to be 1)
```

```
    Null deviance: 71.138  on 5  degrees of freedom
Residual deviance:  4.634  on 4  degrees of freedom
AIC: 22.981
```

```
Number of Fisher Scoring iterations: 5
```

The output from summary() from a logistic regression model resembles the output we saw for linear models in the previous chapters, except that the t-test statistic has been replaced by the Wald test statistic. The final section of the output shows information about the deviance of the model, the over-dispersion parameter (which we will not discuss), and the number of iterations used to obtain the maximum of the likelihood function.

Predicted values for logistic regression models are calculated using the predict() or the fitted() functions. By default, these two functions return different values since predict() returns predicted values on the log odds scale while fitted() gives the predicted probabilities. To obtain predicted probabilities with the predict() function, we should include the type="response" option.

```
> fitted(logreg)      # Predicted probabilities
          1         2         3         4         5         6
0.1637646 0.2086229 0.3232715 0.6106772 0.9441721 0.9994916
> predict(logreg)     # Predicted log odds
           1          2          3          4          5
-1.6304804 -1.3332461 -0.7387774  0.4501599  2.8280345
           6
 7.5837837
> predict(logreg, type="response")
          1         2         3         4         5         6
0.1637646 0.2086229 0.3232715 0.6106772 0.9441721 0.9994916
```

In Example 12.4, we included dose as a categorical explanatory variable for the moth data. We analyze that model using the glm() simply by including dose as a factor, as shown below:

```
> fdose <- factor(dose)
> logreg2 <- glm(malemoths ~ fdose, family=binomial)
> logreg2

Call:  glm(formula = malemoths ~ fdose, family = binomial)

Coefficients:
(Intercept)        fdose2        fdose4        fdose8
     -2.944         1.558         2.744         3.563
    fdose16       fdose32
      5.142        28.696

Degrees of Freedom: 5 Total (i.e. Null);  0 Residual
Null Deviance:        71.14
Residual Deviance: 2.619e-10     AIC: 26.35

> summary(logreg2)
```

```
Call:
glm(formula = malemoths ~ fdose, family = binomial)

Deviance Residuals:
[1]  0  0  0  0  0  0

Coefficients:
            Estimate Std. Error z value Pr(>|z|)
(Intercept)   -2.944      1.026  -2.870  0.00411 **
fdose2         1.558      1.168   1.334  0.18234
fdose4         2.744      1.120   2.450  0.01430 *
fdose8         3.563      1.128   3.159  0.00158 **
fdose16        5.142      1.268   4.054 5.02e-05 ***
fdose32       28.696  52998.405   0.001  0.99957
---
Signif. codes:  0 '***' 0.001 '**' 0.01 '*' 0.05 '.' 0.1 ' ' 1

(Dispersion parameter for binomial family taken to be 1)

    Null deviance: 7.1138e+01  on 5  degrees of freedom
Residual deviance: 2.6194e-10  on 0  degrees of freedom
AIC: 26.347

Number of Fisher Scoring iterations: 22
```

The intercept line of the summary() output corresponds to the reference level (dose 1), while the remaining lines are contrasts on the logit scale relative to the reference level. Thus, we can see that the odds of dying for dose 2 are not significantly different from dose 1 (p-value of 0.18234) when we use the Wald test to test differences. For dose 32 we get a large estimate of the contrast with an extremely large standard error due to quasi-complete separation, so the Wald test is useless for dose 32.

To make a likelihood ratio test for the hypothesis of no difference between doses, we can either use the drop1() function to test for explanatory variables in the model or use the anova() function to compare two nested models. For both functions, we need to include the option test="Chisq" to get R to automatically calculate the p-values for us:

```
> drop1(logreg2, test="Chisq")
Single term deletions

Model:
malemoths ~ fdose
       Df  Deviance    AIC   LRT  Pr(Chi)
<none>      2.619e-10 26.347
```

```
fdose    5    71.138 87.485 71.138 5.94e-14 ***
---
Signif. codes:  0 '***' 0.001 '**' 0.01 '*' 0.05 '.' 0.1 ' ' 1

> # Fit model with no effect of dose
> logreg3 <- glm(malemoths ~ 1, family=binomial)
> anova(logreg3, logreg2, test="Chisq") # Compare the two models
Analysis of Deviance Table

Model 1: malemoths ~ 1
Model 2: malemoths ~ fdose
  Resid. Df Resid. Dev Df Deviance P(>|Chi|)
1         5     71.138
2         0  2.619e-10  5   71.138 5.94e-14 ***
---
Signif. codes:  0 '***' 0.001 '**' 0.01 '*' 0.05 '.' 0.1 ' ' 1
```

We get the same result from both approaches and reject the hypothesis of no effect of dose. The drop1() function lists the likelihood ratio test statistic under LRT, and the same value is found under Deviance for anova().

Alternatively, we can use the logLik() function to extract the log likelihoods and then calculate the test statistic by hand:

```
> logLik(logreg2)
'log Lik.' -7.17349 (df=6)
> logLik(logreg3)
'log Lik.' -42.74228 (df=1)
> lrt <- 2*(logLik(logreg2) - logLik(logreg3))
> 1-pchisq(lrt, df=6-1)      # Calculate p-value
[1] 5.939693e-14
attr(,"df")
[1] 6
attr(,"class")
[1] "logLik"
```

We can also specify a logistic regression model in glm() if we use a dataset with exactly one observation per row — just as we did in the previous chapters. The response variable should then be a factor, and the first level of this factor will denote failures for the binomial distribution and all other levels will denote successes. If the response variable is numeric, it will be converted to a factor, so a numeric vector of zeros and ones will automatically work as a response variable.

In Example 12.3 we looked at urinary tract disease in cats. Assume we have the data saved in a file urinary.txt which has the following format:

```
disease food    exercise
yes     normal  little
```

```
yes       normal      little
yes       normal      little
no        normal      little
no        normal      little
no        normal      little
no        normal      little
no        normal      little
yes       normal      much
 .          .            .
[ More datalines here ]
 .          .            .
no        excessive   much
```

The urinary tract disease data can be analyzed by glm() with the following lines, where "no" is considered failures (since that is the first level of the factor disease) and "yes" is a success:

```
> tract <- read.table("urinary.txt", header=TRUE)
> attach(tract)
> result <- glm(disease ~ food + exercise, family=binomial)
> result

Call:
glm(formula = disease ~ food + exercise, family = binomial)

Coefficients:
 (Intercept)     foodnormal   exercisemuch
       2.274         -2.374         -1.014

Degrees of Freedom: 73 Total (i.e. Null);  71 Residual
Null Deviance:        93.25
Residual Deviance: 68.76         AIC: 74.76

> summary(result)

Call:
glm(formula = disease ~ food + exercise, family = binomial)

Deviance Residuals:
    Min        1Q    Median        3Q       Max
-2.1779   -0.7534    0.4427    0.7069    1.6722

Coefficients:
              Estimate Std. Error z value Pr(>|z|)
(Intercept)     2.2737     0.5400    4.210 2.55e-05 ***
foodnormal     -2.3738     0.6058   -3.919 8.90e-05 ***
```

```
exercisemuch   -1.0142      0.6140   -1.652     0.0986 .
---
Signif. codes:  0 '***' 0.001 '**' 0.01 '*' 0.05 '.' 0.1 ' ' 1

(Dispersion parameter for binomial family taken to be 1)

    Null deviance: 93.253  on 73  degrees of freedom
Residual deviance: 68.763  on 71  degrees of freedom
AIC: 74.763

Number of Fisher Scoring iterations: 4
```

We find that there is a significant effect of food intake but that exercise appears to have no effect on urinary tract disease when we also account for food intake.

12.6.1 Model validation

Pearson residuals are computed in R with the use of the residuals() function and the option type="pearson". There is no built-in function to calculate standardized Pearson residuals, so we have to calculate those by hand — the rstandard() function calculates standardized deviance residuals. The following code shows how to calculate standardized Pearson residuals and make a residual plot. We will not go into detail about the calculations.

```
phi  <- summary(logreg)$dispersion   # Extract dispersion value
hi   <- hatvalues(logreg)            # Estimate influence
rstd <- residuals(logreg, type="pearson")/sqrt(phi * (1 - hi))
plot(fitted(logreg), rstd,
     xlab="Predicted values", ylab="Std. Pearson residuals")
```

We already know from Chapter 11 how to calculate the chi-square test statistic, so here we will just introduce the function cut() that is used to divide a numeric vector into intervals and then convert the intervals into a factor. If we want to divide the range of a numeric vector into n intervals of the same length, we just include n as the second argument to cut(). If we wish to make categories of roughly the same size, we should divide the numeric vector based on its quantiles. In that case, we can use the quantile() function, as shown below:

```
> x <- (1:15) **2
> x
 [1]   1   4   9  16  25  36  49  64  81 100 121 144 169 196 225

> cut(x, 3)  # Cut the range of x into 3 interval of same length
 [1] (0.776,75.6] (0.776,75.6] (0.776,75.6] (0.776,75.6]
 [5] (0.776,75.6] (0.776,75.6] (0.776,75.6] (0.776,75.6]
```

```
 [9] (75.6,150]    (75.6,150]    (75.6,150]    (75.6,150]
[13] (150,225]     (150,225]     (150,225]
Levels: (0.776,75.6] (75.6,150] (150,225]

> # Cut the range into 3 intervals of roughly the same size
> cut(x, quantile(x, probs=seq(0, 1, length=4)),
+      include.lowest=TRUE)
 [1] [1,32.3]     [1,32.3]     [1,32.3]     [1,32.3]     [1,32.3]
 [6] (32.3,107]   (32.3,107]   (32.3,107]   (32.3,107]   (32.3,107]
[11] (107,225]    (107,225]    (107,225]    (107,225]    (107,225]
Levels: [1,32.3] (32.3,107] (107,225]
```

The newly defined factors can be saved as variables and used in a new call to glm().

12.7 Exercises

12.1 Difference between logits. Let p_i and p_j denote the probability for the same event, A, in two different populations. Show that $\text{logit}(p_i) - \text{logit}(p_j)$ is identical to the log odds ratio of A between populations i and j.

12.2 ℝ **Moths.** Consider the moth data from Example 12.2 (p. 325). We shall use the data for female moths.

1. Type the data into R (see Section 12.6, where the data for the male moths were typed in).

2. Make a vector p containing the relative frequencies of dead moths for each dose category. Make a vector logitp containing the values of the logit function, see (12.6), evaluated at p. Why is this not possible for the group with dose equal to 1?

3. Make a plot with dose at the x-axis and the logit-transformed relative frequencies at the y-axis. Discuss how this plot is related to a logistic regression model with dose as explanatory variable.

4. Make a plot with log-dose at the x-axis and the logit-transformed relative frequencies at the y-axis. Use the logarithm with base 2 (log2() in R). Which relationship between dose and logit-transformed probabilities does this figure indicate?

5. Fit the logistic regression model with log-dose as explanatory variable and find the estimate of the slope parameter. What is the interpretation of this estimate in terms of odds ratios?

[Hint: Consider a doubling of dose. What is the increment in log-dose? What is the estimated increment in logit-transformed probability? How is this expressed with odds ratios?]

12.3 ℝ **Pneumoconiosis among coalminers.** Data on the degree of pneumoconiosis in coalface workers were collected in order to examine the relationship between exposure time (years) and degree of disease (Ashford, 1959). Severity of disease was originally rated into three categories, but here we will use only two (normal and diseased):

Exposure time	Normal	Diseased
5.8	98	0
15	51	3
21.5	34	9
27.5	35	13
33.5	32	19
39.5	23	15
46	12	16
51.5	4	7

1. Type the data into R (see Section 12.6).

2. Make a vector p containing the relative frequencies of diseased coalminers for each exposure time category. Make a vector containing the values of the logit function, see (12.6), evaluated at p. Furthermore, make a variable with logarithmic exposure time (use the natural logarithm).

3. Make a plot with logarithmic exposure time at the x-axis and the logit-transformed relative frequencies at the y-axis. Does a logistic regression model with logarithmic exposure time as explanatory variable seem to be reasonable?

4. Fit the logistic regression model with logarithmic exposure time and state the estimates. What is the effect (in terms of odds ratios) of a doubling of exposure time?

5. What is the estimated probability that a random coalminer with an exposure time of 30 years is diseased?

   [Hint: Compute first logit(p). Remember that the explanatory variable is the logarithmic exposure time. How is this value back-transformed to get the probability p?]

6. How many years of exposure gives a 50% risk of having developed the disease?

   [Hint: We are looking for the exposure time corresponding to $p = 0.5$. What is the corresponding value of logit(p)? Use the estimates to compute the (logarithmic) exposure time.]

12.4 Ⓡ **Willingness to pay.** The Danish slaughterhouse association offered visits to breeders if the slaughterhouse repeatedly encountered problems with the meat. The expenses were covered by a mandatory fee payed to the association by all breeders, but data were collected to examine if the breeders were willing to pay a fee for the visit (Andersen, 1990). The breeders have been grouped according to the size of the herd of pigs and whether the breeder has previously been offered a visit.

Herd size	Previous visit	Willing to pay	Not willing to pay
<500	yes	8	10
<500	no	58	131
500–1000	yes	21	30
500–1000	no	15	33
>1000	yes	12	17
>1000	no	6	14

1. Examine how the willingness to pay for the visit depends on the two explanatory variables.

2. Formulate a conclusion of the statistical analysis where you report relevant odds ratios.

12.5 Ⓡ **Low birth weight.** Data from 189 infants were collected in order to study risk factors for low birth weight (Hosmer and Lemeshow, 1989). The dataset is available in the MASS package and is called birthwt. The dataset contains a total of 10 variables, of which we shall use only three:

low is an indicator for low birth weight. It is 1 if the birth weight is less than 2500 grams and 0 otherwise.

race is 1 if the mother is white, 2 if she is black, and 3 otherwise.

smoke is 1 if the mother has been smoking during pregnancy and 0 otherwise.

1. Use the following commands to make the dataset birthwt available:

```
> library(MASS)
> data(birthwt)
```

Use the factor() command to code the variables race and smoke as categorical variables.

2. Fit a logistic regression model where the rate of low birth weights depends on the mother's race, her smoking habits during pregnancy, and their interaction.

3. Examine if the model can be reduced; *i.e.*, remove insignificant terms one at a time and refit the model until all terms are significant. For each test, remember to state the conclusion in words, too.

4. What is the final model? Use summary() to find the estimates in the final model. How much does smoking during pregnancy affect the odds for low birth weight? How does the mother's race affect the risk of low birth weight?

12.6 ⓡ **Moths.** Consider the moth data from Example 12.2 (p. 325) again. In this exercise we shall make an analysis of the complete dataset, including both male and female moths.

1. Type in the following commands to enter the complete dataset:

```
> dose <- c(1,2,4,8,16,32)
> newdose <- c(dose,dose)
> sex <- c(rep("Male", 6), rep("Female", 6))
> sex <- factor(sex)
> mdeath <- c(1,4,9,13,18,20)
> fdeath <- c(0,2,6,10,12,16)
> msurv <- 20 - mdeath
> fsurv <- 20 - fdeath
> response <- matrix(c(mdeath, fdeath, msurv, fsurv),
+                    ncol=2)
```

Make sure you understand the content of each of the variables.

2. Try each of the following commands and explain what you see in the graphs.

```
> colorcode <- c(rep("red", 6), rep("blue",6))
> plot(newdose, logit(c(mdeath,fdeath)/20),
+      col=colorcode)
> plot(log2(newdose), logit(c(mdeath,fdeath)/20),
+      col=colorcode)
```

Which model do you believe would be appropriate for these data?

3. Fit a logistic regression model with model formula sex + newdose + sex*newdose. What are the assumptions of this model?

4. Examine if the effect of dose is the same for male and female moths.
   [Hint: Which term in the model formula is related to this?]

5. Is there a difference in lethality for male and female moths? If so, give an estimate of the difference in terms of odds ratios.

6. Make a residual plot, similar to that of Figure 12.3, for the model from question 3. Discuss the relevance of the residual plot when the data are grouped as in this case.

7. Compute the expected values corresponding to the model from question 3 for each of the 24 combinations of sex and dose and death/survival. Carry out the chi-square test that compares the expected and observed values. What is the conclusion?

12.7 ® **Nematodes in herring fillets.** An experiment was carried out in order to investigate the migration of nematodes in Danish herrings (Roepstorff et al., 1993). The fish were allocated to eight different treatment groups corresponding to different combinations of storage time and storage conditions until filleting. After filleting, it was determined whether nematodes were present in the fillet or not.

The data are listed in the table below. The numbers 14/185 for time 0 and storage condition 0 mean that nematodes were found in the fillet for 14 fish in that group, whereas no nematodes were present in 185 fish; similarly for the other groups.

Time	Storage condition				
	0	1	2	3	4
0	14/185				
36		4/96	3/97	8/92	
132		3/97	3/97	2/93	2/88

The external file herring.txt contains one line per fish in the experiment and four variables:

condi is the storage condition. The possible values are 0–4, where 0 corresponds to filleting immediately after catch.

time is the duration of storage in hours before the fish is filleted. The possible values are 0, 36, and 132, where 0 corresponds to filleting immediately after catch.

group is the combination of storage condition and storage time. Notice that a storage time 0 is equivalent to storage condition 0 and that no fish were stored 132 hours under condition 4. Hence, there are only 8 combinations; *i.e.*, 8 levels of the group variable.

fillet is 1 if nematodes are present in the fillet and 0 otherwise.

1. Read the data into R. Use the factor() function such that the variables condi and group are coded as categorical variables. Make a new variable timefac which is equivalent to time, except that it is coded as a categorical variable.

2. Fit a logistic regression model with the following command:

```
glm(fillet ~ group, family=binomial)
```

3. Examine if the difference between 36 and 132 hours of storage is the same for storage conditions 1–4.

   [Hint: This is a test for interaction. Why? Fit the model with additive effects of `timefac` and `condi` and compare it to the original model with a likelihood ratio test. Use the `anova()` function.]

4. Examine if the storage conditions and the storage time affect the prevalence of nematodes. Formulate a conclusion and quantify any significant effects.

5. Consider a random fish. Compute the estimated odds for presence of nematodes if

   - the fish is filleted immediately after catch.
   - the fish is stored for 36 hours after one of conditions 1–4.
   - the fish is stored for 132 hours after one of conditions 1–4.

   Compute the corresponding probabilities.

   [Hint: How do you back-transform from logit values to probabilities; *i.e.*, what is the inverse function of the logit function?]

6. Make a plot with the storage time (0, 36, 132) on the *x*-axis and the logarithm of the estimated odds on the *y*-axis. Does the relationship seem to be roughly linear?

7. Fit the logistic regression model where time — used as quantitative variable — is used as explanatory variable. Explain why this is a sub-model of the model where time is used as a categorical variable, and carry out a test that compares the two models (use `anova()`). What is the conclusion?

8. Use the new model to estimate the same odds and probabilities as in question 5. Compare the two sets of estimated probabilities.

Chapter 13

Case exercises

The case exercises in this chapter are longer exercises that use R extensively, introduce some additional topics, and put more emphasis on all the steps that comprise a full data analysis than the exercises from each of the preceding chapters. Some topics are dealt with in more detail here than in the main text, and the quite detailed questions should help the reader through the analyses and the choices and considerations that are made during a data analysis. In particular, the case exercises contain hints that should make them useful for self-study.

The topics and theoretical requirements for each case are as listed below:

- Case 1 (linear modeling): Chapters 1 and 2

- Case 2 (data transformations): Chapter 3 and Sections 4.1–4.3

- Case 3 (two sample comparisons): Chapters 4 and 5 and Sections 6.1, 6.2, and 6.6.4

- Case 4 (linear regression with and without intercept): Chapter 5

- Case 5 (analysis of variance and test for linear trend): Chapters 5 and 6

- Case 6 (regression modeling and transformations): Chapter 7

- Case 7 (linear models): Chapter 8

- Case 8 (binary variables): Chapters 9 and 10

- Case 9 (agreement): Chapters 6 and 10

- Case 10 (logistic regression): Chapters 11 and 12

Case 1: Linear modeling

An experiment was undertaken by researchers at the Faculty of Life Sciences at University of Copenhagen to investigate if rice straw could replace wheat straw as a potential feed for slaughter cattle in Tanzania. Rice straw presents a substantial economic advantage, since it is produced locally across Tanzania while the majority of wheat straw is imported. In this particular experiment, we wish to model the daily weight gain of cattle that are fed with rice straw.

Data from this experiment can be found in the external file `ricestraw.txt`. In the file, `time` is the number of days that the calf has been fed rice straw, while `weight` is the weight change (in kg) since the calf was first fed rice straw. Each animal is measured only once after the initial weigh-in.

Part I: Data analysis

1. Read the data into R. Appendix B.2 explains how to input data in R and access variables in the data frame.

2. Try to get a sense of the data and the distribution of the variables by computing summary statistics and graphical methods of the two variables.

3. Make a scatter plot that shows weight change as a function of time. What can you tell about the relationship between weight change and time from the graph?

4. Fit a linear regression model to the data and add the estimated regression line to the scatter plot of the data. Does it appear reasonable to describe the relationship with a straight line? Why/why not?

5. What are the estimated parameters from the linear regression model?

6. What can be concluded about the daily weight gain from the model? Does it increase or decrease, and by how much?

7. What is the biological explanation/interpretation of the estimate for the slope? Does the estimate have a biological relevance? Why/why not?

8. What is the biological explanation/interpretation of the estimate for the intercept? Does the estimate have a biological relevance? Why/why not?

9. Assume that a new calf from the same population is included in the study. How many kilograms would you expect the calf to have gained after 30 days of rice straw feeding?

10. In the previous question we predicted the effect of 30 days of rice straw feed. Are we able to use the model to predict the effect of any number of days (say, 45, 100, or 3)? Why/why not?

11. In this experiment the weight gain and time were measured in kilograms and days, respectively. How will the parameters (the intercept and slope) change if we measure the weight gain in grams instead of kilograms?

12. Cattle breeders state that calves on average gain 1 kg of weight per day when they are fed rice straw. Do the data seem to contradict this statement?

[Hint: To check if the statement from the cattle breeders is true, we need to fit a linear regression model with a fixed slope of 1. Then we can compare the model (the fitted regression line with slope 1) and check if it seems to match the observed data.

One way to fit a regression line with a fixed slope of 1 is to calculate the estimates by hand, as seen in Exercise 2.6.

We cannot fix a regression slope directly in lm(), but it is still possible to use lm() to get the estimate for the intercept with a small "trick". Recall the mathematical representation for the straight line:

$$y = \alpha + \beta \cdot x.$$

We want to set $\beta = 1$. If we do that, we essentially have the model

$$y = \alpha + 1 \cdot x \Leftrightarrow y - x = \alpha.$$

Thus, if we define a new variable, $z = y - x$, then we can use lm() to estimate α based on the model

$$z = \alpha.$$

A model that contains only an intercept can be specified in R by using a 1 on the right-hand side of the model in the call to lm(); *i.e.*, lm(z ~ 1).]

Part II: Model validation

13. Discuss whether you believe the assumptions for the linear regression model listed in Section 2.2 are fulfilled in this case.

14. Calculate the residuals from the regression model. Describe the distribution of the residuals by one or more appropriate graphs and calculate the residual standard deviation.

15. Assume now that you wish to use a linear regression model for prediction, as in question 9. What would be preferable: that the standard deviation of the residuals is small or large? Why?

Case 2: Data transformations

This case is about data transformation and how data transformations can sometimes be useful because assumptions are fulfilled for a transformation of the data, rather than for the original data.

Part I: Malaria parasites

A medical researcher took blood samples from 31 children who were infected with malaria and determined for each child the number of malaria parasites in 1 ml of blood (Williams, 1964; Samuels and Witmer, 2003). The external dataset `malaria.txt` contains the counts in the variable `parasites`.

1. Read the data into R and attach the dataset. Print the `parasites` vector on the screen.

2. Compute the sample mean and the sample standard deviation of the `parasites` variable.

3. Make a boxplot, a histogram, and a QQ-plot of the `parasites` variable. Does the distribution of the data look symmetric? Does it look like a normal distribution?

4. Make a new variable, y, which has the logarithm to the malaria counts as values. Use the natural logarithm, denoted `log()` in R.

5. Make a boxplot, a histogram, and a QQ-plot of the new vector y. Does the distribution of the log-transformed data look symmetric? Does it look like a normal distribution? Compare to question 3.

6. Compute the sample mean and the sample standard deviation of y.

Consider in the following the population of children infected with malaria, and assume that the distribution of logarithmic parasite counts in the population is normally distributed with mean equal to 8.03 and standard deviation equal to 1.86; *cf.* question 6.

7. Compute an interval in which you would expect to find the *logarithmic parasite counts* for the central 95% of the children in the population.

   [Hint: See Sections 4.1.4 and 4.5.1.]

8. Compute an interval in which you would expect to find the parasite counts — not the logarithmic counts — for 95% of children in the population.

   [Hint: How do you "reverse" the results from the logarithmic scale to the original scale? What is the inverse function of the natural logarithm?]

9. Compute the probability that a child drawn at random from the population has a *logarithmic parasite count* that exceeds 10.

10. Compute the probability that a child drawn at random from the population has a parasite count that exceeds 22,000.

    [Hint: Rewrite the condition to a scale where you know the distribution.]

11. Compute the probability that a child drawn at random from the population has a parasite count between 50,000 and 100,000.

12. Compute the probability that a child drawn at random from the sample has a parasite count between 50,000 and 100,000. Explain the difference between this probability and the probability from question 10.

    [Hint: Count!]

Part II: Pillbugs

An experiment on the effect of different stimuli was carried out with 60 pillbugs (Samuels and Witmer, 2003). The bugs were split into three groups: 20 bugs were exposed to strong light, 20 bugs were exposed to moisture, and 20 bugs were used as controls. For each bug it was registered how many seconds it used to move six inches. The external dataset pillbug.txt contains two variables, time and group.

13. Explain what type of setup the data represent.

    [Hint: Is it a linear regression? An ANOVA? A single sample? Two samples? Or something else?]

14. Make sure that you are confident with the notation from Sections 3.3 and 3.4. What is k, n_j, α_j, etc., for these data?

15. Read the data into R and make parallel boxplots. Does the variation seem to be of roughly the same size for the three groups?

16. Consider instead the log-transformed time, logtime <- log(time), and make parallel boxplots for this variable. Does the variation for this variable seem to be of roughly the same size for the three groups? Which variable, time or logtime, would you prefer for future analysis?

17. Use the lm() function to compute the group sample means of the logtime variable. Discuss what types of variation are relevant in a comparison of the three groups.

Case 3: Two sample comparisons

This case provides two data examples where the purpose is that of comparing two samples but where slightly different methods are suitable due to different patterns of variation.

Part I: Vitamin A storage in the liver

In an experiment on the utilization of vitamin A, 20 rats were given vitamin A over a period of three days (Bliss, 1967). Ten rats were fed vitamin A in corn oil and ten rats were fed vitamin A in castor oil. On the fourth day, the liver of each rat was examined and the vitamin A concentration in the liver was determined. The external dataset `oilvit.txt` contains two variables, `type` and `avit`.

1. Read the data into R and make sure you understand the structure of the data. Use the following commands and explain what they do:

   ```
   x <- avit[type=="corn"]
   y <- avit[type=="am"]
   ```

2. Make a QQ-plot for each of the variables x and y. Is it reasonable to assume that they are normally distributed? Discuss how the small size affects the possibilities for validation of the normality assumption.

3. Compute the sample standard deviations of x and y. Are they very different or are they roughly the same?

4. Specify a statistical model for the data. Compute estimates for the expected values of vitamin A concentration for each of the two types and compute the pooled standard deviation. Compare the pooled standard deviation to the standard deviations from question 3.

   [Hint: Use formula (3.5) for the pooled sample standard deviation.]

5. Compute an estimate and a 95% confidence interval for the expected difference in vitamin A concentrations between the two types of oil.

   [Hint: Use the `t.test()` function with the option `var.equal=T`.]

6. Examine with a statistical test if the vitamin A concentration in the liver differs for the two types of oil. Remember to specify the relevant hypothesis and the *p*-value and to make a conclusion in words.

7. Do the conclusions from the confidence interval and the hypothesis test agree?

Assume for a moment that only 10 rats were used in the study, but that each rat was used twice: once where it was fed vitamin A in corn oil and once where it was fed vitamin A in castor oil. (This would not have been possible in the current experiment because the rats were killed before the liver was examined; but just assume that the liver could be examined when the rat is alive.)

8. Is the above analysis still appropriate? Why/why not? How would you analyze the data?

   [Hint: What are the assumptions (stated before question 4)? Does one or more of them fail in the new setup?]

9. Assume that the observations were indeed paired such that, for example, the first observation of x and y came from the same rat. Compute an estimate and a 95% confidence interval for the difference in vitamin A concentration between the two types of oil. Compare to question 5 and explain how and why the results differ.

   [Hint: Use the option `paired=T` to the `t.test()` function.]

Part II: Fish flavor in lamb meat

An experiment was carried out at the Royal Veterinary and Agricultural University in Denmark in order to examine if the addition of fish to lamb feed carries over a flavor of fish to the lamb meat (Skovgaard, 2004). Eleven lambs were used and assigned at random to one of two groups: five lambs were fed a standard feed, whereas six lambs got feed with fish added. After slaughter and a period of storage, the meat was examined chemically for a substance that is related to the taste of fish. The data are as follows:

Feed	Fish flavor					
Standard	3.81	3.00	3.85	3.30	3.78	
With fish	9.42	3.95	7.23	6.86	6.09	3.99

10. Use the following (or similar) commands to read the data into R:

    ```
    > standard <- c(3.81, 3.00, 3.85, 3.30, 3.78)
    > fish <- c(9.42, 3.95, 7.23, 6.86, 6.09, 3.99)
    ```

11. Compute the sample standard deviations of `standard` and `fish`. Are they very different or are they roughly the same?

12. Specify a statistical model for the data.

13. Compute an estimate and a 95% confidence interval for the expected difference in fish flavor between the two types of feed. Examine with a statistical test if the difference is significant.

    [Hint: Use the `t.test()` function.]

14. Repeat question 13, but now with the extra assumption that the standard deviations are the same for the two types of feed. Compare the two analyses and explain how and why the results differ.

Case 4: Linear regression with and without intercept

A classical dataset contains data on body weight and heart weight for 97 male cats and 47 female cats (Fisher, 1947). The dataset cats is available in the MASS package for R. It contains the variables Sex (sex of the cat), Bwt (body weight in kilograms), and Hwt (heart weight in grams).

1. Use the following commands to make and attach a dataset, males, with the data for male cats only:

```
> library(MASS)                        # Load the MASS package
> males <- subset(cats, Sex=="M")   # Only male cats
> attach(males)
```

2. Run the command:

```
> summary(lm(Hwt~Bwt))
```

Specify the corresponding statistical model and estimate the parameters. Give a precise interpretation of the slope parameter in terms of body weight and heart weight.

3. Make a scatter plot of the data and add the estimated regression line with the abline() function.

4. Compute the 95% confidence interval for the intercept parameter and for the slope parameter.

In the following, we shall consider the linear regression model "without intercept"; *i.e.*, with the intercept parameter fixed at zero.

$$y_i = \beta \cdot x_i + e_i,$$

with the usual assumptions on $e_1, \ldots, e_n$. For the cat data, x and y correspond to body weight and heart weight, respectively.

5. Use one of the confidence intervals from question 4 to explain why this is a reasonable model for the cat data.

6. In order to derive the least squares estimate of β, the function

$$Q(\beta) = \sum_{i=1}^{n} (y_i - \beta \cdot x_i)^2$$

should be minimized. Show that $Q(\beta)$ is the smallest possible for

$$\beta = \tilde{\beta} = \frac{\sum_{i=1}^{n} x_i \cdot y_i}{\sum_{i=1}^{n} x_i^2}.$$

7. Compute $\tilde{\beta}$ for the cat data (use the `sum()` function).

8. Run the following command and recognize the value of $\tilde{\beta}$ in the output:

   ```
   > summary(lm(Hwt~Bwt-1))
   ```

9. Add the new fitted regression line to the plot from question 3. You may use another color or another line type; try, for example, one of the options `col="red"` or `lty=2` to `abline()`.

10. Compute an estimate for the expected heart weight for a random cat with body weight 2.5 kilograms.

11. Find $SE(\tilde{\beta})$ in the output from question 8 and use it to compute the 95% confidence interval for β without the use of the `confint()` function.

    [Hint: Use formula (5.22). Which p should you use? Use `qt()` to compute the relevant t quantile.]

12. Use `confint()` to compute the same confidence intervals. Compare it to the confidence interval for the slope parameter from question 4.

13. Finally, run the command:

    ```
    > t.test(Hwt/Bwt)
    ```

    Which statistical model does this correspond to? What is the interpretation of the estimated mean parameter? Explain the difference between this model and the linear regression model without intercept.

Case 5: Analysis of variance and test for linear trend

A large running event is held each year in the early autumn in Copenhagen. In 2006 the event was held over four days (Monday through Thursday), with participation of roughly 100,000 people. The participants enroll as teams consisting of five persons, and each person on the team runs 5 kilometers. The sex distribution in the team classifies the teams into six groups: 5 men and no women, 4 men and 1 woman, *etc.* The total running time for the team (not for each participant) is registered. On average, there are 800 teams per combination of race day and group. In the table below, the median running times, measured in seconds, is listed for each combination:

Men/Women	5/0	4/1	3/2	2/3	1/4	0/5
Monday	7930	8019	8253	8517	8793	9035
Tuesday	7838	8021	8313	8552	8857	9061
Wednesday	7630	7858	8093	8160	8790	8785
Thursday	7580	7766	8069	8349	8620	8672

The external file dhl.txt has six variables with the following content:

day is the day of the race.

men, women are the number of men and women, respectively, on the teams in the corresponding group.

hours, minutes, seconds state the running time.

We first need to arrange the data properly.

1. Read the data into R and attach the dataset. Make a new variable group with the command group <- factor(women). Explain the difference between the variables women and group.

2. Combine the variables hours, minutes, and seconds into a new variable time that contains the running time measured in seconds. Make sure that you get the same values as those listed in the table.

In the next questions we work on the model given by the model formula

$$\text{time} = \text{day} + \text{group} \qquad (13.1)$$

3. What kind of model is this and why might it be reasonable to analyze the data with this model?

   [Hint: Are day and group quantitative or categorical variables?]

4. Fit the model with lm(). Make the residual plot and a QQ-plot for the standardized residuals. Is there anything to worry about?

5. The model contains no interaction term between day and sex group. Explain why it is not possible to include the interaction for these data. Explain in words what it means that there is no interaction.

6. Try the command interaction.plot(group, day, time) and explain what you see. Based on this plot, do you think the assumption of no interaction between day and sex group is reasonable?

7. Use the summary() function to get an estimate for the expected difference in total running time (for 25 kilometers) between teams with women only and teams with men only.

   [Hint: What is the reference group and what is the interpretation of the estimate corresponding to level 5 of group?]

8. Compute an estimate for the expected difference between the 5 kilometer running time for women and men.

   [Hint: How should you transform the estimate from the previous question?]

In the next questions we work on the model given by the model formula

$$\texttt{time = day + women} \qquad (13.2)$$

9. What type of model is this and how does it differ from model (13.1)? How many different slope parameters are included in the model?

   [Hint: Is women a quantitative or categorical variable? What happens when you combine it with day and do not include an interaction between them?]

10. Fit the model with lm() and find the estimate for the slope parameter in the model. Give a precise interpretation of the estimate and compare it to the estimate from question 8. Explain why you do not get the same estimate.

11. Explain why model (13.2) is a sub-model of model (13.1); *cf.* Section 6.4. Which extra restrictions are imposed by model (13.2) compared to model (13.1)?

12. Use anova() to test if model (13.1) describes the data significantly better than (13.2). Which of the models would you use for further analyses?

Finally, we are going to examine the differences between days. It was very rainy during the week of the event, and the organizers had to change the route between Tuesday and Wednesday.

13. Carry out a test to examine if there is an overall effect of day.

14. Make a new factor route with levels 1 and 2, where observations from Monday and Tuesday have level 1 and observations from Wednesday and Thursday have level 2.

   [Hint: You can, for example, use commands

   ```
   route <- c(rep(1,12), rep(2,12))
   route <- factor(route)]
   ```

15. Fit the model time = route + women. Explain why this is a sub-model of (13.2), and compare the two models with anova(). What is the conclusion?

16. Use `summary()` to give an estimate for the difference between the expected running times for the two routes. Which route is the fastest? Is the difference significant?

17. Recap the analysis from this case. What was your starting model? Which model reductions were you able to carry through, and what were the corresponding conclusions? What is the final model? Estimate the sex effect and the route effect in the final model.

Case 6: Regression modeling and transformations

The data consist of height and diameter measurements from 18 pine trees (Jeffers, 1959), and we will use linear regression models to explore the relationship between the variables. We shall pay special attention to transformation issues and try to answer the following questions:

- Is it reasonable to use a linear regression with height as the response variable and diameter as the explanatory variable, or is it perhaps more appropriate to consider transformations of the variables?

- It is much easier to measure the diameter of a tree than the height, so prediction of the height from observation of the diameter is useful. How do we do make predictions if transformations have been necessary?

The external dataset `pine.txt` contains the variables `diam` and `height`.

1. Read the data into R and make a scatter plot of diameter and height. Which variable is natural to use as "x-variable", and which variable is natural to use as "y-variable"?

   [Hint: Recall what kind of predictions are of interest.]

2. Fit the linear regression model

$$H_i = a + b \cdot D_i + e_i, \tag{13.3}$$

   where D_i and H_i denote the diameter and the height of the ith tree. Add the corresponding regression line to the plot.

   [Hint: Section 2.5.]

3. Make the residual plot corresponding to model (13.3). Does the model seem to fit the data appropriately?

   [Hint: Sections 7.1.1 and 7.3.1.]

4. Construct the log-transformed variables

```
logdiam <- log(diam)
logheight <- log(height)
```

and plot `logheight` against `logdiam`.

Recall, for later use, that the `log()` function is the natural logarithm.

5. Fit the linear regression model

$$\log(H_i) = \alpha + \beta \cdot \log(D_i) + e_i \tag{13.4}$$

to the data. Add the fitted regression line to the plot from question 4.

[Hint: What is the new "y-variable" and the new "x-variable"?]

6. Make the residual plot corresponding to model (13.4) as well as a QQ-plot of the standardized residuals. Does the model seem to fit the data appropriately?

7. Model (13.4) says that $\log(H_i) \approx \alpha + \beta \cdot \log(D_i)$. Take exponentials on both sides in order to derive the corresponding relationship between D and H. Insert estimates $\hat{\alpha}$ and $\hat{\beta}$; this gives H as a function on D.

[Hint: Recall the calculus with exponentials: $e^{u+v} = e^u \cdot e^v$ and $e^{u \cdot v} = (e^u)^v$.]

8. Make the (D, H)-plot from question 1 again. This time, add the fitted relationship from question 7. Compare with the straight line fit from question 2.

[Hint: Use a command like `lines(diam, exp(fit2))` or commands like

```
Hhat <- exp(alphahat) * diam^betahat
lines(diam, Hhat)
```

Explain what is going on.]

9. Consider a tree with diameter 27. Use formula (7.3) to compute the predicted value for the log-height of the tree as well as the corresponding 95% prediction interval.

[Hint: Which model is the relevant one? Why? You will need the sample mean and the sum of squares for $\log(D)$. Compute them with the commands

```
mean(logdiam)
17*var(logdiam)]
```

10. Use the `predict()` function to compute the predicted value and the prediction interval from the previous question.

11. Consider again a tree with diameter 27. Compute the predicted value for the height of the tree as well as the corresponding 95% prediction interval. What is the interpretation of the interval?

    [Hint: How do you "reverse" the prediction and the interval endpoints from question 9 to answer this question?]

12. Use model (13.3) to compute a 95% prediction interval for the height of a tree that has diameter equal to 27. Compare to your results from question 11. Which of the intervals do you trust the most? Explain why.

Case 7: Linear models

Tager et al. (1979) were among the first to systematically investigate if respiratory function in children was influenced by smoking or whether the child was exposed to smoking at home. Their primary objective was to examine if children exposed to smoking have lower respiratory function. The dataset contains information on more than 600 children. In this case, the measured outcome of interest is forced expiratory volume (FEV), which is, essentially, the amount of air an individual can exhale in the first second of a forceful breath.

The data recorded in the dataset are found in the file `fev.txt` and include the following variables: FEV (liters), Age (years), the height (Ht, measured in inches), Gender, and exposure to smoking status (Smoke, 0 = no, 1 = yes). Note that all five variables start with an upper-case letter and that R is case-sensitive, so the variable names need to be written in the same way.

In this case we shall analyze the relationship between exposure to smoke and respiratory function, but we shall also account for variables other than exposure to smoking that may influence respiratory function.

Part I: Descriptive statistics and naive analysis

1. Read the data into R.

2. Which of the variables in the dataset are quantitative and which variables are categorical?

3. Make sure that R uses the correct data type for each of the variables; *i.e.*, that categorical variables are coded as factors.

4. Try to get a sense of the data and the distribution of the variables in the dataset by using summary statistics and relevant plots; *e.g.*, plotting fev as a function of each of the four explanatory variables.

5. Try to create parallel boxplots that compare the fev between the two types of smoking status. What do you see? And is this what you would expect? Why/why not?

6. As a naive analysis, try to compare the mean forced expiratory volume of the smokers to the non-smokers by the use of a *t*-test. As always, you should make sure to validate the model before drawing conclusions. What is the conclusion based on this analysis? If any difference is found, be sure to quantify the results.

Part II: Linear model

In this part of the case, we will fit a more complicated and reasonable starting model that uses all of the available explanatory variables simultaneously.

7. Consider the following starting model suggested by an investigator:

```
fev = age + height + height^2 + gender + smoke +
      smoke*gender + gender*age
```

Which of the four variables — age, height, gender, or smoking status — in this starting model would you consider being quantitative and why? Which variables are categorical and why?

In the rest of this case, we will include age and height as quantitative explanatory variables.

8. How would you interpret the interaction gender*age? How would you interpret the interaction gender*smoke? Write down in words how you will interpret height + height^2.

9. Make sure that R considers gender and smoke as categorical variables and age and height as quantitative variables. Fit the starting model in R.

10. Validate the starting model and "correct" the model if necessary (*i.e.*, use a transformation or a more appropriate starting model).

11. Simplify the model as much as possible (*i.e.*, test the significance of the terms in the model and remove non-significant terms). Remember to use the hierarchical principle!

12. Specify the final model; *i.e.*, the model you obtain when it is impossible to reduce the model any further.

13. Based on the final model, what are your conclusions: Does smoking status influence respiratory function? If yes, how much? List all the explanatory variables that influence respiratory function and discuss the results.

Case 8: Binary variables

This case consists of four parts that can be solved independently of each other. Part I discusses interpretation of confidence intervals for binary variables (although the interpretation is relevant for all confidence intervals). Part II considers sample size estimation for binary observations, and part III contains example data from the 2008 US presidential election. In part IV, the first steps of fitting a binomial model are taken.

Part I: Interpretation of confidence intervals

The resilience of a genetically modified plant to a particular disease was examined in an experiment with plants. Out of 104 genetically modified plants, 54 ended up being attacked by the disease. The proportion of susceptible plants in the sample is $\hat{p} = 54/104 = 0.519$, $SE(\hat{p}) = 0.049$, and a 95% confidence interval becomes [0.421; 0.617].

Discuss which of the following statements are correct and explain why/why not. When we write "genetically modified plants", we refer to the same type of plants as in the experiment and not genetically modified plants in general.

1. 51.9% of *all* the genetically modified plants are susceptible to the disease.

2. It is likely that exactly 51.9% of all the genetically modified plants are susceptible to the disease.

3. We do not know exactly what the proportion of susceptible genetically modified plants is, but we know it is between 42.1% and 61.7%.

4. We do not know exactly what the proportion of susceptible genetically modified plants are, but the interval from 42.1% to 61.7% is likely to contain the true proportion.

5. I am 95% confident that between 42.1% and 61.7% of the genetically modified plants in this sample are susceptible to the disease.

6. I am 95% confident that between 42.1% and 61.7% of the genetically modified plants are susceptible to the disease.

7. 95% of all possible datasets that contain the 104 genetically modified plants will show that 51.9% of the plants are susceptible to the disease.

Part II: Sample size calculations

In an experiment with germination, it was found that 360 out of 400 seeds sprouted seedlings under certain conditions.

8. Calculate a 90% confidence interval for the proportion of seedlings under these conditions. You can use the following result from R for the necessary quantile from the normal distribution:

```
> qnorm(.95)
[1] 1.644854
```

9. Discuss if a 95% confidence interval for the proportion of seedlings will be wider or smaller than the confidence interval found in question 8.

10. How much smaller would the confidence interval be if we had observed the same proportion of seedlings for a dataset that was four times as large; *i.e.*, 1440 seedlings out of a total of 1600 seeds?

11. Let us now assume that the original experiment with 400 seeds was a pilot experiment and that we need to plan a new experiment where we wish to determine the proportion of seedlings, p, such that it has a standard error of one percentage point; *i.e.*, 0.01. How many seeds should we use in the new experiment to achieve the desired standard error?

12. Discuss which factors influence the length of a confidence interval for binomially distributed data based on formulas (10.5) and (10.6). How will each of the factors influence the length? Discuss which of them you in practice can control in an experiment.

13. Let us now assume that we wish to plan yet another experiment with new and different growth conditions that cannot be directly compared to the conditions from the original experiment. Thus, we do not have a qualified "guess", $\hat{p}$, for the proportion of seedlings.

Which value of p makes $p(1 - p)$ as large as possible? Argue why we can use this value of p in order not to risk having too few observations to achieve the desired precision.

14. Use the answer from question 13 to calculate the necessary sample size to obtain a standard error of 0.01 when we have no prior knowledge about the true proportion p.

Part III: The 2008 US presidential election

After the first television debate between the presidential candidates Barack Obama and John McCain in 2008, CNN published the following statement on their web page:

> "A national poll of people who watched the first presidential debate suggests that Barack Obama came out on top, but there was overwhelming agreement that both Obama and John McCain would be able to handle the job of president if elected."

The poll was made by Gallup and it found that 57% of the people in the poll said that Obama did better than McCain in the debate. In addition, CNN wrote

> "Poll interviews were conducted with 524 adult Americans who watched the debate and were conducted by telephone on September 26. All interviews were done after the end of the debate. The margin of error for the survey is *plus or minus 4.5 percentage points.*"

15. Verify the calculations made by Gallup. Do you get the same margin of error?

16. In Danish polls, the survey institutes typically aim for a margin of error of ± 2 percentage points. How many individuals should Gallup have polled for the 2008 US presidential debate in order to obtain a margin of error of ± 2 percentage points? (We assume here that the proportion of people who felt that Obama did best is the same.)

Part IV: Gender distribution

In a classical statistical study by Geissler (1889) of over 6000 19th century families from Saxony, it was investigated if "gender runs in families" (*i.e.*, whether some couples are more likely to produce either boys or girls) or whether having a same-gender family is just a statistical chance. The data comprise 6115 families with 12 children from Germany in the 19th century. The families are categorized according to the number of boys and girls in the family and the distribution is shown in Table 13.1.

17. Formulate a binomial model that describes the number of boys in a single family.

18. Assume that the gender of all the children is independent. Estimate the probability of observing a boy, p_{boy}, based on all 6115 families.

19. If we assume that the gender of all children is independent then we can use the estimate from question 18 together with the binomial formula

Table 13.1: Distribution of gender in families with 12 children

Boys	Girls	Families	Expected families	Observed − expected
0	12	3	0.9	+
1	11	24		
2	10	104		
3	9	286		
4	8	670		
5	7	1033		
6	6	1343		
7	5	1112		
8	4	829		
9	3	478		
10	2	181		
11	1	45		
12	0	7		

(10.1) to calculate the expected number of families (out of 6115) that has $0, 1, \ldots, 12$ boys, respectively.

[Hint: Calculate the probability for a single family and use the fact that families are independent. You can check your own calculations since the sum of the expected number of families should be identical to the actual observed number of families; *i.e.*, 6115.]

20. Fill out the last two columns in Table 13.1. The fourth column should contain the expected number of families calculated in the previous question. In the last column you should write the sign of the difference between the observed and expected number of families. Write a "+" if observed is larger than the expected and a "−" otherwise.

How would you expect the distribution of the signs to be in the last column if the sex of the child was completely random? Does that pattern match what you see? Based on the observed distribution of the signs, would you conclude that "gender runs in families" or that "gender does not appear to run in families"?

Exercise 11.7 completes the case by performing a proper statistical goodness-of-fit test to see if a binomial distribution can be used to describe the distribution of gender in families.

Case 9: Agreement

A common situation is to evaluate if two different measurement methods are in *agreement* — *i.e.*, they produce more or less the same results. We can think of two instruments — a mechanical and a digital — and we wish to make sure that the two instruments give the same readings. A similar situation is when two judges or raters are asked to measure a particular variable and we wish to quantify how much homogeneity, or consensus, there is in the ratings given by the judges. In this case exercise, we shall consider only the situation with two raters or two methods.

Part I: Quantitative variables

Mass spectrometry is a technique that is used to separate and identify molecules based on the mass of the chemical compounds that constitute the molecules. The mass spectrometry technique is an important tool to analyze proteins that are active in a cell, and there exist several different methods to separate the chemical compounds in a molecule. Two common methods are GC-MS (gas chromatography-mass spectrometry) and HPLC (high performance liquid chromatography). The biggest difference between the two methods is that one uses gas while the other uses liquid. The table below includes data from the two methods measured on 16 samples. We wish to determine if the two methods measure the same amount of muconic acid in human urine.

Sample	HPLC	GC-MS
1	127	138
2	129	101
3	123	147
4	496	443
5	142	167
6	32	62
7	173	224
8	192	266
9	52	69
10	299	320
11	19	8
12	321	364
13	190	256
14	31	45
15	311	331
16	34	19

We will start by discussing problems we might encounter if we use traditional statistical methods for comparison of methods. We should keep in mind that we wish to evaluate if the two methods are in agreement — that means that we get roughly the same result for the two methods for every observation. It is not sufficient that the two methods are identical on average.

1. Calculate the correlation coefficient between the HPLC and GC-MS methods and discuss why it is problematic to use the correlation coefficient as a measure of agreement.

2. Give two examples of data where the correlation is high but where the data are not in agreement.

3. Test the hypothesis that the average level for GC-MS is the same as the average level of HPLC by using a paired t-test. What is the conclusion? Discuss why it is problematic to use a paired t-test as a test for agreement.

4. Provide an example dataset where the two methods are not in agreement but where a paired t-test fails to reject the hypothesis that the difference in means is zero.

Altman and Bland (1983) presented the following method to evaluate if two quantitative measurement methods are in agreement. In the following, x_1 are the measurements from the first method and x_2 are the measurements from the second method. The steps in the Bland-Altman method are

- Plot the difference $d = x_1 - x_2$ against the mean $a = (x_1 + x_2)/2$.

- Use the graph to check that there are no systematic changes in the difference as the average value increases.

- Test the hypothesis $H_0 : \beta = 0$, where β is the slope from a linear regression analysis where d is modeled as a function of a.

- If H_0 is rejected, then we need to find a transformation of x_1 or x_2 or both and restart by checking the graph.

- If H_0 is *not* rejected, then we can calculate the *limits of agreement*,

$$\bar{d} \pm 1.96 \cdot s_d,$$

where s_d is the standard deviation of the differences. Roughly 95% of the observed difference should fall within the limits of agreement.

The two methods are said to be in agreement if the differences within these limits are not clinically relevant (*i.e.*, differences within the limits of agreement are acceptable from a biological point of view).

5. Why is it interesting to test the hypothesis $H_0 : \beta = 0$ that the regression slope is equal to zero?

6. Analyze the data and determine if the two methods can be said to be in agreement. From the laboratory, a maximum difference of 50 is deemed acceptable.

7. Discuss how the Bland-Altman method tries to overcome the problems that exist with using the correlation coefficient or the pairwise t-test to determine agreement.

Part II: Categorical variables

The method presented in part I cannot be used for categorical variables since the calculation of the limits of agreement are based on the normal assumption. If we have categorical response data, we can use Cohen's kappa to measure agreement (Cohen, 1960).

Consider the following data, where we wish to examine if two raters are in agreement on how to classify 48 horses according to lameness:

	Rater 1	
Rater 2	Not lame	Lame
Not lame	11	20
Lame	5	12

8. Calculate the estimated probability of agreement; *i.e.*, the number of times the same rating is assigned by each rater divided by the total number of ratings.

The estimated probability from the previous question can be used as a measure of agreement. It assumes that the data categories are nominal and it does not take into account that agreement may happen solely based on chance. We will now study a measure of agreement that takes agreement by chance into account.

Cohen's kappa is defined as

$$\kappa = \frac{P(\text{agreement}) - P(\text{chance agreement})}{1 - P(\text{chance agreement})}, \tag{13.5}$$

where $P(\text{agreement})$ is the estimated probability of agreement calculated in the previous question. If the raters are in complete agreement, then $\kappa = 1$, and if there is no agreement among the raters (other than what would be expected by chance), then $\kappa \leq 0$.

The probability of a chance agreement is calculated as the sum over all categories of the probabilities that both raters by chance will classify a random observation as belonging to that category. For example, in the horse data above, then rater 1 has probability $16/48 = 0.333$ of classifying a random horse as "not lame" while the same probability for rater 2 is $31/48 = 0.646$. For the "lame" category, the probabilities are 0.667 and 0.354. Thus,

$$P(\text{chance agreement}) = 0.333 \cdot 0.646 + 0.667 \cdot 0.354 = 0.451.$$

9. Calculate Cohen's kappa for the horse lameness data.

10. Construct a dataset containing 100 observations rated on a scale with three categories such that Cohen's kappa lies somewhere between 0.6 and 0.8.

11. Construct a dataset containing 100 observations rated on a scale with three categories such that Cohen's kappa lies somewhere between 0.1 and 0.2.

Case 10: Logistic regression

The data for this case come from Finney (1952) and concern the comparison of three insecticides (rotenone, deguelin, and a mixture of those). A total of 818 insects were exposed to different doses of one of the three insecticides. After exposure, it was recorded if the insect died or not. Data is available in the external file `poison.txt`. There are three variables in the dataset: `status` (dead=1, alive=0), `poison` (the insecticide), and `logdose` — the natural logarithm of the dose.

1. Read the data into R.

2. Try to get a sense of the data. For example, use summary statistics or graphs that show the relationship between `status` and `logdose`. Another option is to tabulate two categorical variables against each other using the `table()` function.

We start with a simple analysis to get an understanding of how to run the logistic regression model in R. Normally we would start with a larger and more proper start model. In the simple analysis, we wish to describe the influence of dose on the log odds ratio; *i.e.*, the model described with the model formula

```
status = logdose
```

Note that we specify the model in the same way as we would for a linear model, so it is only from the context that we know that the response variable in this situation is binary.

3. Fit the simple logistic regression model in R with the following command:

```
> glm(status ~ logdose, family="binomial")
```

4. Make a residual plot to validate the logistic regression model. What do you see and why do you think the plot looks like this? Discuss how useful residual plots are for logistic regression models.

5. Use a goodness-of-fit test to validate the model. What is your conclusion about the model fit?

   [Hint: Look at the data and identify some obvious groups for log-dose.]

6. Give a precise interpretation of the parameter estimate corresponding to log-dose. Is the mortality increasing or decreasing with increased dose? Make sure to back-transform the estimate for log-dose, so you can give the interpretation in terms of odds instead of log odds.

7. Use summary() to evaluate the effect of log-dose and to test if the effect of log-dose is significantly different from zero. What is the corresponding p-value and what is your conclusion about the effect of log-dose?

8. Use drop1() to test the hypothesis that log-dose has no influence on the survival of insects. State the p-value and the conclusion in words.

We shall now make a proper statistical analysis, where we include both explanatory variables (poison and logdose) and where log-dose is included both linearly and quadratically.

9. What is the model formula corresponding to the logistic regression model that allows for both a linear and a quadratic effect of log-dose and for an effect of insecticide? Fit the model with glm().

10. Are there any differences among the three insecticides? If yes, which one is most effective and how many times greater odds are there to kill an insect if you use the most effective insecticide compared to the least effective insecticide?

11. Is there any effect of dose? If yes, what is the fold increase of odds if the dose is doubled?

    [Hint: Recall that $\log(2) = 0.6931$.]

Appendix A

Summary of inference methods

This appendix brings a non-technical overview of important aspects of statistical reasoning and statistical analysis. The considerations are general in nature and thus not restricted to the models we have discussed in this book. We start with an overview of the basic statistical concepts (Section A.1) and proceed with an outline of a typical statistical analysis (Section A.2). The hardest part of a statistical analysis is often that of selecting a good statistical model, and we go into detail about the considerations in that respect in Section A.3.

A.1 Statistical concepts

Statistical model. A statistical model describes the distribution of the response in terms of explanatory variables and random variation. We have been concerned with models based on the normal distribution for continuous response variables (linear models) and the binomial distribution for binary variables (logistic regression). The mean of the response depends on the explanatory variables through unknown parameters, and the distribution type determines the random variation.

Independence. Variables are independent if they do not share information; *i.e.*, knowing the value of one (or more variables) does not change our knowledge about the others.

Model validation. Model validation is important because the confidence intervals and hypothesis tests are valid only if the statistical model is appropriate for the data. For models based on the normal distribution, model validation is carried out as an investigation of the residuals. Model validation for logistic regression models is more subtle but may involve Pearson residuals or goodness-of-fit tests.

Estimates and standard errors. The estimate of a parameter is the value that fits the best with the observed data. This may be understood in terms of least squares or likelihood. The standard error is the (estimated) standard deviation of the parameter estimate and thus measures the variability of the estimate.

Confidence intervals. A confidence interval for a parameter is an interval, computed from the data, that contains values of the parameter that are in accordance with the data at a certain confidence level. It is computed as

$$\text{estimate} \pm \text{quantile} \cdot \text{SE}(\text{estimate}),$$

where the quantile is either a t quantile (for models based on the normal distribution) or a quantile in the normal distribution (for models based on the binomial distribution). The 97.5% quantile is used for a 95% confidence interval, the 95% quantile for a 90% confidence interval, *etc.*

Hypothesis tests. Hypotheses are restrictions on the parameters of the model. A test statistic is a function of the data that measures how well the data and the hypothesis agree. The corresponding p-value is the probability — if the hypothesis is true — of sampling data that agree as little or even less with the hypothesis as our observed data does, where agreement is measured with the test statistic. The hypothesis is rejected if the p-value is smaller than or equal to the significance level, which is often taken to be 0.05.

Prediction. Prediction is about forecasting the value of new observations. The corresponding 95% prediction interval includes the central 95% most likely values of such a new observation.

A.2   Statistical analysis

Statistical analysis may of course take many forms depending on the data type and the purpose of the analysis, but a typical statistical analysis involves the following steps:

Graphical investigation. It is important to explore the data graphically in order to "get a feeling" for the data. Graphical exploration of the data also works as a tool to determine a reasonable statistical model and to understand associations between variables. Relationships that cannot be recognized in well-considered figures are rarely worth looking for in numerical analyses. On the other hand, the graphical investigation cannot stand alone — actual statistical analyses are necessary to identify and quantify significant effects.

Model selection and model validation. Model selection is an iterative process where models are suggested and validated until a satisfactory description of the data is obtained. A good model fits the data and at the same time makes sense from a biological point of view. Finding such a model is not always an easy task. The model selection process includes selection of response variable, distribution type, and explanatory variables and possibly also involves transformation of one or more variables. We go into more detail about model selection in Section A.3.

Model reduction. The selected (initial) model possibly includes effects which are not supported by the data. The model reduction procedure "trims" the model for such insignificant terms. The hierarchical principle (Infobox 8.2) should be obeyed; so a test for the main effect of a categorical variable is not meaningful as long as the variable is included in an interaction term with other categorical variables. There may still be several possible orders in which to test the relevant hypotheses. The basic rule is to test effects of particular interest as late in the process as possible such that they are tested in the simplest possible model that is still in accordance with the data. The model reduction step produces a model where all terms are significant. This is called the final model.

Quantification of significant effects. Just because an effect is statistically significant it does not mean it is important from a biological point of view, so it is extremely important to quantify relevant effects in the final model. This involves computation of estimates and confidence intervals for biologically relevant parameters. Relevant parameters may, for example, be differences between treatments or an effect of a quantitative variable (a slope parameter). It is important to report the results on a scale where the interpretations make sense; in particular, to "back-transform" the estimates if some of the variables have been transformed during the analysis. The results of the analysis may very well be illustrated graphically rather than as numbers in a table since this is usually easier to grasp.

Prediction. Sometimes one purpose of the analysis is to be able to predict future values of the response. Then relevant predictions and prediction intervals should be computed.

Conclusion. The conclusion of the analysis includes a description of the initial model, the results from the model reduction step, and the quantifications of significant effects (in figures and/or tables). It is important that all results are stated in biological terms as well as in terms of the statistical model. Notice that p-values should be listed both for terms that are removed during the model reductions and for the terms that remain in the final model.

A.3 Model selection

Selection of a meaningful statistical model is often the hardest part of a statistical analysis. The model should fit the data in the sense that it describes the variation in the data reasonably well. Moreover, the model should describe the phenomenon under investigation in a way that can be interpreted biologically. In particular, the parameters should generally be biologically meaningful and it should be possible to formulate relevant hypotheses as restrictions on the parameters in the statistical model. This means that the model selection process involves statistical (mathematical) as well as biological reasoning.

For complex data, there will often be several possible models, each of them having advantages and drawbacks. This makes us think of model selection — or model building — as an "art" rather than as a problem with one correct solution. Model selection is an iterative process. A model is suggested and it is investigated to see whether it fits the data and whether it can answer our questions. Perhaps it fulfills both criteria, and so we continue the analysis with this model. In other cases it turns out that the suggested model is not appropriate, and we have to refine it some way or the other; for example, by introducing extra explanatory variables or by transformation of the data. Graphical exploration of the data and model validation tools are essential parts of this process.

Some of the considerations can be summarized as follows:

Response variable. The first thing to determine is which variable (or variables) should be used as response; *i.e.*, as the variable for which we would like to make a model.

Distribution type. The response variable determines the distribution type, at least to some extent. In this book have discussed only the normal distribution and the binomial distribution as building blocks for statistical models. If the response is binary, then a model based on the binomial distribution is appropriate. If the response is quantitative and continuous, then the normal distribution might be appropriate (but this has to be tested during model validation). Perhaps the response should be transformed in order for the normal distribution to be reasonable.

Independence. The statistical models in this book have assumptions on independence among the observations, and it is necessary to think about whether this assumption is reasonable or not. For paired observations, for example, the assumption is usually not reasonable for the original observations — but rather for the differences computed for each subject. Hence, we might have to reconsider our choice of response variable.

Explanatory variables. We need to decide which explanatory variables to include in the model — and how. This is generally a matter of biological reasoning: Which variables possibly affect the response? On the other hand, the type of explanatory variable determines how it should be included in the model. A categorical variable imposes a grouping structure on the data, whereas a quantitative variable introduces a linear relationship between this variable and the response. Moreover, we need to decide which interactions to include in the model.

Transformation. Transformations of quantitative variables are sometimes useful in the model selection process. Transformations change the relationship between explanatory variables and the response, and furthermore, the assumptions of variance homogeneity and normality are sometimes more reasonable for transformation of the response variable. Transformations change the biological interpretations of the parameters, and it is important to back-transform parameter estimates and confidence intervals when they are reported and interpreted at the end of the analysis.

Appendix B

Introduction to R

R is a programming language and software environment for statistical computing and graphics. R has become the *lingua franca* in computational statistics and is used in many applied fields. The flexibility and large number of user-contributed packages make it an extremely useful statistical toolbox for almost any data analysis and visualization problem.

This appendix will provide only a very brief introduction to R. The introduction in this section should be sufficient for readers to use R for the exercises and statistical analyses in the rest of the book, as explained in the R sections of each chapter. Readers interested in a broader introduction to R can consult the books by Crawley (2007) or Dalgaard (2008) or the documents found on the R project homepage, www.r-project.org. A quick guide to installation of R is given in Section B.6.

R uses a command line interface, and although several graphical user interfaces are available, we will here use the basic command line interface.

B.1 Working with R

R usually works interactively, using a "question-and-answer" model. R provides a command prompt, >, and waits for input. The user enters commands at the command prompt and presses enter; R will process the commands, print any output and wait for further input. For example, if you wish to add 3 and 2, you write 3 + 2:

```
> 3+2
[1] 5
```

and R responds with the second line (the answer 5) when you press enter.

The help() function provides information on the various functions and their syntax in R. help(log) will give information on logarithms and exponentials, and help(mean) will provide information about the function to calculate the mean. If you want an overview of all functions, you can start the help page with the help.start() command. Then you can browse around the available reference pages. To search for functions related to a specific topic, use help.search(). For example,

```
> help.search("chi-square")    # search for help on installed
> help.search("linear model") # functions with certain keywords
```

but note that help.search() searches only through installed packages. Thus, you cannot use help.search() to look for functions in add-on packages that are not installed or loaded.

Note how everything after # is read as a comment and is disregarded by R. This is very useful when commands are saved in files for later use; see Section B.5 about how to save commands and output.

You quit R by calling the function q().

B.1.1   Using R as a pocket calculator

R can be used as a simple pocket calculator. All standard mathematical functions (power, exponential, square-root, logarithm, *etc.*) are built-in. For example,

```
> 2+4                 # Add 2 and 4
[1] 6
> log(8)              # Take natural logarithm of 8
[1] 2.079442
> exp(1)              # Exponential function
[1] 2.718282
> 2*(3+5)-log(12)     # Combine expressions
[1] 13.51509
> log(8, base=2)      # Base 2 logarithm of 8
[1] 3
> 2^4                 # 2 to the power 4
[1] 16
```

In R, you can also assign results to variables and use them for later computations. Assignment is done by the "arrow operator", <-, which is formed by first typing < and then -.

```
> x <- 2              # Assign the value of 2 to x
> 5 + x               # Can use x in computations
[1] 7
> 4*x-1
[1] 7
> y <- 17             # Assign the value 17 to y
> x + y
[1] 19
> exp(x)              # Corresponds to exp(2)
[1] 7.389056
> exp(x + (y-15))
[1] 54.59815
> s <- "this is a string"  # Assigns a string to the variable s
```

```
> s                        # Prints the string
[1] "this is a string"
```

R is quite flexible with the names that are allowed for variables, with only very few caveats:

- variable names cannot start with a digit.

- variable names cannot contain a hyphen (-).

- variable names are case sensitive. BMI, Bmi, and bmi are three different variables.

- short/common names are already used for some functions: c q t C D F I T, so you should refrain from using these names.

Note that when you write just the variable name and press enter, R prints its value. The function print() can also be used to print the contents of a variable.

```
> x <- 2.5
> x
[1] 2.5
> print(x)
[1] 2.5
```

B.1.2 Vectors and matrices

All elementary data types in R are *vectors*. The combine function, c(), is used to construct a vector from several elements:

```
> c(11, 3, 8, 6)    # Create a vector with four elements 11 3 8 6
[1] 11 3 8 6
```

Note how the first output line starts with [1]. R shows the index of the first element on that line in square brackets. Thus, in the example above we can see that "11" must be the first element, "3" the second, and so forth. In the example below we can see that the first "30" shown on the second output line is the 9th element of the dayspermonth vector.

```
> dayspermonth <- c(31, 28, 31, 30, 31, 30, 31, 31,
+ 30, 31, 30, 31)
> dayspermonth
[1] 31 28 31 30 31 30 31 31
[9] 30 31 30 31
```

The sequence function, seq(), is used to generate a sequence of numbers, and the replicate function, rep(), replicates a vector a given number of times:

```
> seq(2, 5)            # Sequence from 2 to 5 with steps of 1
[1] 2 3 4 5
> seq(1, 7, 2)         # Sequence from 1 to 7 with steps of 2
[1] 1 3 5 7
> rep(c(1, 2, 3), 2) # Replicates the vector 1 2 3 twice
[1] 1 2 3 1 2 3
> 1:6                  # Identical to seq(1, 6)
[1] 1 2 3 4 5 6
```

Standard arithmetic in R is *vectorized*: mathematical functions work element-wise on vectors; *i.e.*, x + y adds each element of x to the corresponding element of y.

```
> weight <- c(65, 77, 84, 82, 93)      # Create vector with
> weight                               # weights in kg
[1] 65 77 84 82 93
> height <- c(158, 180, 186, 179, 182) # Vector with heights
> height                               # (in cm)
[1] 158 180 186 179 182
> height/100                           # Divide elements by 100
[1] 1.58 1.80 1.86 1.79 1.82
> bmi <- weight/((height/100)^2)       # Calculate BMI values
> bmi
[1] 26.03749 23.76543 24.28026 25.59221 28.07632
```

When calculating the body mass index in the example above, the calculations are carried out element-wise; *i.e.*, the BMI for the first individual is $\frac{65}{1.58^2} = 26.04$.

If operating on two vectors of different length, the shorter one is replicated to the length of the longer. If the length of the shorter vector is a divisor of the length of the longer vector, then R will silently replicate the shorter vector. If the length of the shorter vector is not a divisor of the longer vector, then R will still replicate the shorter vector to the length of the longer vector but will produce a warning message:

```
# Add vectors 1 2 3 4 5 and 1. Short vector is replicated to
# 1 1 1 1 1 before addition
> seq(1, 5) + 1
 [1]  2  3  4  5  6

# Add vectors 1 2 3 4 and 2 3. Short vector is replicated to
# 2 3 2 3 before addition
> seq(1, 4) + 2:3
[1] 3 5 5 7

# Add vectors 1 2 3 4 5 and 0 1. Short vector is replicated to
# 0 1 0 1 0 before addition
```

```
> seq(1, 5) + c(0, 1)
[1] 1 3 3 5 5
Warning message:
In seq(1, 5) + c(0, 1) :
  longer object length is not a multiple of shorter object length
```

Matrices are constructed using, for example, the function matrix(), as shown below:

```
> A <- matrix(1:6, 2, 3)     # Matrix with 2 rows and 3 columns
> B <- matrix(1:6, 3, 2)     # Matrix with 3 rows and 2 columns
> A
     [,1] [,2] [,3]
[1,]   1    3    5
[2,]   2    4    6
> B
     [,1] [,2]
[1,]   1    4
[2,]   2    5
[3,]   3    6
> A %*% B                    # Matrix multiplication of A and B
     [,1] [,2]
[1,]  22   49
[2,]  28   64
```

Note that by default the elements are inserted column-wise in the matrix. If you wish to insert elements row-wise, then you should use the option byrow=TRUE:

```
> A <- matrix(1:6, 2, 3, byrow=TRUE)
> A
     [,1] [,2] [,3]
[1,]   1    2    3
[2,]   4    5    6
```

R has several useful indexing mechanisms to extract specific elements from a vector or matrix:

```
> a <- 1:7
> b <- c(110, 200, 230, 60, 70, 210, 160)
> a
[1] 1 2 3 4 5 6 7
> b
[1] 110 200 230  60  70 210 160
> a[5]              # The 5th element of a
[1] 5
> a[5:7]            # Elements 5, 6 and 7 of a
```

```
[1] 5 6 7
> a[c(1, 4, 6, 2)]    # Elements 1, 4, 6 and 2 (in that order)
[1] 1 4 6 2
> a[-6]               # All but the 6th element of a
[1] 1 2 3 4 5 7
> a[b>200]            # Elements of a for which the corresponding
[1] 3 6               # elements of b are greater than 200
> a[3] <- 10          # Replace 3rd element of a with value 10
> a
[1]  1  2 10  4  5  6  7
```

Elements from a matrix are accessed in the same manner:

```
> A <- matrix(1:6, 2, 3)
> A
     [,1] [,2] [,3]
[1,]    1    3    5
[2,]    2    4    6
> A[2, 1]                 # The element on row 2 and in column 1
[1] 2
> A[1,]                   # First row of A
[1] 1 3 5
> A[,3]                   # 3rd column of matrix A
[1] 5 6
> A[,c(2,3)]              # 2nd and 3rd columns of A
     [,1] [,2]
[1,]    3    5
[2,]    4    6
> A[1,1] <- 10           # Replace top left element with the
> A                      # value 10
     [,1] [,2] [,3]
[1,]   10    3    5
[2,]    2    4    6
```

B.2   Data frames and reading data into R

The simplest way to get small datasets into R is to use the combine function to enter the data directly into R. For example, to enter the digestibility data used in Example 2.1, we could type

```
> stearic.acid <- c(29.8, 30.3, 22.6, 18.7, 14.8, 4.1, 4.4,
+ 2.8, 3.8)
> digest <- c(67.5, 70.6, 72.0, 78.2, 87.0, 89.9, 91.2,
```

```
+ 93.1, 96.7)
```

This method is useful only for very small datasets and is prone to data entry errors. In most situations datasets are stored in an external file, and we should use R to read that file directly.

B.2.1 Data frames

A data frame is a collection of vectors of the same length. This is like a matrix, except that each column can contain a different data type and not just numbers like a matrix. The data.frame() function constructs a data frame in R by grouping variables together.

Suppose we have already entered the two vectors for the digestibility data, as described above. We can then combine the two variables into a data frame if we specify the labels and the variables that should constitute the data frame:

```
> indata <- data.frame(digest = digest, acid = stearic.acid)
> indata
  digest acid
1   67.5 29.8
2   70.6 30.3
3   72.0 22.6
4   78.2 18.7
5   87.0 14.8
6   89.9  4.1
7   91.2  4.4
8   93.1  2.8
9   96.7  3.8
```

Note that we specify the label names to be used in the data frame (digest and acid) and for each of the labels which vector is used for that name. When we print the data frame, we see the complete contents and the corresponding row numbers.

Variables inside a data frame are *not* directly accessible from R. For example, if we ask to print the acid variable stored in the indata data frame, we get

```
> acid
Error: object "acid" not found
```

We can extract or access a vector from a data frame by using the $-operator.

```
> indata$acid
[1] 29.8 30.3 22.6 18.7 14.8  4.1  4.4  2.8  3.8
```

All variables in a data frame can be made directly accessible with the attach() function.

```
> attach(indata)
> acid
[1] 29.8 30.3 22.6 18.7 14.8  4.1  4.4  2.8  3.8
> digest
[1] 67.5 70.6 72.0 78.2 87.0 89.9 91.2 93.1 96.7
```

Note that attach() overwrites existing objects. Thus, if there already exists an object called acid, then the acid variable from indata will replace the original acid object when attach(indata) is called. Thus, one has to be careful when using the attach() command. The command detach(indata) can be used to undo the attachment of the data frame.

Indexing for data frames is done in the same way as for matrices:

```
> indata[3,2] <- 25    # Replace obs. 3 for 2nd variable with 25
> indata[3,]           # Extract row 3
  digest acid
3     72   25
> indata[c(1, 5, 6),]  # Extract rows 1, 5 and 6
  digest acid
1   67.5 29.8
5   87.0 14.8
6   89.9  4.1
> indata[,1]           # Extract vector/column 1
[1] 67.5 70.6 72.0 78.2 87.0 89.9 91.2 93.1 96.7
> indata["acid"]       # Extract the acid variable
  acid
1 29.8
2 30.3
3 25.0
4 18.7
5 14.8
6  4.1
7  4.4
8  2.8
9  3.8
```

B.2.2   Reading text files

Suppose the digestibility data from Example 2.1 was stored in a plain text file, digest.txt, with the following contents:

```
acid        digest
29.8        67.5
30.3        70.6
22.6        72.0
18.7        78.2
```

14.8	87.0
4.1	89.9
4.4	91.2
2.8	93.1
3.8	96.7

The observations are written in columns with blanks or tabs between them. The columns do not need to be aligned, but multi-word observations like high income need to be put in quotes or combined into a single word, or they will be considered as two different columns. Data stored in simple text files can be read into R using the read.table() function:

```
> indata <- read.table("digest.txt", header=TRUE)
```

where the first argument is the name of the data file, and the second argument (header=TRUE) is optional and should be used only if the first line of the text file provides the variable names. If the first line does not contain the column names, the variables will be labeled consecutively V1, V2, V3, *etc.* Data read with read.table() are stored as a data frame within R.

R looks for the file digest.txt in the current working directory, but the full path* can be specified in the call to read.table(); *e.g.*,

```
> indata <- read.table("d:/digest.txt", header=TRUE)
```

The current working directory of R can be seen with the getwd() function, and the function setwd() is used to set the current working directory:

```
> getwd()           # Get the current working directory
[1] "C:/R"
> setwd("d:/")      # Set the current working directory to d:/
```

If you are working in windows, the working directory can also be changed from the File menu (choose Change dir).

B.2.3   Reading spreadsheet files

The easiest way to read data from spreadsheets into R is to export the spreadsheet to a delimited file like a comma-separated file, .csv. Use read.csv()[†] to read in the delimited file just like you would use read.table.

```
> indata <- read.csv("digest.csv")
```

Both read.csv() and read.csv2() assume that a header line is present in the .csv (*i.e.*, header=TRUE is the default). If you do not have a header line in the .csv file, you need to specify the header=FALSE argument.

---

*Note that when specifying paths in R you should generally use the forward slash, '/'. This is standard on unix-like systems but different from windows. For example, you should use something like c:/My Documents/mydata.txt under windows.

[†]Use read.csv2() to read in semi-colon-separated files where ',' is the decimal point character, which is the default .csv format for some language installations.

B.2.4 Reading SAS and SPSS files

It is possible to read datasets from other statistical packages directly into R. These methods require the use of the `foreign` package (see Section B.6.1 on how to install R packages). Use

```
> library(foreign)
```

to load the add-on package `foreign`. The foreign package provides functionality on how to read data stored by other statistical programs (Minitab, S, SAS, SPSS, Stata, and Systat) as well as dBase database files. Here we will give examples on how to read SAS and SPSS files.

SAS datasets are stored in different formats that depend on the operating system and the version of SAS. To read SAS datasets, it is necessary to save the SAS dataset as a SAS transport (XPORT) file, since that can be read on any platform. The following code can be used from inside SAS to store a SAS dataset `sasdata` in the SAS XPORT format:

```
library mydata xport "somefile.xpt";

DATA mydata.thisdata;
  SET sasdata;
RUN;
```

To read the `sasdata` stored in the file `somefile.xpt`, we can use the `read.xport()` function, also from the foreign package:

```
> read.xport("somefile.xpt")
```

Datasets stored by the SPSS "save" and "export" commands can be read by the `read.spss()` function from the foreign package:

```
> read.spss("spssfilename.sav", to.data.frame = TRUE)
```

The option `to.data.frame=TRUE` makes sure that the SPSS data file is stored as a data frame in R. If that option is not included, the dataset is stored as a list.

B.3 Manipulating data

Often it is necessary to do some data manipulation before starting the statistical analysis. For example, we may want to analyze only a subset of the original data or we may want to consider transformations of the original variables.

Consider the cucumber disease spread experiment described in Example 8.4 on p. 217. Assume the data are stored in a file, `cucumber.txt`, that contains the following data:

```
disease climate dose
51.5573   A       2.0
51.6001   A       2.0
47.9937   A       3.5
48.3387   A       3.5
57.9171   A       4.0
51.3147   A       4.0
48.8981   B       2.0
60.1747   B       2.0
48.2108   B       3.5
51.0017   B       3.5
55.4369   B       4.0
51.1251   B       4.0
```

and we read in that file using

```
> indata <- read.table("cucumber.txt", header=TRUE)
```

We can use the subset() function to extract a subset of the data frame based on some criteria. For example, to extract the subset of the indata data frame that concerns climate "A", we can type

```
> climateA <- subset(indata, climate=="A")
> climateA
   disease climate dose
1 51.5573        A  2.0
2 51.6001        A  2.0
3 47.9937        A  3.5
4 48.3387        A  3.5
5 57.9171        A  4.0
6 51.3147        A  4.0
```

Note that we need to use two equality signs, ==, to tell R that climate=="A" is a test for the rows of indata, where the vector climate is equal to A. If we use only a single equality sign, then R will interpret that as an argument/assignment to one of the parameters in the function call to subset.

The transform() function can be used to transform or create new variables within a data frame. If we want to create a new vector that contains the logarithm of the dose, we can type

```
> indata <- transform(indata, logdose=log(dose))
```

This command takes the logarithm of the dose variable inside the indata data frame and creates a new variable, logdose. transform() returns a new data frame, which we assign to the same name as the original data frame, so we overwrite indata. It now contains

```
> indata
```

```
   disease climate dose   logdose
1  51.5573       A  2.0 0.6931472
2  51.6001       A  2.0 0.6931472
3  47.9937       A  3.5 1.2527630
4  48.3387       A  3.5 1.2527630
5  57.9171       A  4.0 1.3862944
6  51.3147       A  4.0 1.3862944
7  48.8981       B  2.0 0.6931472
8  60.1747       B  2.0 0.6931472
9  48.2108       B  3.5 1.2527630
10 51.0017       B  3.5 1.2527630
11 55.4369       B  4.0 1.3862944
12 51.1251       B  4.0 1.3862944
```

Instead of using the `transform()` function, we can create a new variable inside a data frame directly by using the $ operator. The following line creates the variable `mylog` inside the `indata` data frame by taking logarithms of the original dose values:

```
> indata$mylog <- log(indata$dose)
```

B.4   Graphics with R

R is an extremely useful tool for producing graphics. The default graphs are most often pretty and of publication quality, and you can modify them more or less as you like. Various plots like scatter plots, boxplots, and histograms are introduced in Section 1.6 and in the text when we need them for the analyses. In this section we will briefly cover a few ways to modify plots in R and discuss how we can export and save graphs as pdf files so they can be included in other documents.

By default, R uses an "ink on paper" approach: once something is plotted, you cannot remove it. That means that we sometimes have to consider the way we construct a plot, since we can only add to it once we have begun the graph.

The high-level plotting functions, like `plot()`, `hist()`, `barplot()`, `boxplot()`, and `qqnorm()`, all generate a new plot, and any modifications to this initial plot need to be changed by adding options to these functions. Four of the most frequently used options are `col=`, which changes the plotting color, `pch=`, which changes the plotting symbol for points, and `lty=` and `lwd=`, which change the line type and line width when plotting lines. Figure B.1 shows the different plotting symbols and line types. There are many more options available to change every aspect of a graph, and `help(par)` will list and explain all the options.

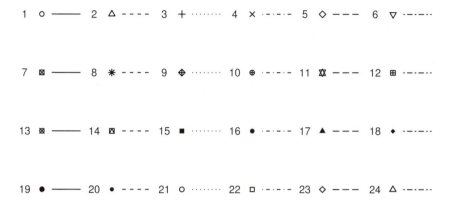

Figure B.1: Plotting symbols (pch=) and line types (lty=) that can be used with the R plotting functions.

Once a plot is started, we can use the functions points() and lines() to add points and lines to the existing plot, respectively.

The title() , text(), and mtext() functions are used to add text to the current plot. title() adds a title, text() adds text inside the plot, and mtext() adds text to one of the margins. The following lines show how these functions can be applied:

```
> plot(1:10, pch=1:10, col="red")  # Plot first 10 symbols in red
> title("Plot title")              # Add a title
> text(3, 2, "Some text here")      # Add text at position (3,2)
> text(2, 7, "Other text here")     # Add text at position (2,7)
> mtext("Margin text", 2)           # Put margin text along y-axis
```

The text() function requires the coordinates on the plot where the text should be inserted. mtext() works slightly different, since the region where the text should be added is specified by the second option: 1=bottom/below the figure, 2=left, 3=top, 4=right.

If we wish to save the graph so we can include it in other documents we can use one of the following methods. Note that R also has the opportunity to save graphs in formats other than pdf, but we include only the pdf file format here because it is easily used under most operating systems:

```
> plot(x,y)                         # Scatter plot
> dev.print(pdf, "my-plot.pdf")     # Copy graph to pdf-file
```

The plot() command produces a scatter plot of two variables x and y on the screen. The dev.print() command copies the graph to the file my-plot.pdf. Note that R should also know what type of file to create; there are other possibilities (try help(dev.print)).

As an alternative, the graph may be produced directly to a file, without making it on the screen at all:

```
> pdf("my-file.pdf")   # Opens a pdf-file
> plot(x,y)            # Scatter plot produced in the file
> dev.off()            # File is closed
```

The pdf() opens a file, the second line makes a scatter plot in the file, and the final dev.off() closes the file once again.

Notice that the files are saved in the working directory in any case.

B.5   Reproducible research

When making a statistical data analysis it is critical to be able to document the analyses and to be able to reproduce them. We will describe two ways to store the analysis: saving commands and saving the complete history.

It is a good idea to make a directory (or perhaps several directories) for your work with R. You should then save data files, R-scripts, and history files in this directory and start every R session by changing the working directory to this directory (see Section B.2.2).

B.5.1   Writing R-scripts

Instead of writing commands directly at the prompt, you may write the commands in a separate file, a so-called R-script, and then transfer the commands to the R-prompt. When you quit R, the file with the commands still exists (if you saved it) and you can run the exact same commands in another R session.

You can use your favorite editor for writing R-scripts. In windows, R has a simple built-in editor that you can start from the File menu. Choose New script in order to open a new file/script and Open script to open a file/script from an earlier session. Commands are very easily transferred from the script editor to the R command prompt with this editor: use Control-r and the current line is transferred to the prompt and evaluated by R. You may also choose to run larger parts of the script at once by selecting a region of the file. Alternatively, you may run all the commands in a file with the source() command:

```
> source("my-analysis.r")
```

where my-analysis.r is a file with R commands.

You may of course switch between writing commands directly at the prompt and in the script editor. One suggestion is to write commands at the prompt as long as you "play around" and copy the commands that you will need for later use to the script editor. In this way you avoid saving the commands that — for some reason — were not appropriate anyway.

B.5.2 Saving the complete history

As a slightly different alternative, R allows the user to save (and reload) the complete session history as a text file. This allows the user to save the analysis for a later time (and possibly to use it again) and keep each and every step in the analysis as documentation for how the dataset was analyzed.

The savehistory() function saves the complete R session history:

```
> savehistory("cucumber-analysis.txt")  # Save complete session
```

Notice that savehistory() saves only the commands used during the session and not the actual vectors and data frames that were created. That means that you still need to save a copy of the external dataset file if data were read from an external file. To read a saved history you use the loadhistory() function:

```
> loadhistory("cucumber-analysis.txt")
```

All the commands from the session are saved, including all the commands with errors and the commands that you, during the analysis, decided were not useful or appropriate.

B.6 Installing R

To download R go to CRAN (the Comprehensive R Archive Network) from the R homepage (http://www.r-project.org). On the CRAN homepage, you find links to pre-compiled binary distributions for Windows, Mac, and several Linux distributions. Follow the link for your operating system and download and install the base file. CRAN is also the place to find extra packages with additional functions and/or datasets for R.

B.6.1 R packages

Apart from the basic functionalities in R, there exist a huge number of *R packages*. An R package is a collection of R functions (and datasets). The functions cannot be used before the package is *installed* and *loaded*.

Assume that we are interested in functions or datasets from a package called "foreign". The command

```
> install.packages("foreign")  # Install the foreign package
```

downloads and installs the package from one of the CRAN mirrors. This needs to be done only once. On most Windows installations, the foreign package is already installed. The package should be loaded in order to use its functions. This is done with

```
> library("foreign")  # Loads the foreign package
```

which should be run in every R session for which you want to use functions from the package.

Alternatively, a package can be installed from the `Packages -> Install package(s)` menu item in the Windows setup.

B.7 Exercises

B.1 ⓡ **Simple calculations in R. Variables.** Start R and do the following:

1. Calculate $3 \cdot 4$.

2. Calculate 0.95^{10}.

3. Calculate $(2 \cdot 4) - \frac{2}{7}$ and $2 \cdot (4 - \frac{2}{7})$. Notice if and when you need to place parentheses in the two expressions.

4. Assign the value $\frac{2}{7}$ to a variable named x. Print the value of x on the screen.

5. Assign the value of the logarithm of x to the variable y. (Use the natural logarithm.)

6. Calculate the exponential function of -y.

7. Calculate by a single expression the square root of the exponential function taken in 2*x. (This expression can be reduced using pen and paper, but that is not the intention.)

B.2 ⓡ **Simple statistical calculations. Vectors.** The length of 20 cones from a conifer (*Picea abies*) are shown below (in mm):

125.1	114.6	99.3	119.1	109.6
102.0	104.9	109.6	134.0	108.6
120.3	98.7	104.2	91.4	115.3
107.7	97.8	126.4	104.8	118.8

1. Read the data into a vector named `conelen`. Use `c()` followed by the data.

2. Calculate the mean (use the `mean()` function) of the cone lengths. Calculate the sum (use the `sum()` function) of the cone lengths and divide by the number of observations (the `length()` of the vector).

3. Find the minimum, maximum, and median of the observations (functions `min()`, `max()`, and `median()`, respectively). Try to sort the `conelen` vector by using the `sort()` function. Can you figure out how R defines the median?

4. Which numbers are printed when you type `summary(conelen)`?

B.3 ℝ **Indices and logical variables.** Use the conifer data from exercise B.2 for this exercise. R has a logical data type that has two possible values, `TRUE` and `FALSE` (or `T` and `F`, for short). The `TRUE` and `FALSE` values can be used for indexing vectors and data frames.

1. Print the first 5 lengths on the screen. (Use the vector `1:5` and index using `[ ]`.) Print out all cone lengths except observation 8.

2. What happens when you write `conelen > 100`? Explain what happens when you write `conelen[conelen > 100]`. Try also `long <- (conelen > 100)` followed by `conelen[long]`.

3. What does `short <- !long` do? Print the lengths of the short cones.

B.4 ℝ **Missing values.** Use the data from exercise B.2. A common problem with a real dataset is "missing values", where some observations are missing for various reasons. Missing values are scored in R as `NA` (not available).

1. Type in `conelen[8] <- NA` and print the vector `conelen`. What has happened?

2. What happens when a variable that contains a missing value is used in a calculation? For example, try `x <- NA` and then `x + 5`.

3. Calculate the sum and the mean of the cone lengths. Is there a problem?

4. Construct a new logical vector, for example, with the name `notmiss`. The vector `notmiss` should take the value `FALSE` if the corresponding value in `conelen` is missing and the value `TRUE` if the value is present in `conelen`. Print out `notmiss` so you are certain it is correct. (You can use the function `is.na()` here.)

5. Type `conelen[notmiss]`. What happens?

6. Calculate the mean of the 19 non-missing cone lengths.

B.5 ℞ **Data manipulation.** The chickwts dataset from R contains information on the growth rate of chickens for various feed supplements.

1. Type data(chickwts) to make the chickwts available in R. print(chickwts) will display the data frame.

2. Use the subset() function to create a subset, mini, of the chickwts data that contains only the two feed supplements "soybean" and "sunflower".

3. The weight variable in the mini dataset contains the weight in grams of the chickens at 6 weeks. Create a new variable, logweight, in the dataset that contains the logarithm of the weights.

4. Create a new subset, mini2, that contains the subset of chickwts corresponding to the "meatmeal" and "horsebean" supplements.

   Use the rbind() function to join the two data frames mini and mini2. The rbind() function joins rows from data frames that contain the same variables.

Appendix C

Statistical tables

C.1 The χ^2 distribution

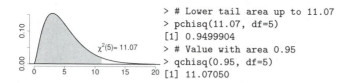

```
> # Lower tail area up to 11.07
> pchisq(11.07, df=5)
[1] 0.9499904
> # Value with area 0.95
> qchisq(0.95, df=5)
[1] 11.07050
```

df	\multicolumn{7}{c}{Lower tail probability}
	0.8	0.9	0.95	0.98	0.99	0.999	0.9999
1	1.642	2.706	3.841	5.412	6.635	10.828	15.137
2	3.219	4.605	5.991	7.824	9.210	13.816	18.421
3	4.642	6.251	7.815	9.837	11.345	16.266	21.108
4	5.989	7.779	9.488	11.668	13.277	18.467	23.513
5	7.289	9.236	11.070	13.388	15.086	20.515	25.745
6	8.558	10.645	12.592	15.033	16.812	22.458	27.856
7	9.803	12.017	14.067	16.622	18.475	24.322	29.878
8	11.030	13.362	15.507	18.168	20.090	26.124	31.828
9	12.242	14.684	16.919	19.679	21.666	27.877	33.720
10	13.442	15.987	18.307	21.161	23.209	29.588	35.564
11	14.631	17.275	19.675	22.618	24.725	31.264	37.367
12	15.812	18.549	21.026	24.054	26.217	32.909	39.134
13	16.985	19.812	22.362	25.472	27.688	34.528	40.871
14	18.151	21.064	23.685	26.873	29.141	36.123	42.579
15	19.311	22.307	24.996	28.259	30.578	37.697	44.263
16	20.465	23.542	26.296	29.633	32.000	39.252	45.925
17	21.615	24.769	27.587	30.995	33.409	40.790	47.566
18	22.760	25.989	28.869	32.346	34.805	42.312	49.189
19	23.900	27.204	30.144	33.687	36.191	43.820	50.795
20	25.038	28.412	31.410	35.020	37.566	45.315	52.386
21	26.171	29.615	32.671	36.343	38.932	46.797	53.962
22	27.301	30.813	33.924	37.659	40.289	48.268	55.525
23	28.429	32.007	35.172	38.968	41.638	49.728	57.075
24	29.553	33.196	36.415	40.270	42.980	51.179	58.613
25	30.675	34.382	37.652	41.566	44.314	52.620	60.140
30	36.250	40.256	43.773	47.962	50.892	59.703	67.633

The table shows values for the cumulative distribution function of a χ^2 distribution with df degrees of freedom (see p. 298). The rows show various degrees of freedom while the column headers show the area under the curve. For example, for a $\chi^2(4)$ distribution with 4 degrees of freedom, the 0.95 quantile is 9.488.

C.2   The normal distribution

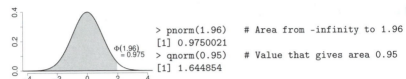

```
> pnorm(1.96)     # Area from -infinity to 1.96
[1] 0.9750021
> qnorm(0.95)     # Value that gives area 0.95
[1] 1.644854
```

z	0	0.01	0.02	0.03	0.04	0.05	0.06	0.07	0.08	0.09
-3.4	0.000	0.000	0.000	0.000	0.000	0.000	0.000	0.000	0.000	0.000
-3.3	0.000	0.000	0.000	0.000	0.000	0.000	0.000	0.000	0.000	0.000
-3.2	0.001	0.001	0.001	0.001	0.001	0.001	0.001	0.001	0.001	0.001
-3.1	0.001	0.001	0.001	0.001	0.001	0.001	0.001	0.001	0.001	0.001
-3.0	0.001	0.001	0.001	0.001	0.001	0.001	0.001	0.001	0.001	0.001
-2.9	0.002	0.002	0.002	0.002	0.002	0.002	0.002	0.001	0.001	0.001
-2.8	0.003	0.002	0.002	0.002	0.002	0.002	0.002	0.002	0.002	0.002
-2.7	0.003	0.003	0.003	0.003	0.003	0.003	0.003	0.003	0.003	0.003
-2.6	0.005	0.005	0.004	0.004	0.004	0.004	0.004	0.004	0.004	0.004
-2.5	0.006	0.006	0.006	0.006	0.006	0.005	0.005	0.005	0.005	0.005
-2.4	0.008	0.008	0.008	0.008	0.007	0.007	0.007	0.007	0.007	0.006
-2.3	0.011	0.010	0.010	0.010	0.010	0.009	0.009	0.009	0.009	0.008
-2.2	0.014	0.014	0.013	0.013	0.013	0.012	0.012	0.012	0.011	0.011
-2.1	0.018	0.017	0.017	0.017	0.016	0.016	0.015	0.015	0.015	0.014
-2.0	0.023	0.022	0.022	0.021	0.021	0.020	0.020	0.019	0.019	0.018
-1.9	0.029	0.028	0.027	0.027	0.026	0.026	0.025	0.024	0.024	0.023
-1.8	0.036	0.035	0.034	0.034	0.033	0.032	0.031	0.031	0.030	0.029
-1.7	0.045	0.044	0.043	0.042	0.041	0.040	0.039	0.038	0.038	0.037
-1.6	0.055	0.054	0.053	0.052	0.051	0.049	0.048	0.047	0.046	0.046
-1.5	0.067	0.066	0.064	0.063	0.062	0.061	0.059	0.058	0.057	0.056
-1.4	0.081	0.079	0.078	0.076	0.075	0.074	0.072	0.071	0.069	0.068
-1.3	0.097	0.095	0.093	0.092	0.090	0.089	0.087	0.085	0.084	0.082
-1.2	0.115	0.113	0.111	0.109	0.107	0.106	0.104	0.102	0.100	0.099
-1.1	0.136	0.133	0.131	0.129	0.127	0.125	0.123	0.121	0.119	0.117
-1.0	0.159	0.156	0.154	0.152	0.149	0.147	0.145	0.142	0.140	0.138
-0.9	0.184	0.181	0.179	0.176	0.174	0.171	0.169	0.166	0.164	0.161
-0.8	0.212	0.209	0.206	0.203	0.200	0.198	0.195	0.192	0.189	0.187
-0.7	0.242	0.239	0.236	0.233	0.230	0.227	0.224	0.221	0.218	0.215
-0.6	0.274	0.271	0.268	0.264	0.261	0.258	0.255	0.251	0.248	0.245
-0.5	0.309	0.305	0.302	0.298	0.295	0.291	0.288	0.284	0.281	0.278
-0.4	0.345	0.341	0.337	0.334	0.330	0.326	0.323	0.319	0.316	0.312
-0.3	0.382	0.378	0.374	0.371	0.367	0.363	0.359	0.356	0.352	0.348
-0.2	0.421	0.417	0.413	0.409	0.405	0.401	0.397	0.394	0.390	0.386
-0.1	0.460	0.456	0.452	0.448	0.444	0.440	0.436	0.433	0.429	0.425
-0.0	0.500	0.496	0.492	0.488	0.484	0.480	0.476	0.472	0.468	0.464

The table shows values for the cumulative distribution function of $N(0,1)$ (see p. 74). To find the value of $\Phi(z)$ for a given value of z, you should cross-index the first two digits (found in the left-most column) with the third and last digit (shown in the column headers). For example, $\Phi(-0.57) = 0.284$. Negative values of z are found on this page; positive values of z are found on the next page.

z	0	0.01	0.02	0.03	0.04	0.05	0.06	0.07	0.08	0.09
0.0	0.500	0.504	0.508	0.512	0.516	0.520	0.524	0.528	0.532	0.536
0.1	0.540	0.544	0.548	0.552	0.556	0.560	0.564	0.567	0.571	0.575
0.2	0.579	0.583	0.587	0.591	0.595	0.599	0.603	0.606	0.610	0.614
0.3	0.618	0.622	0.626	0.629	0.633	0.637	0.641	0.644	0.648	0.652
0.4	0.655	0.659	0.663	0.666	0.670	0.674	0.677	0.681	0.684	0.688
0.5	0.691	0.695	0.698	0.702	0.705	0.709	0.712	0.716	0.719	0.722
0.6	0.726	0.729	0.732	0.736	0.739	0.742	0.745	0.749	0.752	0.755
0.7	0.758	0.761	0.764	0.767	0.770	0.773	0.776	0.779	0.782	0.785
0.8	0.788	0.791	0.794	0.797	0.800	0.802	0.805	0.808	0.811	0.813
0.9	0.816	0.819	0.821	0.824	0.826	0.829	0.831	0.834	0.836	0.839
1.0	0.841	0.844	0.846	0.848	0.851	0.853	0.855	0.858	0.860	0.862
1.1	0.864	0.867	0.869	0.871	0.873	0.875	0.877	0.879	0.881	0.883
1.2	0.885	0.887	0.889	0.891	0.893	0.894	0.896	0.898	0.900	0.901
1.3	0.903	0.905	0.907	0.908	0.910	0.911	0.913	0.915	0.916	0.918
1.4	0.919	0.921	0.922	0.924	0.925	0.926	0.928	0.929	0.931	0.932
1.5	0.933	0.934	0.936	0.937	0.938	0.939	0.941	0.942	0.943	0.944
1.6	0.945	0.946	0.947	0.948	0.949	0.951	0.952	0.953	0.954	0.954
1.7	0.955	0.956	0.957	0.958	0.959	0.960	0.961	0.962	0.962	0.963
1.8	0.964	0.965	0.966	0.966	0.967	0.968	0.969	0.969	0.970	0.971
1.9	0.971	0.972	0.973	0.973	0.974	0.974	0.975	0.976	0.976	0.977
2.0	0.977	0.978	0.978	0.979	0.979	0.980	0.980	0.981	0.981	0.982
2.1	0.982	0.983	0.983	0.983	0.984	0.984	0.985	0.985	0.985	0.986
2.2	0.986	0.986	0.987	0.987	0.987	0.988	0.988	0.988	0.989	0.989
2.3	0.989	0.990	0.990	0.990	0.990	0.991	0.991	0.991	0.991	0.992
2.4	0.992	0.992	0.992	0.992	0.993	0.993	0.993	0.993	0.993	0.994
2.5	0.994	0.994	0.994	0.994	0.994	0.995	0.995	0.995	0.995	0.995
2.6	0.995	0.995	0.996	0.996	0.996	0.996	0.996	0.996	0.996	0.996
2.7	0.997	0.997	0.997	0.997	0.997	0.997	0.997	0.997	0.997	0.997
2.8	0.997	0.998	0.998	0.998	0.998	0.998	0.998	0.998	0.998	0.998
2.9	0.998	0.998	0.998	0.998	0.998	0.998	0.998	0.999	0.999	0.999
3.0	0.999	0.999	0.999	0.999	0.999	0.999	0.999	0.999	0.999	0.999
3.1	0.999	0.999	0.999	0.999	0.999	0.999	0.999	0.999	0.999	0.999
3.2	0.999	0.999	0.999	0.999	0.999	0.999	0.999	0.999	0.999	0.999
3.3	1.000	1.000	1.000	1.000	1.000	1.000	1.000	1.000	1.000	1.000
3.4	1.000	1.000	1.000	1.000	1.000	1.000	1.000	1.000	1.000	1.000

C.3   The *t* distribution

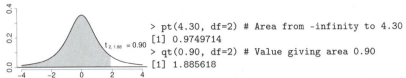

```
> pt(4.30, df=2) # Area from -infinity to 4.30
[1] 0.9749714
> qt(0.90, df=2) # Value giving area 0.90
[1] 1.885618
```

df	0.8	0.9	0.95	0.975	0.98	0.99	0.995	0.999	0.9999
					Lower tail probability				
1	1.376	3.078	6.314	12.706	15.895	31.821	63.657	318.309	3183.099
2	1.061	1.886	2.920	4.303	4.849	6.965	9.925	22.327	70.700
3	0.978	1.638	2.353	3.182	3.482	4.541	5.841	10.215	22.204
4	0.941	1.533	2.132	2.776	2.999	3.747	4.604	7.173	13.034
5	0.920	1.476	2.015	2.571	2.757	3.365	4.032	5.893	9.678
6	0.906	1.440	1.943	2.447	2.612	3.143	3.707	5.208	8.025
7	0.896	1.415	1.895	2.365	2.517	2.998	3.499	4.785	7.063
8	0.889	1.397	1.860	2.306	2.449	2.896	3.355	4.501	6.442
9	0.883	1.383	1.833	2.262	2.398	2.821	3.250	4.297	6.010
10	0.879	1.372	1.812	2.228	2.359	2.764	3.169	4.144	5.694
11	0.876	1.363	1.796	2.201	2.328	2.718	3.106	4.025	5.453
12	0.873	1.356	1.782	2.179	2.303	2.681	3.055	3.930	5.263
13	0.870	1.350	1.771	2.160	2.282	2.650	3.012	3.852	5.111
14	0.868	1.345	1.761	2.145	2.264	2.624	2.977	3.787	4.985
15	0.866	1.341	1.753	2.131	2.249	2.602	2.947	3.733	4.880
16	0.865	1.337	1.746	2.120	2.235	2.583	2.921	3.686	4.791
17	0.863	1.333	1.740	2.110	2.224	2.567	2.898	3.646	4.714
18	0.862	1.330	1.734	2.101	2.214	2.552	2.878	3.610	4.648
19	0.861	1.328	1.729	2.093	2.205	2.539	2.861	3.579	4.590
20	0.860	1.325	1.725	2.086	2.197	2.528	2.845	3.552	4.539
21	0.859	1.323	1.721	2.080	2.189	2.518	2.831	3.527	4.493
22	0.858	1.321	1.717	2.074	2.183	2.508	2.819	3.505	4.452
23	0.858	1.319	1.714	2.069	2.177	2.500	2.807	3.485	4.415
24	0.857	1.318	1.711	2.064	2.172	2.492	2.797	3.467	4.382
25	0.856	1.316	1.708	2.060	2.167	2.485	2.787	3.450	4.352
26	0.856	1.315	1.706	2.056	2.162	2.479	2.779	3.435	4.324
27	0.855	1.314	1.703	2.052	2.158	2.473	2.771	3.421	4.299
28	0.855	1.313	1.701	2.048	2.154	2.467	2.763	3.408	4.275
29	0.854	1.311	1.699	2.045	2.150	2.462	2.756	3.396	4.254
30	0.854	1.310	1.697	2.042	2.147	2.457	2.750	3.385	4.234
40	0.851	1.303	1.684	2.021	2.123	2.423	2.704	3.307	4.094
50	0.849	1.299	1.676	2.009	2.109	2.403	2.678	3.261	4.014
60	0.848	1.296	1.671	2.000	2.099	2.390	2.660	3.232	3.962
80	0.846	1.292	1.664	1.990	2.088	2.374	2.639	3.195	3.899
100	0.845	1.290	1.660	1.984	2.081	2.364	2.626	3.174	3.862
150	0.844	1.287	1.655	1.976	2.072	2.351	2.609	3.145	3.813
1000	0.842	1.282	1.646	1.962	2.056	2.330	2.581	3.098	3.733
∞	0.842	1.282	1.645	1.960	2.054	2.326	2.576	3.090	3.719

The table shows values for the cumulative distribution function of a *t* distribution with df degrees of freedom (see p. 116). The rows show various degrees of freedom while the column headers show the area under the curve. For example, for a *t* distribution with 4 degrees of freedom, the 0.975 quantile is 2.776.

C.4 The *F* distribution

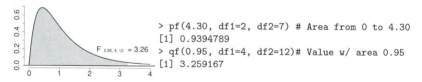

```
> pf(4.30, df1=2, df2=7) # Area from 0 to 4.30
[1] 0.9394789
> qf(0.95, df1=4, df2=12)# Value w/ area 0.95
[1] 3.259167
```

$F_{0.95, 4, 12} = 3.26$

	Numerator degrees of freedom									
	1	2	3	4	5	6	7	8	9	10
1	161.45	199.50	215.71	224.58	230.16	233.99	236.77	238.88	240.54	241.88
2	18.513	19.000	19.164	19.247	19.296	19.330	19.353	19.371	19.385	19.396
3	10.128	9.552	9.277	9.117	9.013	8.941	8.887	8.845	8.812	8.786
4	7.709	6.944	6.591	6.388	6.256	6.163	6.094	6.041	5.999	5.964
5	6.608	5.786	5.409	5.192	5.050	4.950	4.876	4.818	4.772	4.735
6	5.987	5.143	4.757	4.534	4.387	4.284	4.207	4.147	4.099	4.060
7	5.591	4.737	4.347	4.120	3.972	3.866	3.787	3.726	3.677	3.637
8	5.318	4.459	4.066	3.838	3.687	3.581	3.500	3.438	3.388	3.347
9	5.117	4.256	3.863	3.633	3.482	3.374	3.293	3.230	3.179	3.137
10	4.965	4.103	3.708	3.478	3.326	3.217	3.135	3.072	3.020	2.978
11	4.844	3.982	3.587	3.357	3.204	3.095	3.012	2.948	2.896	2.854
12	4.747	3.885	3.490	3.259	3.106	2.996	2.913	2.849	2.796	2.753
13	4.667	3.806	3.411	3.179	3.025	2.915	2.832	2.767	2.714	2.671
14	4.600	3.739	3.344	3.112	2.958	2.848	2.764	2.699	2.646	2.602
15	4.543	3.682	3.287	3.056	2.901	2.790	2.707	2.641	2.588	2.544
16	4.494	3.634	3.239	3.007	2.852	2.741	2.657	2.591	2.538	2.494
17	4.451	3.592	3.197	2.965	2.810	2.699	2.614	2.548	2.494	2.450
18	4.414	3.555	3.160	2.928	2.773	2.661	2.577	2.510	2.456	2.412
19	4.381	3.522	3.127	2.895	2.740	2.628	2.544	2.477	2.423	2.378
20	4.351	3.493	3.098	2.866	2.711	2.599	2.514	2.447	2.393	2.348
25	4.242	3.385	2.991	2.759	2.603	2.490	2.405	2.337	2.282	2.236
30	4.171	3.316	2.922	2.690	2.534	2.421	2.334	2.266	2.211	2.165
35	4.121	3.267	2.874	2.641	2.485	2.372	2.285	2.217	2.161	2.114
40	4.085	3.232	2.839	2.606	2.449	2.336	2.249	2.180	2.124	2.077
45	4.057	3.204	2.812	2.579	2.422	2.308	2.221	2.152	2.096	2.049
50	4.034	3.183	2.790	2.557	2.400	2.286	2.199	2.130	2.073	2.026
55	4.016	3.165	2.773	2.540	2.383	2.269	2.181	2.112	2.055	2.008
60	4.001	3.150	2.758	2.525	2.368	2.254	2.167	2.097	2.040	1.993
65	3.989	3.138	2.746	2.513	2.356	2.242	2.154	2.084	2.027	1.980
70	3.978	3.128	2.736	2.503	2.346	2.231	2.143	2.074	2.017	1.969
75	3.968	3.119	2.727	2.494	2.337	2.222	2.134	2.064	2.007	1.959
100	3.936	3.087	2.696	2.463	2.305	2.191	2.103	2.032	1.975	1.927
200	3.888	3.041	2.650	2.417	2.259	2.144	2.056	1.985	1.927	1.878
500	3.860	3.014	2.623	2.390	2.232	2.117	2.028	1.957	1.899	1.850

The table shows values for the 0.95 quantile for the *F* distribution with df1 degrees of freedom in the numerator and df2 degrees of freedom in the denominator (see p. 160). The columns show the numerator degrees of freedom while the rows are the denominator degrees of freedom. For example, for a $F(4, 12)$ distribution with $(4, 12)$ degrees of freedom, the 0.95 quantile is 3.259.

Bibliography

Agresti, A. and Coull, B. A. (1998). Approximate is better than "exact" for interval estimation of binomial proportions. *The American Statistician*, 52:119–126.

Altman, D. G. and Bland, J. M. (1983). Measurement in medicine: the analysis of method comparison studies. *Statistician*, 32:307–317.

Andersen, S. (1990). *Statistisk analyse af tælledata*. Lecture notes, Landbohøjskolen, Denmark.

Anonymous (1949). Query 70. *Biometrics*, pages 250–251.

Ashby, E. and Oxley, T. A. (1935). The interactions of factors in the growth of *Lemna*. *Annals of Botany*, 49:309–336.

Ashford, J. R. (1959). An approach to the analysis of data for semi-quantal responses in biological assay. *Biometrics*, 15:573–581.

Ashton, K. G., Burke, R. L., and Layne, J. N. (2007). Geographic variation in body and clutch size of gopher tortoises. *Copeia*, 49:355–363.

Azzalini, A. (1996). *Statistical Inference Based on the Likelihood*. Chapman & Hall, London.

Bateson, W., Saunders, E. R., and Punnett, R. C. (1906). Experimental studies in the physiology of heredity. *Reports to the Evolution Committee of the Royal Society*, III:2–11.

Blæsild, P. and Granfeldt, J. (2003). *Statistics with Applications in Biology and Geology*. Chapman & Hall, London.

Bliss, C. I. (1967). *Statistics in Biology*. McGraw-Hill, New York.

Cohen, J. (1960). A coefficient of agreement for nominal scales. *Educational and Psychological Measurement*, 20:37–46.

Collett, D. (1991). *Modelling Binary Data*. Chapman & Hall, London.

Courtney, N. and Wells, D. L. (2002). The discrimination of cat odours by humans. *Perception*, 31:511–512.

405

Crawley, M. (2007). *The R Book*. John Wiley & Sons, New York.

Dalgaard, P. (2008). *Introductory Statistics with R (2nd ed.)*. Springer, New York.

Daniel, W. W. (1995). *Biostatistics: A Foundation for Analysis in the Health Sciences*. Wiley, New York.

Danscher, A. M., Enemark, J. M. D., Telezhenko, E., Capion, N., Ekstrøm, C. T., and Thoefner, M. B. (2009). Oligofructose overload induces lameness in cattle. *Journal of Dairy Science*, 92:607–616.

de Neergaard, E., Haupt, G., and Rasmussen, K. (1993). Studies of *Didymella bryoniae*: the influence of nutrition and cultural practices on the occurrence of stem lesions and internal and external fruit rot on different cultivars of cucumber. *Netherlands Journal of Plant Pathology*, 99:335–343.

Dobson, A. (2001). *An Introduction to Generalized Linear Models (2nd ed.)*. Chapman & Hall, London.

Finney, D. J. (1952). *Probit analysis*. Cambridge University Press, England.

Fisher, R. A. (1947). The analysis of covariance method for the relation between a pair and the whole. *Biometrics*, 3:65–68.

Geissler, A. (1889). Beitrage zur frage des geschlechts verhaltnisses der geborenen. *Zeitschrift des Königlichen Sächsischen Statistischen Bureaus*, 35:1–24.

Hand, D. J., Daly, F., McConway, K., Lunn, D., and Ostrowski, E. (1993). *A Handbook of Small Data Sets*. Chapman & Hall, London.

Haraldsdottir, J., Jensen, J. H., and Møller, A. (1985). Danskernes kostvaner 1985, I. Hovedresultater. *Levnedsmiddelstyrelsen*, page Publikation nr. 138.

Heinecke, R. D., Martinussen, T., and Buchmann, K. (2007). Microhabitat selection of *Gyrodactylus salaris* Malmberg on different salmonids. *Journal of Fish Diseases*, 30:733–743.

Hosmer, D. W. and Lemeshow, S. (1989). *Applied Logistic Regression*. John Wiley & Sons, New York.

Innes, J. R. M., Ulland, B. M., Valerio, M. G., Petrucelli, L., Fishbein, L., Hart, E. R., Pallota, A. J., Bates, R. R., Falk, H. L., Gart, J. J., Klein, M., Mitchell, I., and Peters, J. (1969). Bioassay of pesticides and industrial chemicals for tumorigenicity in mice: A preliminary note. *Journal of the National Cancer Institute*, 42:1101–1114.

Jeffers, J. N. R. (1959). *Experimental Design and Analysis in Forest Research*. Almqvist & Wiksell, Stockholm.

Jensen, A. T., Sørensen, H., Thomsen, M. H., and Andersen, P. H. (2009). Quantification of symmetry for functional data with application to equine lameness classification. Submitted.

Joglekar, G., Schuenemeyer, J. H., and LaRiccia, V. (1989). Lack-of-fit testing when replicates are not available. *The American Statistician*, 43:135–143.

Jørgensen, G. and Hansen, N. G. (1973). Fedtsyresammensætningens indflydelse på fedtstoffernes fordøjelighed. *Landøkonomisk Forsøgslaboratorium. Årbog 1973*, pages 250–259.

Juncker, D., Andersen, H., Skibsted, L., Erichsen, L., and Bertelsen, G. (1996). Meat colour of Danish landrace pigs anno 1973 and 1995. II. colour stability of pork chops during chill storage. *42nd International Congress of Meat Science and Technology*, pages 82–83.

Kleinbaum, D. G. and Klein, M. (2002). *Logistic Regression (2nd ed.)*. Springer, New York.

Lansky, P., Sanda, P., and He, J. (2006). The parameters of the stochastic leaky integrate-and-fire neuronal model. *Journal of Computational Neuroscience*, 21:211–223.

Latter, O. H. (1905). The egg of *Cuculus Canorus*: An attempt to ascertain from the dimensions of the cuckoo's egg if the species is tending to break up into sub-species, each exhibiting a preference for some one foster-parent. *Biometrika*, 4:363–373.

McCullagh, P. and Nelder, J. A. (1989). *Generalized Linear Models*. Chapman & Hall, London.

Mendel, G. (1866). Versuch über Pflanzenhybriden. *Verhandlungen des naturforschenden Vereines in Brünn, Bd. IV für das Jahr 1865*, pages 3–47.

Møller, A. J., Kirkegaard, E., and Vestergaard, T. (1987). Tenderness of pork muscles as influenced by chilling rate and altered carcass suspension. *Meat Science*, 21:275–286.

Neter, J., Kutner, M., Nachtsheim, C., and Wasserman, W. (1996). *Applied Linear Statistical Models*. McGraw-Hill, New York.

Ortego, J. D., Aminabhavi, T. M., Harlapur, S. F., and Balundgi, R. H. (1995). A review of polymeric geosynthetics used in hazardous waste facilities. *Journal of Hazardous Materials*, 42:115–156.

Petersen, H. L. (1954). Pollination and seed setting in lucerne. *Kgl. Veterinær og Landbohøjskole, Årsskrift 1954*, pages 138–169.

Roepstorff, A., Karl, H., Bloemsma, B., and Huss, H. H. (1993). Catch handling and the possible migration of Anisakis larvae in herring, Clupea harengus. *Journal of Food Protection*, 56:783–787.

Rudemo, M. (1979). *Statistik og sandsynlighedslære med biologiske anvendelser. Del 1: Grundbegreber.* DSR Forlag.

Ryan Jr., T. A., Joiner, B. L., and Ryan, B. F. (1985). *The Minitab Student Handbook.* Duxbury Press, Boston.

Samuels, M. L. and Witmer, J. A. (2003). *Statistics for the Life Sciences (3rd ed.).* Pearson Education, Inc., New Jersey.

Shin, J. H., Chung, J., Kang, J. H., Lee, Y. J., Kim, K. I., So, Y., Jeong, J. M., Lee, D. S., and Lee, M. C. (2004). Noninvasive imaging for monitoring of viable cencer cells using a dual-imaging reporter gene. *The Journal of Nuclear Medicine*, 45:2109–2115.

Skovgaard, I. (2004). *Basal Biostatistik 2.* Samfundslitteratur.

Skovgaard, I., Stryhn, H., and Rudemo, M. (1999). *Basal Biostatistik 1.* DSR Forlag.

Sommer, C. and Bibby, B. M. (2002). The influence of veterinary medicines on the decomposition of dung organic matter in soil. *European Journal of Soil Biology*, 38:155–159.

Tager, I., Weiss, S., Rosner, B., and Speizer, F. (1979). Effect of paternal cigarette smoking on the pulmonary function of children. *American Journal of Epidemiology*, 110:15–26.

Verzani, J. (2005). *Using R for Introductory Statistics.* Chapman & Hall/CRC, London.

Weisberg, S. (1985). *Applied Linear Regression (2nd ed.).* John Wiley & Sons, New York.

Wells, D. L. and Hepper, P. G. (2000). The discrimination of dog odours by humans. *Perception*, 29:111–115.

Willeberg, P. (1976). Interaction effects of epidemiologic factors in the feline urological syndrome. *Nordisk Veterinær Medicin*, 28:193–200.

Williams, C. B. (1964). *Patterns in the Balance of Nature.* Academic Press, London.

Zar, J. H. (1999). *Biostatistical Analysis (4th ed.).* Prentice Hall, Upper Saddle River, New Jersey.

Ziv, G. and Sulman, F. G. (1972). Binding of antibiotics to bovine and ovine serum. *Antimicrobial Agents and Chemotherapy*, 2:206–213.

Index

Statistical Data Analysis for the Life Sciences